D0207824

Text Algorithms

Text Algorithms

Maxime Crochemore
Wojciech Rytter

Oxford University Press
New York Oxford

Oxford University Press
Oxford New York Toronto
Delhi Bombay Calcutta Madras Karachi
Petaling Jaya Singapore Hong Kong Tokyo
Nairobi Dar es Salaam Cape Town
Melbourne Auckland
and associated companies in
Berlin Ibadan

Published by Oxford University Press, Inc.
200 Madison Ave., New York, New York 10016

Library of Congress Cataloging-in-Publication Data

Crochemore, Maxime, 1947–
Text algorithms / Maxime Crochemore and Wojciech Rytter.
p. cm.
Includes bibliographic references and index.
ISBN 0-19-50809-0
1. Text processing (Computer science) 2. Computer algorithms.
I. Rytter, Wojciech. II. Title.
QA76.9.T48C76 1994
005.1—dc20 94-20649
CIP

9 8 7 6 5 4 3 2 1

Printed in the United States on acid-free paper.

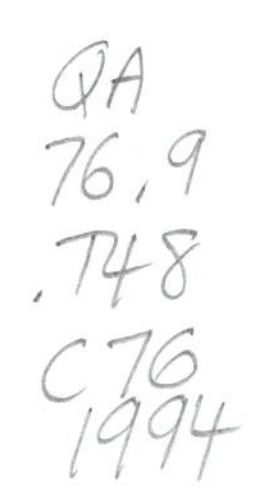

Contents

Preface

The design of algorithms that process strings and texts goes back at least twenty-five years. In particular, the last ten of those years have produced an explosion of new results. This progress is due in part to the human genome effort, to which string algorithms can make an important contribution.

While text algorithms can be viewed as part of the general field of algorithmic research, it has developed into a respectable subfield of its own. This subfield combines theory and practice nicely. The theory deals with symmetries and periodicities in strings, which in practice may lead to the development of faster, newer algorithms.

One measure of the vibrance of this new subfield is the ongoing success of a conference devoted to its study. The Conference on Combinatorial Pattern Matching will be holding its fifth meeting in the summer of 1994 to present theoretical results, new algorithms, and new applications of these algorithms.

Following the remarkable progress in this new field, Maxime Crochemore and Wojciech Rytter embarked on the right project at the right time—writing a textbook on text algorithms. Both authors have made important contributions to the field and, therefore, are excellent choices for the job.

About ten years ago, in a workshop that preceded the conference (called Combinatorial Algorithms on Words), I gave a lecture entitled "Open Problems in Stringology" [Ga 85a]. (Stringology has become the nickname of this new subfield.) In that lecture I described one dozen open problems. It gives me great pleasure to report that about half of these problems have either been solved or significant progress has been made toward their complete solution. Some of these new developments appear in Crochemore and Rytter's new book. The recent *Handbook of Theoretical*

Computer Science includes an excellent chapter entitled "Algorithms for Finding Patterns in Strings" by A.V. Aho, one of the earliest contributors to text algorithms. Each time I have met with Aho in the last fifteen years he has encouraged me to try to find a better algorithm for the membership problem for regular expressions, one of my twelve open problems. An improved solution to this problem would have significant practical consequences; in particular, it would improve the performance of the popular EGREP™ program used by UNIX®. See Chapter 7 of Crochemore and Rytter's book for more details.

Since the research on text algorithms continues, it is not possible to have a book that covers the area completely. Looking at the table of contents of this book, I am convinced that its fifteen chapters cover many of the major developments in the field nicely. Crochemore and Rytter have succeeded in producing a textbook that is as thorough as it is timely.

Zvi Galil

Acknowledgements

We have worked on this book during our visits to each other at LIPN (University of Paris-Nord), Instytut Informatyk (University of Warsaw), LITP (University Paris 7), and Institut Gaspard Monge (University of Marne la Vallée). We thank our colleagues of all these laboratories warmly for the kind and stimulating atmosphere that has favored the creation of this book. We especially thank the people who helped us produce the text by reading preliminary versions of some chapters, or by their comments: Véronique Bruyère, Richard Cole, Artur Czumaj, Zvi Galil, Leszek Gasieniec, Christophe Hancart, Thierry Lecroq, Larry Larmore, Dominique Perrin, Giuseppina Rindone, Laura Toniolo, Marc Zipstein.

Maxime Crochemore,
Wojciech Rytter

Text Algorithms

1

Introduction

1.1 Texts and their processing

One of the simplest, natural types of information representation is by means of written texts. Data to be processed often does not decompose into independent records. This type of data is characterized by the fact that it can be written down as a long sequence of characters. Such linear sequence is called a *text*. The texts are central in "word processing" systems, which provide facilities for the manipulation of texts. Such systems usually process objects that are quite large. For example, this book probably contains more than a million characters. Text algorithms occur in many areas of science and information processing. Many text editors and programming languages have facilities for processing texts. In biology, text algorithms arise in the study of molecular sequences. The complexity of text algorithms is also one of the central and most studied problems in theoretical computer science. It could be said that it is the domain in which practice and theory are very close together.

The basic textual problem is the problem called *pattern matching*. It is used to access information and, no doubt, at this moment many computers are solving this problem as a frequently used operation in some application system. Pattern matching is comparable in this sense to sorting, or to basic arithmetic operations.

Consider the problem of a reader of the French dictionary "Grand Larousse," who wants all entries related to the name "Marie-Curie-Sklodowska." This is an example of a pattern matching problem, or string matching. In this case, the name "Marie-Curie-Sklodowska" is the pattern. Generally we may want to find a string called a *pattern* of length m inside a text of length n where n is greater than m. The pattern can be described in a more complex way to denote a set of strings and not just a single word. In many cases m is very large. In genetics the pattern can correspond to a genome that can be very long; in image processing the digitized images sent serially take millions of characters each. The string-matching problem is the basic question considered in this book, together with its variations. String matching is also the basic subproblem in other algorithmic problems on texts. Following is a (not exclusive) list of basic groups of problems discussed in this book:

- variations on the string-matching problem
- problems related to the structure of the segments of a text
- data compression
- approximation problems
- finding regularities
- extensions to two-dimensional images
- extensions to trees
- optimal time-space implementations
- optimal parallel implementations.

The formal definition of string matching and many other problems is given in the next chapter. We now introduce some of them informally in the context of applications.

1.2 Text file facilities

The UNIX system uses text files for exchanging information as a main feature. The user can get information from the files and transform them through different existing commands. The tools often behave as filters that read their input once and produce the output simultaneously. These tools can easily be connected with each other, particularly through the pipelining facility. This often reduces the creation of new commands to a few lines of already existing commands.

One of these useful commands is *grep*, acronym of "general regular expression print." An example of the format of *grep* is

```
grep Marie-Curie-Sklodowska Grand-Larousse
```

provided "Grand-Larousse" is a file on your computer. The output of this command is the list of lines from the file that contains an occurrence of the name "Marie-Curie-Sklodowska."

This is an instance of the **string-matching** problem. Another example with a more complex pattern can be

```
grep '^Chapter [0-9]' Book
```

to list the titles of a book assuming titles begin with "Chapter" followed by a number. In this case the pattern denotes a set of strings (even potentially infinite), and not simply one string. The notation to specify patterns is known as *regular expressions*. This is an instance of the **regular-expression-matching** problem.

The indispensable complement of *grep* is *sed* (stream editor). It is designed to transform its input. It can replace patterns of the input with specific strings. Regular expressions are also available with *sed*. But the editor contains an even more powerful notation. This allows, for example, the action on a line of the input text containing the same word twice. It can be applied to delete two consecutive occurrences of a same word in a text. This is simultaneously an example of the **repetition-finding** problem, **pattern-matching** problem and, more generally, the problem of finding **regularities in strings**.

The very helpful matching device based on regular expressions is omnipresent in the UNIX system. It can be used inside text editors such as *ed* and *vi*, and generally in almost all UNIX commands. The above tools, *grep* and *sed*, are based on this mechanism. There is even a programming language based on pattern-matching actions. It is the *awk* language, where the name *awk* comes from the initials of the authors, Aho, Weinberger, and Kernighan. A simple *awk* program is a sequence of pattern-action statements:

```
pattern1 { action1 }
pattern2 { action2 }
pattern3 { action3 }
```

The basic components of this program are patterns to be found inside the lines of the current file. When a pattern is found, the corresponding action is applied to the line. Therefore, several actions may be applied sequentially to a same line. This is an example of the **multi-pattern matching** problem. The language *awk* is meant for converting data from one form to another form, counting things, adding up numbers, and extracting information for reports. It contains an implicit input loop, and the pattern-action paradigm often eliminates control flow. This also frequently reduces the size of a program to a few statements.

For instance, the following *awk* program prints the number of lines of the input that contain the word "abracadabra":

```
abracadabra     { count++ }
END             { print count }
```

The pattern "END" matches the end of input file, so that the result is printed after the input has been entirely processed. The language contains attractive features that strengthen the simplicity of the pattern-matching mechanism, such as default initialization for variables, implicit declarations, and associative arrays providing arbitrary kinds of subscripts. All this makes *awk* a convenient tool for rapid prototyping.

The *awk* language can be considered as a generalization of another UNIX tool, *lex*, aimed at producing lexical analyzers. The input of a *lex* program is the specification of a lexical analyzer by means of regular expressions (and a few other possibilities). The output is the source of the specified lexical analyzer in the C programming language. A specification in *lex* is mainly a sequence of pattern-action statements as in *awk*. Actions are pieces of C code to be inserted in the lexical analyzer. At run time, these pieces of code execute the action corresponding to the associated pattern, when found.

The following line is a typical statement of a *lex* program:

```
[A-Za-z]+([A-Za-z0-9])*    { yylval = Install(); return(ID); }
```

The pattern specifies identifiers, that is, strings of characters starting with one letter and containing only letters and digits. This action leads the generated lexical analyzer to store the identifier and to return the string type "ID" to the calling parser. It is another instance of the regular expression-matching problem. The question of constructing **pattern-matching automata** is an important component having a practical application in the *lex* software.

Lex is often combined with *yacc* and *twig* for the development of parsers, translators, compilers, or their prototypes. *Yacc* allows the specification of a parser by grammar rules. It produces the source of the specified parser in C language. *Twig* is a tree-manipulation language that helps to design efficient code generators. It transforms a tree-translation scheme into a code generator based on a fast top-down tree-pattern-matching algorithm. The output generator also uses dynamic programming techniques for optimization purposes, but the essential feature is the tree-pattern-matching action for tree rewriting. It has been shown, by experimental results on few significant applications, that *twig*-generated code generators are often faster than hand-generated compilers even when the latter incorporate some tricky optimizations. One of the basic problems considered for the design of such a tool is the **tree-pattern-matching** problem.

Texts such as books or programs are likely to be changed during elaboration. Even after their completion they often support periodic upgrades. These questions are related to **text comparisons**. Sometimes we also wish to find a string, and do not completely remember it. The search has to be performed with an entirely non-specified pattern. This is an instance of the **approximate pattern matching**.

Keeping track of all consecutive versions of a text may not be helpful because the text can be very long and changes may be hard to find. The reasonable way to control the process is to have an easy access to differences between the various versions. There is no universal notion as to what the differences are, or conversely, what the similarities are, between the two texts. However, it can be agreed that the intersection of the two texts is the longest common subtext of both. In our book this is called the **longest common subsequence** problem, so that the differences between the two texts are the respective complements of the common part. The UNIX command *diff* builds on this notion.

Consider the texts A and B as follows (Molière's joke, 1670):

Text A	Belle Marquise, vos beaux yeux me font mourir d'amour
Text B	D'amour mourir me font, Belle Marquise, vos beaux yeux

The command diff A B produces

```
0a1
> D'amour mourir me font,
3d3
< me font mourir d'amour
```

An option of the command *diff* produces a sequence of *ed* instructions to transform one text into the other. The similarity of texts can be measured as the minimal number of edit operations to transform one text into the other. The computation of such a measure is an instance of the **edit distance** problem.

1.3 Dictionaries

The search for words or patterns in static texts is quite a different question than the previous pattern-matching mechanism. Dictionaries, for example, are organized in order to speed up the access to entries. Another example of the same question is given by indexes. Technical books often contain an index of chosen terms that gives pointers to parts of the text related to words in the index. The algorithms involved in the creation of an index form a specific group.

The use of dictionaries or lexicons is often related to natural language processing. Lexicons of programming languages are small, and their representation is not a difficult problem during the development of a compiler. To the contrary, English contains approximately 100,000 words, and even twice that if inflected forms are considered. In French, inflected forms produce more than 500,000 words. The representation of lexicons of this size makes the problem a bit more challenging.

A simple use of dictionaries is illustrated by spelling checkers. The UNIX command, *spell*, reports the words in its input that are not stored in the lexicon. This rough approach does not yield a pertinent checker, but, practically, it helps to find typing errors. The lexicon used by *spell* contains approximately 70,000 entries stored within less than 60 kilobytes of random-access memory.

Quick access to lexicons is a necessary condition for producing good parsers. The data structure useful for such access is called an index. In our book indexes correspond to data structures representing all factors of a given (presumably long) text. We consider problems of the construction of such structures: **suffix trees**, **directed**

acyclic word graphs, **factor automata**, **suffix arrays**. The *PAT* tool developed at the NOED Center (Waterloo, Canada) is an implementation of one of these structures tailored to work on large texts. There are several applications that effectively require some understanding of phrases in natural languages, such as data retrieval systems, interactive softwares, and character recognition.

An image scanner is a kind of photocopier. It is used to give a digitized version of an image. When the image is a page of text, the natural output of the scanner must be in a digital form available to a text editor. The transformation of a digitized image of a text into a usual computer representation of the text is realized by an Optical Character Reader (OCR). Scanning a text with an OCR can be 50 times faster than retyping the text on a keyboard. Thus, OCR softwares are likely to become more common. But they still suffer from a high degree of imprecision. The average rate of error in the recognition of characters is approximately one percent. Even if this may happen to be rather small, this means that scanning a book produces approximately one error per line. This is compared with the usually very high quality of texts checked by specialists. Technical improvements on the hardware can help to eliminate certain kinds of errors occurring on scanned texts in printed forms. But this cannot alleviate the problem associated with recognizing texts written by hand. In this case, isolation of characters is much more difficult, and ambiguous forms can be encountered at the recognition stage. Reduction of the number of errors can thus only be achieved by considering the context of the characters, which assumes some understanding of the text. Image processing is related to the problem of **two-dimensional pattern matching**. Another related problem is the data structure for all subimages which is discussed in this book in the context of the **dictionary of basic factors (subimages)**.

No system is presently able to parse natural languages. Indeed, it is questionable whether such an effective system can exist. Words are certainly units of syntax in languages, but they are not units of meaning. This is because words can have several meanings, and also because they can be parts of compound phrases which carry a meaning only as a whole. It appears that the proportion in the lexicon (of French and several other European languages) of idiomatic sentences—of metaphoric and technical sentences whose meaning cannot be inferred from their elements—is very high. The natural consequence is that compound expressions and the so-called frozen phrases must be included in computerized dictionaries. The data base must also incorporate at least the syntactic properties of the elements. At present, no complete dictionary established according to these principles exists. This certainly represents a huge amount of data to be collected, but this seems an indispensable work. Data bases for the European languages—French, English, Italian, and Spanish—are presently being built. And studies of Arabic, Chinese, German, Korean and Madagascan have been performed. These questions are studied, for instance, at LADL (Paris), and the reader can refer to [Gr 91] for more information on the specific aspects of computational linguistics.

The theoretical approach to the representation of lexicons is either by means of trees or finite state automata. It appears that both approaches are equally efficient.

This shows the practical importance of the **automata theoretic approach** to text problems. At LITP (Paris) we have shown that the use of automata to represent lexicons is particularly efficient. Experiments have been done on a 500,000 word lexicon of LADL (Paris). The representation supports direct access to any word of the lexicon and takes only 300 kbytes of random-access memory.

1.4 Data compression

One of the basic problems in storing a large amount of textual information is the **text compression** problem. Text compression means reducing the representation of a text. It is assumed that the original text can be recovered from its compressed form. No loss of information is allowed. Text compression is related to the **Huffman coding problem** and the **factorization problem**. This kind of compression contrasts with other kinds of compression techniques applied to sounds or images, in which approximation is acceptable. Availability of large mass storage does not decrease the interest for compressing data. Indeed, users always take advantage of extra available space to store more data or new kinds of data. Moreover, the question remains important for storing data on secondary storage devices.

Examples of implementations of dictionaries reported above show that **data compression** is important in several domains related to natural language analysis. Text compression is also useful for telecommunications. It actually reduces the time to transmit documents via telephone networks, for example. The success of Facsimile is perhaps to be credited to compression techniques.

General compression methods often adapt themselves to the data. This phenomenon is central in achieving high compression ratios. However, it appears, in practice, that methods tailored for specific data lead to the best results. We have experimented with this fact on data sent by "géostationnaires" satellites. The data have been compressed to seven percent without any loss of information. The compression is very successful if there are redundancies and regularities in the information message. The analysis of data is related to the problem of **detecting regularities in texts**. Efficient algorithms are particularly useful to expertize the data.

1.5 Applications of text algorithms in genetics

Molecules of nucleic acids carry a large segment of information about the fundamental determinants of life, and, in particular, about the reproduction of cells. There are two types of nucleic acids known as deoxyrybonucleic acid (DNA) and ribonucleic acid (RNA). DNA is usually found as double-stranded molecules. In vivo, the molecule is folded up like a ball of string. The skeleton of a DNA molecule is a sequence on the four-letter alphabet of nucleotides: adenine (A), guanine (G), cytosine (C), and thymine (T). RNA molecules are usually single-stranded composed of ribonucleotides: A, G, C, and uracil (U). Processes of "transcription" and "translation" lead to the production of proteins, which also have a string composed of 20 amino acids as a primary structure. In a first approach all these molecules can be viewed as texts. The discovery fifteen years ago of powerful sequencing techniques has led to a rapid accumulation of sequence data (more than 10 million nucleotides of sequences). From the collection of sequences up to their analysis many algorithms on texts are implied. Moreover, only fast algorithms are often feasible because of the huge amount of data involved.

Collecting sequences is accomplished through gel audioradiographs. The automatic transcription of these gels into sequences is a typical **two-dimensional pattern-matching** problem in two dimensions. The reconstruction of a whole sequence from small segments is another example of a problem that occurs during this step. This problem is called the **shortest common superstring problem:** construction of the shortest text containing several given smaller texts.

Once a new sequence is obtained, the first important question to ask is whether it resembles any other sequence already stored in data banks. Before adding a new molecular sequence into an existing data base one needs to know whether or not the sequence is already present. The comparison of several sequences is usually realized by writing one over another. The result is known as an alignment of the set of sequences. This is a common display in molecular biology that can give an idea of the evolution of the sequences through the mutations, insertions, and deletions of nucleotides. Alignment of two sequences is the **edit distance problem:** compute the minimal number of edit operations to transform one string into another. It is realized by algorithms based on dynamic programming techniques similar to the one used by the UNIX command *diff*. The problem of the **longest common subsequence** is a variation of the alignment of sequences. A tool, called *agrep*, developed at the University of Arizona, is devoted to these questions, related to **approximate string matching**.

Further questions about molecular sequences are related to their analysis. The aim is to discover the functions of all parts of the sequence. For example, DNA sequences contain important regions (coding sequences) for the production of proteins. However, no good answer is presently known for finding all coding sequences of a DNA sequence. Another question about sequences is the reconstruction of their three-dimensional structure. It seems that a part of the information resides in the sequence itself. This is because, during the folding process of DNA, for example, nucleotides match pairwise (A with T, and C with G). This produces approximate palindromic symmetries (as TTAGCGGCTAA). Involved in all these questions are **approximate searches** for specific patterns, for **repetitions**, for **palindromes,** or any other **regularities**.

1.6 Structure of this book

The structure of this book reflects both the relationships between applications of text algorithmic techniques and the classification of algorithms according to the measures of complexity considered. The book can be viewed as a "parade" of algorithms in which our main aim is to discuss the foundations of these algorithms and their interconnections.

Chapter 2 begins with the algorithmic and theoretical foundations: machine models, discussion of algorithmic efficiency, review of basic problems and their interaction, and some combinatorics of words.

The topic of Chapters 3 and 4 is *string matching*. The chapters deal with two famous algorithms, Knuth-Morris-Pratt and Boyer-Moore algorithms, respectively. The recentmost improvements and results of these algorithms are incorporated. We discuss those string-matching algorithms which are efficient with respect to only one measure of complexity—sequential time. Hence, the chapter can be treated as a "warm-up" for other chapters devoted to the same topic. We later discuss more difficult algorithms in Chapter 13—*time-space optimal string-matching* algorithms—that work in linear time and simultaneously use only a very small additional memory. The consideration of two complexity measures together usually leads to more complicated algorithms. Chapters 3 and 4 can also be treated as a preparation to Chapters 9 and 14 where we present advanced *parallel string-matching* algorithms, and to Chapter 12 that deals with two-dimensional pattern matching.

Chapters 5 and 6 are the basic chapters of the book. They cover data structures succinctly representing all subwords of a text, namely, *suffix trees* and *subword graphs*. The considerations are typical from the point of view of data structures manipulations: trees, links, vectors, lists, etc. The prepared data structures are used later in the discussion of other problems; particularly in the computation of the longest common factor of texts, factorization problems, and finding squares in words. It is also used to construct certain kind of automata related to string-matching in Chapter 7.

Chapter 7 presents an automata theoretical approach to string matching that leads to a natural solution of the *multi-pattern-matching* problem. It covers Aho-Corasick automata for multi-pattern matching, automata for the set of factors and suffixes of a given text, and their applications.

Chapter 8 considers *regularities in texts: symmetries and repetitions*. It covers problems related to symmetries of words as well as problems related to repetitions. Surprisingly it occurs that finding a palindrome in a word in linear time is much easier than finding a square. Generally, problems related to symmetries are easier to solve than those related to repetitions. The basic component of this chapter is a new data structure, called the dictionary of basic factors, which is introduced as a basic technique for the development of efficient, but not necessarily optimal, algorithms.

Chapter 9 deals with *parallel algorithms on texts*. Only one, very general model of parallel computations is considered. The parallel implementation and applications of the dictionary of basic factors, suffix trees, and subword graphs are presented. We discuss a parallel version of the Karp-Miller-Rosenberg algorithm (Chapter 8) and some computations related to trees.

Chapter 10 is about data compression. It discusses only some of the *compression techniques for texts*. This area is exceptionally broad as compression of data is a vital part of many real-world systems. We focus on a few important algorithms. Two of the subjects here are the Ziv-Lempel algorithm, and applications of data structures to the text factorization from Chapters 5 and 6.

Questions related to algorithms involving the notion of *approximate patterns* are presented in Chapter 11. It opens with algorithms to compute edit distance of sequences (alignment of sequences). Two specific classical notions on approximate patterns are then considered, the first for string matching with errors, and the second for patterns with "don't care" symbols.

Two-dimensional pattern matching, in Chapter 12, introduces the problem of identifying a subimage inside a larger one. The notion naturally extends that of pattern matching in texts. Basic techniques on the subject show the importance of algorithms presented in Chapter 7.

Chapter 13 presents three different *time-space optimal string-matching* algorithms. None of them has ever appeared in a textbook, and the third one is presented in a completely new version. This is advanced material as the algorithms rely heavily on the combinatorial properties of words, mostly related to periodicities.

The discussion of *optimal parallel algorithm on texts* is the subject of Chapter 14. It includes more technical considerations than Chapter 9. The well-known Vishkin's algorithm and a recent optimal parallel algorithm for string matching, based on suffix versus prefix computations, designed by Kedem, Landau and Palem are presented here.

The last chapter considers several miscellaneous problems such as tree-pattern-matching problems, or the computation shortest common superstrings, maximal suffixes, and Lyndon factorizations.

One can partition algorithmic problems discussed in this book into practical and theoretical problems. Certainly string matching and data compression are in the first class, while most problems related to symmetries and repetitions are in the second. However, we believe that all the problems are interesting from an algorithmic point of view and enable the reader to appreciate the importance of combinatorics on words. In most textbooks on algorithms and data structures the presentation of efficient algorithms on words is quite short as compared to issues in graph theory, sorting, searching, and some other areas. At the same time, there are many presentations of interesting algorithms on words accessible only in journals and in a form directed mainly at specialists. We hope that this book will cover a gap on algorithms on words in book literature, and bring together the many results presently dispersed in the masses of journal articles.

Selected references

[AKW 88] Aho, A.V., Kernighan, B.W., & Weinberger, P.J., *The AWK Programming Language*, Addison-Wesley, Reading, Mass., 1986.

[ASU 86] Aho, A.V., Sethi, R., & Ullman, J., *Compilers—Principles, Techniques, and Tools*, Addison-Wesley, Reading, Mass., 1988, Chapter 3.

[BCW 90] Bell, T.C., Cleary, J.C., & Witten, I.H., *Text Compression*, Prentice Hall, Englewood Cliffs, New Jersey, 1990.

[BR 87] Bishop, M.J., & Rawlings, C.J., *Nucleic Acid and Protein Sequence Analysis: A Practical Approach*, IRL Press Limited, Oxford, England, 1987.

[Gr 91] Gross, M., Constructing lexicon-grammars, in B.T.S. Atkins & A. Zampolli eds, *Computational Approaches to the Lexicon*, Oxford University Press, 1991.

[Sa 89] Salton, G., *Automatic Text Processing*, Addison-Wesley, Reading, Mass., 1989.

[Sm 90] Smith, P.D., *An Introduction to Text Processing*, The MIT Press, Cambridge, Mass., 1990.

[Wa 89] Waterman, M.S., *Mathematical Methods for DNA Sequences*, CRC Press, Boca Raton, Florida, 1989.

2

Foundations

This chapter gives algorithmic and theoretical foundations required in further chapters. The models of computations are discussed, and basic problems on texts are exposed in a precise way. The chapter ends with some combinatorial properties of texts.

2.1 Machine models and presentation of algorithms

The model of sequential computations is a Random Access Machine (RAM). Evaluation of the time complexity of algorithms is done according to the uniform cost criterion, i.e., one time unit per basic operation. However, to avoid "abuses" of the model, the sizes of all integers occurring in algorithms will be "reasonable." The integers will be within a polynomial range of input size and can be represented by a *logarithmic number* of bits. We refer the reader to *The Design and Analysis of Computer Algorithms* by A.V. Aho, J.E. Hopcroft, and J.D. Ullman for more details on the RAM model.

The sequential algorithms in this book are mainly presented in Pascal programming language. A few more complex algorithms are written in an informal language, in a style and a semantic similar to that of Pascal. Sometimes, the initial algorithm for a given problem is usually more easily understood using a kind of a pseudo-language. In this language we can use a terminology appropriate to the natural model for the problem. The flow-of-control constructs can be the same as in Pascal, but with the advantage of having informal names for objects. For example, in algorithms

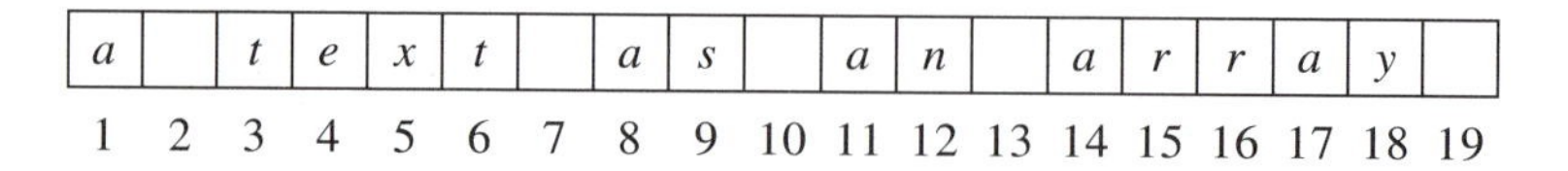

Figure 2.1: Array representation of a text.

on trees we can use notions such as "next son," or "father of the father," etc. In this situation, an informal algorithm is given to derive the main ideas of the construction. We generally attempt to convert such an informal algorithm into a Pascal program.

We consider texts as arrays of characters. We stick to the convention that indices of characters of a text run from 1 to the length of the text (see Figure 2.1). The symbols of a text called *text* of length n are stored in *text*[1], *text*[2], ..., *text*[n]. We may assume that location *text*[n+1] contains a special character.

Throughout this book, we consider two global variables *pat* and *text* that are declared by

```
var text: text_type; pat: pattern_type;
```

The types themselves can be defined by

```
type text_type = array[1..n] of char ;
     pattern_type = array[1..m] of char,
```

assuming that n and m are Pascal integer constants declared in a proper way.

We assume a lazy evaluation of Boolean expressions. For instance, the expression

```
(i>0) and (t[i]=c)
```

is false when i=0, even if t[0] does not exist. And the expression

```
(i=0) or (t[i]=c)
```

is true under the same conditions. Replacing all Boolean expression to avoid the convention is a straightforward exercise. Doing this in the presentation of algorithms would transform them into an unreadable form.

Pascal functions are presented in this book with the nonstandard (but common) "return" statement. Such instruction exists in almost all modern languages and aids in writing more structured programs. The meaning of the statement "**return**(*expression*)" is that the function returns the value of the expression to the calling procedure and then stops. The following function shows an application of the return statement.

```
function text_inequality(text1,text2 : text_type) : boolean;
begin
```

```
{ left-to-right scan of text1 and text2 }
    m := length(text1); n := length(text2);
    j := 0;
    while j <= min(m,n) do
    { the prefixes of size j are the same }
      if text1[j+1]=text2[j+1] then j := j+1
      else return(true);
    return(false);
end;
```

The "**return**" statement is very useful to shorten the presentation of algorithms. It is also easy to implement. It can be replaced by

```
text_inequality := expression; goto L
```

where L is the label of the last **end** of the function. The following description of the function *text_inequality* is equivalent to the above text.

```
function text_inequality(text1,text2 : text_type) : boolean;
label 1992;
begin
{ left-to-right scan of text1 and text2 }
    m := length(text1); n := length(text2);
    j := 0;
    while j <= min(m,n) do
    { the prefixes of size j are the same }
      if text1[j+1]=text2[j+1] then j := j+1
      else begin text_inequality:=true; goto 1992 end;
    text_inequality:= false;
1992 : end;
```

Several models of sequential computations more theoretical than the RAM are relevant to text processing: Turing machines, multihead finite automata, etc. For example, it is known that string matching can be done in real time on a Turing machine or even on a multihead finite automaton (without any memory but with a constant number of heads). Generally, we do not discuss algorithms on such highly theoretical models; in a few cases we give only some brief remarks related to the computations on these models.

Concerning parallel computations, a very general model is assumed, since we are interested mainly in exposing the parallel nature of some problems without going into the details of the parallel hardware. The parallel random access machine (PRAM), a parallel version of the random access machine, is used as a standard model for presentation of parallel algorithms (see Figure 2.2).

The PRAM consists of a number of processors working synchronously and communicating through a common random access memory. Each processor is a ran-

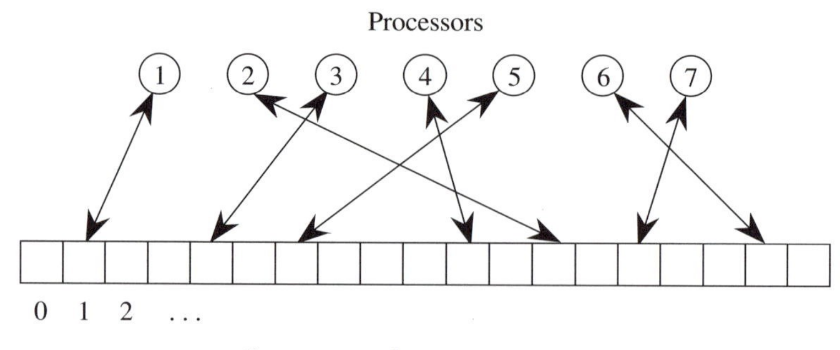

Figure 2.2: PRAM machine.

dom access machine with the usual operations. The processors are indexed by the consecutive natural numbers, and synchronously execute the same central program; however, the action of a given processor also depends on its number (known to the processor). In one step, a processor can access one memory location. The models differ with respect to simultaneous access of the same memory location by more than one processor. For the CREW (concurrent read, exclusive write) variety of PRAM machine (see Figure 2.3), any number of processors can read from the same memory location simultaneously, but write conflicts are not allowed: no two processors can attempt to write simultaneously into the same location.

CRCW (concurrent read, concurrent write) denotes the PRAM model in which, in addition to concurrent read, write conflicts are allowed: many processors can attempt to write into the same location simultaneously but only if they all attempt to write the same value (see Figure 2.4).

Parallel algorithms are presented in a similar way as in *Efficient parallel algorithms* by A. Gibbons and W. Rytter. There is no generally accepted universal language for the presentation of parallel algorithms. The PRAM is a rather idealized model. We have chosen this model as the best one suitable for the presentation of

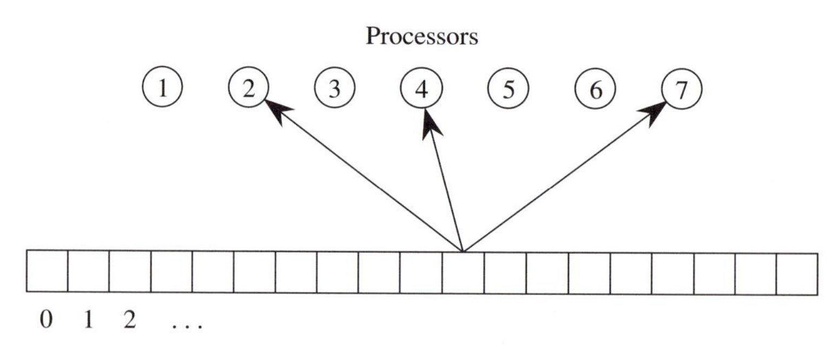

Figure 2.3: A concurrent read.

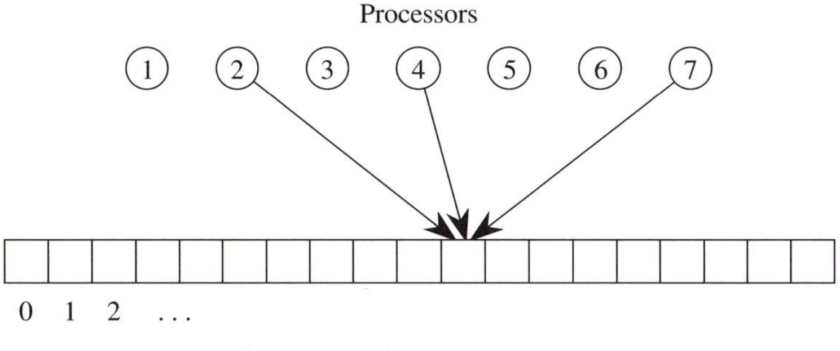

Figure 2.4: A concurrent write.

algorithms, and especially for the presentation of the inherent parallelism of some problems. It would be difficult to adequately present these algorithms with languages oriented toward concrete existing hardware of parallel computers. Moreover, the PRAM model is widely accepted in the literature on parallel computation on texts.

Parallelism will be expressed by the following type of parallel statement:

for all *i* in *X* in parallel do action(*i*).

Execution of this statement consists of

- assigning a processor to each element of X;
- executing in parallel by assigned processors the operations specified by action(i).

Usually the part "i in X" looks like "$1 \leq i \leq n$" if X is an interval of integers. We discuss this more in depth in Chapter 9.

2.2 Efficiency of algorithms

Efficient algorithms can be classified according to what is meant by efficiency. There exist different notions of efficiency depending on the complexity measure involved. Several such measures are discussed in this book: sequential time, memory space, parallel time, and number of processors. This book deals with "feasible" problems. We can define them as problems having efficient algorithms, or as solvable in time bounded by a small-degree polynomial. In the case of sequential computations we are interested in lowering the degree of the polynomial corresponding to time complexity. The most efficient algorithms usually solve a problem in linear time complexity. We are also interested in space complexity. Optimal space complexity often means a constant number of (small integer) registers in addition to input data. Therefore, we say that an algorithm is time-space optimal iff it works simultaneously in linear time and in constant extra space. These are the most advanced sequential algorithms, and also the most interesting, both from a practical and theoretical point of view.

In the case of parallel computations we are generally interested in the parallel time $T(n)$ as well as in the number of processors $P(n)$ required for the execution of the parallel algorithm. The total number of elementary operations performed by the parallel algorithm is not greater than the product $T(n)P(n)$. Efficient parallel algorithms are those that operate in no more than polylogarithmic (a polynomial of logs of input size) time with a polynomial number of processors. The class of problems solvable by such algorithms is denoted by NC and hence we call the related algorithms NC-algorithms. An NC-algorithm is optimal iff the total number of operations $T(n)P(n)$ is linear. Another possible definition is that this number is essentially the same as the time complexity of the best known sequential algorithm solving the given problem. However, we adopt the first option here because algorithms on strings usually have a time complexity which is at least linear.

Precisely evaluating the complexity of an algorithm according to some measure is often difficult, and, moreover, it is unlikely to be of much use. The "big O" notation clarifies what the important terms of a complexity expression are. It estimates the asymptotic order of the complexity of an algorithm and helps to compare algorithms to each other. Recall that if f and g are two functions from and to integers, then we say that $f = O(g)$ if $f(n) \leq C.g(n)$ when $n>N$, for some constants C and N. We write $f = \Theta(g)$ when the functions f and g are of the same order, which means that both equalities $f = O(g)$ and $g = O(f)$ hold.

Comparing functions through their asymptotic orders leads to these kinds of inequalities: $O(n^{0.7}) < O(n) < O(n.\log(n))$, or $O(n^{\log(n)}) < O(\log(n)^n) < O(n!)$.

Within sequential models of machines one can distinguish further types of computations: off-line, on-line and real-time. These computations are also related to efficiency. It is understood that real-time computations are more efficient than general on-line, and that on-line computations are more efficient than off-line. Each algorithm is an off-line algorithm: "off-line" conceptually means that the whole input data can be put into the memory before the actual computation starts. We are not interested then in the intermediate results computed by the algorithm, but only in the final result (though this final result can be a sequence or a vector). The time complexity is measured by the total time from the moment the computation starts (with all input data previously memorized) up to the final termination. In contrast, an on-line algorithm is like a sequential transducer. The portions of the input data are "swallowed" by the algorithm step after step, and after each step an intermediate result is expected (related to the input data read so far). It then reads the next portion of the input, and so on. In on-line algorithms the input can be treated as an infinite stream of data, consequently we are not interested mainly in the termination of the algorithm for all such data. The main interest for us is the total time $T(n)$ for which we have to wait to get the n-th first outputs. The time $T(n)$ is measured starting at the beginning of the whole computation (activation of the transducer). Suppose that the input data is a sequence and that after reading the n-th symbol we want to print "1" if the text read to this moment contains a given pattern as a suffix, otherwise we print "0". Hence we have two streams of data: the stream of input symbols and an output stream of answers "1" or "0". The main feature of the on-line algorithm is that we have to give it an output value before reading the next input symbol. The real-time computations are these on-line algorithms that are in a certain sense optimal; the elapsing time between reading two consecutive input symbols (the time spent for computing only the last output value) should be bounded by a constant. Most linear on-line algorithms are in fact real-time.

We are primarily interested in off-line computations in which the worst case time is linear, however, on-line and real-time computations, as well as average complexities will also be discussed briefly in this book.

2.3 Notation and formal definitions of basic problems on texts

Let A be an input *alphabet*—a finite set of symbols. Elements of A are called the *letters*, the *characters*, or the *symbols*. Typical examples of alphabets are: the set of all ordinary letters, or the set of binary digits. The texts (also called words or strings) over A are sequences of elements of A. The length (size) of a text is the number of its elements (with repetitions). Therefore, the length of *aba* is 3. The length of a word x is denoted by $|x|$. The input data for our problems will be words, and the size n of the input problem will usually be the length of the input word. In some situations, n will denote the maximum length or the total length of several words if the input of the problem consists of several words.

The i-th element of the word x is denoted by $x[i]$ and i is its position in x. We denote by $x[i...j]$ the subword $x[i]x[i+1]...x[j]$ of x. If $i>j$, by convention, the word $x[i...j]$ is the empty word (the sequence of length zero).

We say that the word x of length n is a *factor* (also called a subword) of the word y if $x=y[i+1...i+n]$ for some integer i. We also say that *x occurs in y at position i* or that the position i is a *match* for x in y.

We also define the notion of a *subsequence* (sometimes called a subword). The word x is a subsequence of y if x can be obtained from y by removing zero or more (not necessarily adjacent) letters from it. Likewise x is a subsequence of y if $x=y[i_1]y[i_2]...y[i_m]$, where $i_1,i_2, ...,i_m$ is an increasing sequence of indices in y.

Next we define basic problems and groups of problems covered in this book. We often consider two texts *pat* (the pattern) and *text* of respective lengths m and n.

1. String matching (the basic problem) Given texts *pat* and *text*, verify if *pat* occurs in *text*. Hence it is a decision problem: the output is a Boolean value. It is usually assumed that $m \leq n$. Therefore, the size of the problem is n. A slightly advanced version entails searching for all occurrences of *pat* in *text*, that is, computing the set of positions of *pat* in *text*. Denote by *MATCH(pat,text)* this set. In most cases an algorithm computing *MATCH(pat,text)* is a trivial modification of the decision algorithm, hence, we sometimes present only the decision algorithm for string matching.

Instead of just one pattern, one can consider a finite set of patterns and ask if a given text *text* contains a pattern from this set. The size of the problem is now the total length of all patterns plus the length of *text*.

2. Construction of string-matching automata For a given pattern *pat*, construct the minimal deterministic finite automaton accepting all words containing *pat* as a suffix. Denote such an automaton by *SMA*(*pat*). A more general problem is to construct a similar automaton *SMA*(Π) accepting the set of all words containing one of the patterns from the set $\Pi=\{pat_1, \ldots, pat_k\}$ as a suffix.

3. Approximate string matching We are again given a pattern *pat* and a text *text*. The problem consists of finding an approximate occurrence of *pat* in *text*, that is, a position i such that $dist(pat, text[i+1\ldots i+m])$ is minimal. Another instance of the problem is to find all positions i such that $dist(pat, text[i+1\ldots i+m])$ is less than a given constant. Here, *dist* is a function defining a distance between two words. There are two standard distances: the Hamming distance (number of positions at which two strings differ), and the edit distance (see Problem 7).

4. String matching with "don't care" symbols We are given a special universal symbol ∅ that matches any other symbol (including itself). It is called the "don't care symbol." In fact, ∅ can be a common name for several unimportant symbols of the alphabet. For two symbols a and b, we write $a \approx b$ iff they are equal or if at least one of them is a don't care symbol. For two words of the same length we write $x \approx y$ if, for each i, $x[i] \approx y[i]$.

The problem consists of verifying whether $pat \approx text[i+1\ldots i+m]$ for some i, or more generally, to compute the set of all such positions i.

5. Two-dimensional pattern matching For this problem, the pattern *pat* and the text *text* are two-dimensional arrays (or matrices) in which elements are symbols. We are to answer whether *pat* occurs in *text* as a subarray, or to find all of the positions of *pat* in *text* . The approximate and don't care versions of the problem can be defined similarly as for the (one-dimensional) string-matching problem.

Instead of arrays one can also consider some other "shapes," e.g., trees.

6. Longest common factor Compute the maximal length of longest common factors of two words x, y (denoted by $LCF(x, y)$). One can also consider longest common factors for more than two words.

7. Edit distance Compute the edit distance, $EDIT(x, y)$, between two words. It is the minimal number of edit operations needed to transform one string into the other. Edit operations are: insertion of one character, deletion of one character, and change of one character. The problem is closely related to the approximate string matching and to the problem defined below. It is similar to the notion of alignment of DNA sequences in Molecular Biology.

8. Longest common subsequence Compute the maximal length of common subsequences of two words x, y. The length is denoted by $LCS(x, y)$. Note that several subsequences may have this length and that we may also want to compute one or all of these words.

This problem is related to the edit distance problem as follows: if $l{=}LCS(x, y)$, then one can transform x to y by first deleting $m\text{-}l$ symbols of x (all but those of a longest common subsequence) and then inserting $n\text{-}l$ symbols to get y. Therefore, the computation of *LCS*s is a particular case of the computation of edit distances.

9. Longest repeated factor We are to compute a longest factor x of the text *text* that occurs at least twice in *text*. Its length is denoted by *LRF*(*text*). Similarly we can consider a longest factor that occurs at least k times in *text*. Its length is $LRF_k(text)$.

10. Finding squares and powers A much more difficult problem is that of computing the consecutive repetitions of a same factor. A square word (or simply a square) is a non-empty word of the form xx. Let us define the predicate *Square-free*(*text*): its value is "true" iff *text* does not contain any square word factor. Similarly one can define cubes and the predicate *Cube-free*(*text*). Generally one can define the predicate $Power\text{-}free_k(text)$: its value is "true" iff *text* does not contain any factor of the form x^k (with x non-empty).

These problems have seemingly no practical application. However, they are related to the most classical mathematical problems on words (the investigation of which started at the beginning of this century).

11. Computing symmetries in texts The word x is symmetrical if x is equal to its reversed x^R. A palindrome is any symmetrical word containing at least two letters (a one-letter word has an uninteresting obvious symmetry!). Even palindromes are palindromes of an even length. Denote by *Pal* and *P* the set of all palindromes and the set of even length palindromes, respectively.

There are several interesting problems related to symmetries. Compute *PREFPAL*(*text*), the length of the shortest prefix of *text* that is a palindrome. Another interesting problem is to find all factors of the text that are palindromes. Or, compute *MAXPAL*(*text*), the length of a longest symmetrical factor of *text*.

Yet another question is to count the number of all palindromes which are factors of *text*. A word is called a *palstar* if it is a composition of palindromes (it is contained in the language Pal^*); it is an *even palstar* if it is a composition of even palindromes. Two algorithmic problems related to palstars are: verify if a word is a palstar, or an even palstar.

Instead of asking about the composition of some number of palindromes we turn our interest to compositions of a fixed number k of palindromes. Let $P_k(text)$ be the predicate whose value is true if *text* is a composition of k even palindromes; analogously we can define $Pal_k(text)$ as a predicate related to compositions of any k palindromes (odd or even).

12. Number of all distinct factors The problem is to compute the number of all distinct factors of *text*. Observe that there is a quadratic number of factors, while linear time algorithm is possible. Such a linear time counting of a quadratic number of objects is possible due to succinct (linear-sized) representations of the set of all factors of a given text.

13. Maximal suffix Compute the lexicographically maximal suffix of the word *text* (it is denoted by *maxsuf*(*text*)). Observe that the lexicographically maximal suffix of a text is the same as its lexicographically maximal factor. The lexicographic order is the same order commonly used, for example, in encyclopedias or dictionaries (if we omit the question of accents, uppercase letters, hyphens, etc.).

14. Cyclic equivalence of words Check whether two words x and y are cyclically equivalent, i.e., whether $x=uv$, $y=vu$ for some words u, v. A straightforward linear time algorithm for this problem can be constructed by applying a string-matching procedure to the pattern $pat=x$ and the text $text=yy$. However, it is possible to design a much simpler, more efficient algorithm.

15. Lyndon factorization Compute the decomposition of the text into a non-increasing sequence of Lyndon words. Lyndon words are lexicographically (strictly) minimal within the set of all their cyclic shifts. It is known that any word can be uniquely factorized into a sequence of Lyndon words.

16. Compression of texts For a given word *text* we would like to have a most succinct representation of *text*. The solution depends on what we mean by a representation. If, by representation, we mean an encoding of single symbols by binary strings, then we have the problem of the efficient construction of Huffman codes and Huffman trees. Much more complicated problems arise if some parts of the text *text* are encoded using other parts of the text. This leads to some factorization and textual substitution problems. We devote an entire chapter to this subject.

17. Unique decipherability problem We have a code which is a function h assigning to each symbol a of the alphabet K a word over an alphabet A. The function can be extended in a natural way to all words over the alphabet K using the equation $h(xy)=h(x)h(y)$. We ask whether the so-extended function is one-to-one (denote by $UD(h)$ the corresponding predicate). The size n of the problem is the total length of all code words $h(a)$ over all symbols a in K.

The next three problems are related to the construction of a succinct (but useful) representation of sets of all factors or all suffixes of a given word *text*. Denote by *Fac*(*text*), *Suf*(*text*) the set of all respective factors, suffixes of *text*. There may be a quadratic number of elements to be represented but the size of the representation must be linear. Moreover, we also require that the construction time is linear.

18. Suffix trees The set *Fac*(*text*) is prefix closed. This means that prefixes of words in *Fac*(*text*) are also elements of the set. The natural representation for each prefix-closed set *F* of words is a tree *T1* whose edges are labeled with symbols of the alphabet. The set *F* equals the set of labels of all paths from the root down to nodes of the tree. However, such a tree *T1* corresponding to *Fac*(*text*) can be too large. It can have a quadratic number of internal nodes. Fortunately, the number of leaves is always linear and, hence, the number of internal nodes having at least two sons (out-degree bigger than one) is also linear. Therefore, we can construct a tree *T* whose nodes are all "essential" internal nodes and leaves of *T1*. The edges of *T* correspond to maximal paths (chains) in *T1* consisting only of nodes of out-degree one; the label of each such edge is the composite label of all edges of the corresponding chain. Denote the constructed tree *T* by *T*(*text*). It is called the suffix (or the factor) tree of *text*. The size of the suffix tree is linear and we also present a linear time algorithm to construct it.

The suffix tree is a very useful representation of all factors; the range of applications is essentially the same as that of the two following data structures: suffix and factor automata.

19. Smallest suffix automata and directed acyclic word graphs Compute the smallest deterministic finite automaton *DAWG*(*text*) accepting the set *Suf*(*text*) of all suffixes of *text*. The size of the automaton happens to be linear and we prove later that its size does not depend on the size of the alphabet (if transitions to dead state are not counted). The automaton *DAWG*(*text*) is essentially the same data structure as the so-called directed acyclic word graph (dawg, for short). Dawgs are directed graphs in which their edges are labeled by symbols, and such a graph represents the set of all words "spelled" by its paths from the root to the sink.

Dawgs are (in a certain sense) more convenient than suffix trees because each edge of the dawg is labeled by a single symbol, while edges of factor trees are labeled by words of varying lengths. However, it is possible to consider compacted versions of dawgs.

20. Smallest factor automata Construct the smallest deterministic automaton *FA*(*text*) accepting the set *Fac*(*text*) of all factors of *text*. This is approximately the same problem as the previous one. It is known that the size of *FA*(*text*) does not exceed the size of *DAWG*(*text*). In fact the suffix automaton also accepts (after a minor modification) the set *Fac*(*text*) of all factors. In most applications *DAWG*(*text*) can be used instead of *FA*(*text*), and the order of magnitude of the complexity does not change significantly. However, the factor automaton is the most optimal representation of the set of all subwords.

21. Equivalence of two words over a partially commutative alphabet Let C be a binary relation on the alphabet. This relation represents the commutativity between pair of symbols. The only necessary property of this relation is that it should be symmetric. Two words are equivalent according to C ($x \approx y$) iff one of them can be obtained from the other by commuting certain symbols in the word several times. Formally, $ubav \approx uabv$ if $(a, b) \in C$ (symbols a, b commute). The relation $\approx$ is transitive because we can apply operations of commuting two symbols many times. One can reformulate many of the problems defined earlier in the framework of partially commutative alphabets. Instead of one word, one can also consider the whole equivalence class containing this word (called a *trace*). We can ask if such an equivalence class for a given word contains a square or a palindrome or a given factor (the latter is the string-matching problem for partially commutative alphabets).

However, we consider only the complexity of one basic problem: checking equivalence of two strings. It is one of the simplest problems in this area encompassing many sophisticated questions.

22. Two-dpdas The acronym "2dpda" is an abbreviation for "two-way deterministic pushdown automaton." This is essentially the same device as a pushdown automaton described in many standard textbooks on automata and formal languages; the only difference is its ability to move input heads in two directions. We will demonstrate that any language accepted by such an abstract machine has a linear time recognition procedure on random access machines, which is of algorithmic interest. The notion of 2dpdas is historically interesting. Indeed, one of the first approaches to obtaining a linear time algorithm for string matching used 2dpdas.

The list below shows some relations between chapters and problems. The first introductory chapters are naturally related to all problems.

Chapter 3: Basic string-matching algorithms—1, 2, 5, 10, 11, 13

Chapter 4: The Boyer-Moore algorithm and its variations—1, 5

Chapter 5: Suffix trees—1, 2, 6, 9, 10, 12, 13, 17, 18

Chapter 6: Subword graphs—1, 2, 6, 9, 10, 12, 13, 17, 19, 20

Chapter 7: Automata-theoretic approach—1, 2, 3, 5, 6, 9, 10, 12, 13, 19, 20, 22

Chapter 8: Regularities in words: symmetries and repetitions—9, 10, 11

Chapter 9: Almost optimal parallel algorithms—1, 2, 9, 10, 11, 15, 18, 19, 20

Chapter 10: Text compression techniques—10, 16, 19

Chapter 11: Approximate pattern matching—3, 4, 7, 8, 17

Chapter 12: Two-dimensional pattern matching—5

Chapter 13: Time-space optimal string matching—1, 10, 13

Chapter 14: Time-processor optimal string matching—1

Chapter 15: Miscellaneous—1, 5, 13, 14, 15, 17, 19, 21

This book is about efficient algorithms for textual problems. By efficient, we mean algorithms working in polynomial sequential time (usually linear time or $n.\log(n)$ time), or in polylogarithmic parallel time with a polynomial number of processors (usually linear). However, we have to warn the reader that many textual problems have no efficient algorithms (unless the main question in complexity theory "is P=NP?" holds true, which seems unlikely). We refer the reader to [GJ 79] for the definition and discussion on NP-completeness. Roughly speaking, an NP-complete problem is a problem that has (with high probability) no efficient exact algorithm.

Below are some examples of NP-complete problems on texts. Usually, NP-complete problems are stated as decision problems, but we present some problems in their more natural version where outputs are not necessarily Boolean. Essentially, this does not affect the hardness of the problem.

Shortest common superstring

The general problem is to reconstruct a text from a set of its factors. There is, of course, no unique solution, so the problem is usually constrained to a minimal length condition. Given a finite set S of words, find a shortest word x that contains all words of S as factors, i.e., such that S is included in *Fac*(*text*). The length of text is denoted *SCS*(S) (see [MS 77]). The size of the problem is the total length of all words in S.

Shortest common supersequence

Given a finite set S of words, find a shortest word x such that every word in S is a subsequence of x (see [Ma 78]). The word x is called a supersequence of S.

Longest common subsequence

Given a finite set S of words, find a longest word x that is a subsequence of every word in S (see [Ma 78]). It is worth observing that the problem is solvable in polynomial time if $|S|$ is a constant. The seemingly similar problem of computing the longest common factor of words of S is also easily solvable in polynomial time (number of distinct factors in quadratic, as opposed to the exponential, number of distinct subsequences).

String-to-string correction

Given words x and y, and an integer k, is it possible to transform x into y by a sequence of at most k operations of single symbol deletion or adjacent-symbol interchange? (See [Wa 75]). The problem becomes polynomial-time solvable if we also allow the operations of changing a single symbol and inserting a symbol (see [WF 74]).

The rank of a finite set of words

Given a finite set X of words and an integer k, does there exist a set Y of at most k words such that each word in X is a composition of words in Y, see [Ne 88]? The smallest k for which Y exists is called the rank of X.

There are several other NP-complete textual problems: hitting string [Fa 74], grouping by swapping [Ho 77], and problems concerning data compression [St 77], [SS 78]. We refer the reader to *Computers and Intractability: A Guide to the Theory of NP-Completeness* by M.R. Garey and D.S. Johnson.

2.4 Combinatorics of texts

As already seen in the previous section, a text x is simply a sequence of letters. The first natural operation on sequences is concatenation. The concatenation of x and y, also called their product, is denoted by a mere juxtaposition: xy, or sometimes with an extra dot, $x.y$, to make the decomposition of the resulting word apparent. This is coherent with the notation of sequences as juxtaposition of their elements. The concatenation of k copies of the same word x is denoted by x^k.

There is no need to bracket a sequence representing a text because concatenation is associative: $(xy)z = x(yz)$. The empty sequence (of zero length) is denoted by ε. The set of all sequences on the alphabet A is denoted by A^*, and the set of non-empty words is denoted by A^+. In the algebraic terminology, sequences are called words, and concatenation provides to A^* a structure of monoid (associativity and ε). Moreover, this monoid is free, which essentially means that two words are equal iff their letters coincide at the same positions.

Consider a word x that decomposes into uvw (for three words u, v and w). The words u, v and w are called *factors* of x. Moreover, u is called a *prefix* of x, and w a *suffix* of x. The occurrence of factor v in x can be formally considered as the triplet (u, v, w). The position of this occurrence is the length of u denoted by $|u|$. This notion is intrinsic; it does not depend on the representation of the word x. However, it coincides with our convention of representing texts by arrays, and with the definition of positions already given in the previous section.

We define the notion of a period of a word, which is central in almost all string-matching algorithms. A *period* of a word x is an integer p, $0<p \leq |x|$, such that

$\forall i \in \{1, \ldots, |x|-p\}\ x[i] = x[i+p]$.

This is the usual definition of a period for a function defined on integers, as x can be viewed. Note that the length of a word is always a period of it, so that any word has at least one period. We denote by *period*(x) the smallest period of x.

Example The periods of *aabaaabaa* (of length 9) are 4, 7, 8 and 9. ♦

Proposition 2.1 Let x be a non-empty word and p be an integer such that $0<p \leq |x|$. Then each of the following conditions equally defines p as a period of x:

1. x is a factor of some y^k with $|y| = p$ and $k>0$,
2. x may be written $(uv)^k u$ with $|uv| = p$, v non-empty and $k>0$,
3. for some words y, z and w, $x = yw = wz$ and $|y| = |z| = p$.

Proof Condition 1 is equivalent to saying that p is a period of x if one considers $y = x[1 \ldots p]$. 1 implies 2: if x is a factor of y^k it can be written ry^is with r a suffix of y and s a prefix of y. Let t be such that $y = tr$. Then $x = r(tr)^is = (rt)^irs$. Since t and s are both prefixes of y, one of them is a prefix of the other. If s is a proper prefix of t, then $t = sv$ for some non-empty word v and, setting $u = rs$, $x = (rsv)^irs = (uv)^iu$. Furthermore $|uv| = |rsv| = |rt| = |tr| = |y| = p$. Also note that $i>0$ because $|rs|<|y|=p \leq |x|$. In the other situation, t is a prefix of s and the word u such that $s = tu$ is also a prefix of r. Then, for some word w, $r = uw$ and $x = (uwt)^{i+1}u$ or $x = (uv)^{i+1}u$, if we set $v = wt$. Moreover, $|uv| = |uwt| = |rt| = |tr| = |y| = p$ and $i+1>0$.

Condition 2 implies 3: if $x = (uv)^k u$ with $|uv| = p$, let y, z and w be defined by

$$y = uv,\ z = vu,\ w = (uv)^{k-1}u.$$

The definition of w is valid because $k>0$. The conclusion follows: $x = yw = wz$. 3 implies 1: first note that, if $yw = wz$, we also have $y^iw = wz^i$ for any integer i. Since y is non-empty ($|y| = p$ and $p>0$), we can choose an integer k such that $|y^k|>|x|$. The equation $y^kw = wz^k$ shows that x $(=wz)$ is a factor and even a prefix of y^k with $|y| = p$. The proof is complete. ♦

A *border* of a word x is any word that is both a prefix and a suffix of x. Borders and periods are dual notions, as shown by condition 3 of the previous proposition (see Figure 2.5). We can state it more precisely, defining *Border*(x) as the longest proper border of a non-empty word x. The border of a word x is the longest (nontrivial) overlap when we try to match x with itself. Since *Border*(x) is (strictly) shorter than x, iterating the function from any non-empty word eventually leads to the empty word. We also say that x is *border-free* if *Border*(x) is the empty word.

Example The borders of *aabaaabaa* are *aabaa*, *aa*, *a* and ε. ♦

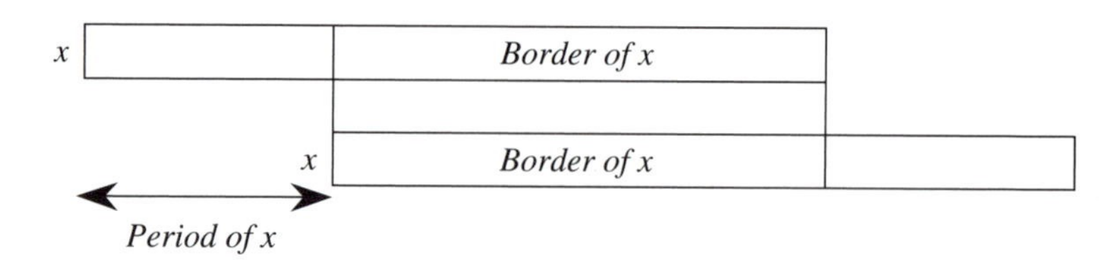

Figure 2.5: Duality between periods and borders of texts.

Proposition 2.2 Let x be a word and let k (≥ 0) be the smallest integer such that $Border^k(x)$ is empty. Then

$$(x, Border(x), Border^2(x), \ldots, Border^k(x))$$

is the sequence of all borders of x in order of decreasing length. Moreover

$$(|x|-|Border(x)|, |x|-|Border^2(x)|, \ldots, |x|-|Border^k(x)|)$$

is the sequence of all periods of x in increasing order.

Proof Both properties hold if x is empty. In this case, the sequence of borders reduces to (x) and the sequence of periods is empty since $k = 0$. Consider a non-empty word x and let $u = Border(x)$. Let w be a border of x. Then it is either $Border(x)$ itself or a border of $Border(x)$. By induction it belongs to the sequence

$$(Border(x), Border^2(x), \ldots, Border^k(x))$$

and this proves the first point.

If p is a period of the non-empty word x, by condition 3 of Proposition 2.1 we have $x = yw = wz$ with $|y|=p$. Then w is a proper border of x and $p = |x|-|w|$. The conclusion follows from the fact that $(Border(x), Border^2(x), \ldots, Border^k(x))$ is the sequence of proper borders of x. ♦

Note that the period $period(x)$ (the smallest period of x) corresponds to the longest proper border of x and is precisely $|x|-|Border(x)|$.

The next theorem provides an important property on periods of a word. It is often used in combinatorial proofs on words. We first give a weak version of the theorem which is usually sufficient for most applications.

Lemma 2.3 (Weak Periodicity Lemma) Let p and q be two periods of a word x. If $p+q \leq |x|$, then $\gcd(p, q)$ is also a period of x.

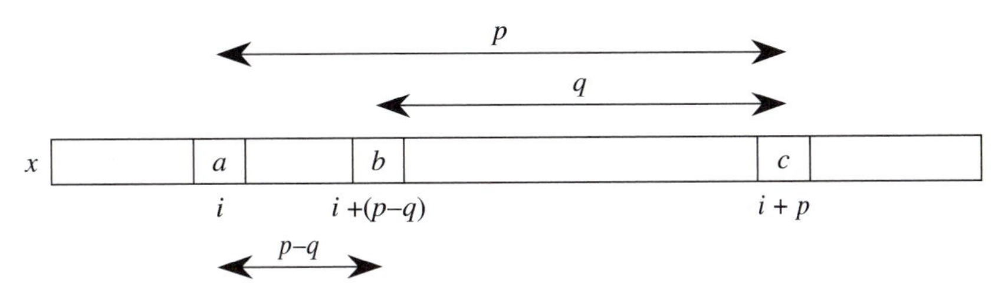

Figure 2.6: $p-q$ is also a period: letters a and b are both equal to letter c.

Proof The conclusion trivially holds if $p = q$. Assume now that $p > q$. First we show that the condition $p+q \leq |x|$ *implies that* $p-q$ is a period of x. Let $x = x[1]x[2]\ldots x[n]$ ($x[i]$'s are letters). Given $x[i]$ the i-th letter of x, the condition implies that either $i-q \geq 1$ or $i+p \leq n$. In the first case, q and p being periods of x, $x[i] = x[i-q] = x[i-q+p]$. In the second case, for the same reason, $x[i] = x[i+p] = x[i+p-q]$. Thus $p-q$ is a period of x. This situation is shown in Figure 2.6. The rest of the proof, left to the reader, is by induction on the integer $\max(p, q)$, after noting that $\gcd(p, q)$ equals $\gcd(p-q, q)$. ♦

Lemma 2.4 (Periodicity Lemma) Let p and q be two periods of a word x. If $p+q-\gcd(p, q) \leq |x|$, then $\gcd(p, q)$ is also a period of x.

Proof As for the weak periodicity lemma we prove the conclusion by induction on $\max(p, q)$. The conclusion trivially holds if $p = q$, therefore, we assume now that $p > q$ and let $d = \gcd(p, q)$. The word x can be written yv with $|y| = p$ and v a border of x. It can also be written zw with $|z| = q$ and w a border of x.

First we show that p-q is a period of w. Since $p > q$, z is a prefix of y. Let z' be such that $y = zz'$. From $x = zw = yv$, we get $w = z'v$. But since v is a border of x shorter than the border w, it is also a border of w. This proves that p-$q = |z'|$ is a period of w, and thus w is a prefix of a power of z'. To prove that q is a period of w, we have only to show that $q \leq |w|$. Indeed, since $d \leq p-q$ (because $p > q$), we have $q \leq p-d = p+q-d-q \leq |x|-q = |w|$.

This also shows that $(p-q)+q-\gcd((p-q), q)$ (which is $p-d$) $\leq |w|$. By induction hypothesis, we deduce that d is a period of w. Thus z and z', whose lengths are multiples of d, are powers of a same word u of length d. The word w is a prefix of a power of u. Thus, the word x has period $\gcd(p, q)$ as expected. ♦

Fibonacci words

In one sense, the inequality that appears in the statement of the periodicity lemma is optimal. The aim of the following example is to exhibit a word x having two periods p and q satisfying $p+q-\gcd(p, q) = |x|+1$ and that does not satisfy the conclusion of the lemma. Fibonacci words are defined on the alphabet $A = \{a, b\}$. Let Fib_n be the n-th Fibonacci word ($n \geq 0$). It is defined by

$$Fib_0 = \varepsilon,\ Fib_1 = b,\ Fib_2 = a, \text{ and } Fib_n = Fib_{n-1}Fib_{n-2} \text{ for } n>2.$$

The lengths of Fibonacci words are the well-known Fibonacci numbers, $f_0 = 0, f_1 = 1, f_2 = 1, \ldots$ The first Fibonacci words of $(Fib_n, n>2)$ are

$Fib_3 = ab,$ $|Fib_3| = 2,$
$Fib_4 = aba,$ $|Fib_4| = 3,$
$Fib_5 = abaab,$ $|Fib_5| = 5,$
$Fib_6 = abaababa,$ $|Fib_6| = 8,$
$Fib_7 = abaababaabaab,$ $|Fib_7| = 13,$
$Fib_8 = abaababaabaababaababa,$ $|Fib_8| = 21,$
$Fib_9 = abaababaabaababaababaabaababaabaab,$ $|Fib_9| = 34.$

Fibonacci words satisfy a large number of interesting properties related to periods and repetitions. Note that Fibonacci words (except the two first words of the sequence) are prefixes of their successors. Indeed, there is an even stronger property: the square of any Fibonacci word is a prefix of its succeeding Fibonacci words of high enough rank.

To prove the extreme property of Fibonacci words related to the Periodicity Lemma, we consider, for $n>2$, the prefix of Fib_n of length $|Fib_n|-2$, denoted g_n. It can be shown that, for $n>5$, g_n is a prefix of both Fib_{n-1}^2 and Fib_{n-2}^3. Then $|Fib_{n-1}|$ and $|Fib_{n-2}|$ are periods of g_n. It can also be shown that $\gcd(|Fib_{n-1}|, |Fib_{n-2}|) = 1$, a basic property of Fibonacci numbers. Thus, we get $|Fib_{n-1}|+|Fib_{n-2}|-\gcd(|Fib_{n-1}|, |Fib_{n-2}|) = |Fib_{n-1}|+|Fib_{n-2}|-1 = |Fib_n|-1 = |g_n|+1$. Moreover, the conclusion of the periodicity theorem does not hold on g_n because this would imply that g_n was a power of a single letter which is obviously false.

Among other properties of Fibonacci words, it must be noted that they have no factor of the form u^4 (u not empty). Therefore, Fibonacci words contain a large number of periodicities, but none with an exponent higher than 3.

Fibonacci words come up in the analysis of some algorithms. A common expression of their length is used: $|Fib_n|$ is $F^n/\sqrt{5}$ rounded to the nearest integer. It follows that logs are based Φ, the golden ratio ($=(1+\sqrt{5})/2$).

Selected references

[AHO 74] Aho, A.V., Hopcroft, J. E., & Ullman, J. D., *The Design and Analysis of Computer Algorithms*, Addison-Wesley, Reading, Mass., 1974.

[CLR 89] Cormen, T.H., Leirserson, C.E., & Rivest, R.L., *Introduction to Algorithms*, The MIT Press, Cambridge, Mass., 1989.

[GJ 79] Garey, M.R., & Johnson, D.S., *Computers and Intractability: A Guide to the Theory of NP-Completeness*, Freeman, New York, 1979.

[GR 88] Gibbons, A., & Rytter, W., *Efficient Parallel Algorithms*, Cambridge University Press, Cambridge, England, 1988.

[GB 91] Gonnet, G.H., & Baeza-Yates, R., *Handbook of Algorithms and Data Structures*, Addison-Wesley, Reading, Mass., 1991, second edition.

[Lo 83] Lothaire, M., *Combinatorics on Words*, Addison-Wesley, Reading, Mass., 1983.

[Se 88] Sedgewick, R., *Algorithms*, Addison-Wesley, Reading, Mass., 1988, second edition.

Basic string-matching algorithms

The string-matching problem is the most studied problem in algorithmics on words, and there are many algorithms for solving this problem efficiently. Recall that we assumed that the pattern *pat* is of length m, and that the text *text* has length n. If the only access to text *text* and pattern *pat* is by comparisons of symbols, then the lower bound on the maximum number of comparisons required by a string-matching algorithm is $n-m$. The best algorithms presented in this book will make no more than $2.n-m$ comparisons in the worst case. Recently, the lower bound has been improved, to approximately $4/3.n$ on a two-letter alphabet. However, algorithms can become quite sophisticated and as such are beyond the scope of this book.

We begin with a brute-force algorithm that uses quadratic time. Such a naive algorithm is, in fact, an origin of a series of more and more complicated yet more efficient algorithms. The informal scheme of such a naive algorithm is:

for *i* := 0 **to** *n-m* **do** check if *pat*=*text* [*i*+1...*i*+*m*].

The actual implementation differs with respect to how we implement the checking operation: scanning the pattern from the left or scanning the pattern from the right. We then get two brute-force algorithms. Both algorithms have quadratic worst-case complexity. In this chapter we discuss the first of these (left-to-right scanning of the pattern).

To shorten the presentation of some algorithms we assume that *pat* and *text,* together with their respective lengths, m and n, are global variables.

```
function brute_force1 : boolean;
    var i, j : integer;
begin
    i := 0;
    while i ≤ n-m do begin
        { left-to-right scan of pat }
        j := 0;
        while j<m and pat [j+1]=text [i+j+1] do j := j+1;
        if j=m then return(true);
        { inv1(i, j) }
        i := i+1; { length of shift = 1 }
    end;
    return(false);
end;
```

If $pat=a^{n/2}b$ and $text=a^{n-1}b$ then a quadratic number of symbol comparisons takes place. However, the average complexity is not so bad. Assume that the alphabet has two symbols and that each symbol appears with the same probability. Then the probability that the test "$pat[j+1]=text[i+j+1]$" is successful is at most 1/2. It is now quite easy to calculate that the average number of such successful tests for a fixed i does not exceed 1. The number of unsuccessful tests does not exceed n. Hence, we have the following:

Fact Assume $|A|=2$. The expected number of all symbol comparisons done by algorithm *brute_force1* does not exceed $2.n$.

In the following, we are primarily interested in worst-case time complexity, and particularly in linear-time algorithms.

3.1 The Knuth-Morris-Pratt algorithm

Our first linear-time algorithm is a natural, improved version of the naive algorithm discussed earlier. We present a constructive proof of the following.

Theorem 3.1 The string-matching problem can be solved in $O(|text|+|pat|)$ time using $O(|pat|)$ space. The constants involved in "O" notation are independent of the size of the alphabet.

Remark We are disappointed with this theorem on one point. The size of additional memory is rather large though linear in the size of the pattern. In Chapter 13, we show that a constant number of registers suffices to achieve linear-time complexity (again the size of the alphabet does not intervene).

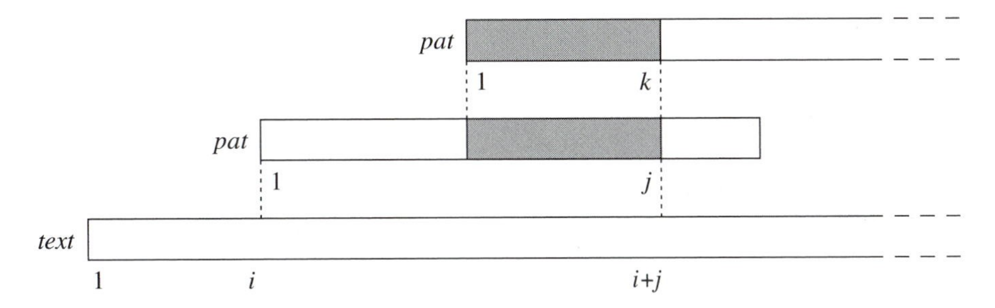

Figure 3.1: Shifting the pattern to the next safe position.

Let us look more closely at the algorithm *brute_force1* and at its main invariant *inv*1(*i*, *j*): *pat*[1...*j*]=*text*[*i*+1...*i*+*j*] and *pat*[*j*+1]...*text*[*i*+*j*+1]. In fact, we first use the slightly weaker invariant

*inv*1'(*i*, *j*) : *pat*[1...*j*]=*text*[*i*+1...*i*+*j*].

The invariant essentially says that the value of *j* gives us a lot of information about the last part of the text scanned up to this point.

Using the invariant *inv*1'(*i*, *j*), we are able to make longer shifts of the pattern. Present shifts in algorithm *brute_force1* always have a length of 1. Let *s* denote (the length of) a "safe" shift, where "safe shift *s*" means that based on the invariant we know that there is no occurrence of the pattern at positions between *i* and *i*+*s*, but there may be one at position *i*+*s*.

Assume *j*>0, let *k*=*j*-*s*, and suppose an occurrence of the pattern starts at position *i*+*s*. Then, *pat*[1...*k*] and *pat*[1...*j*] are suffixes of the same text *text*[1...*i*+*j*] (see Figure 3.1). Hence, the following condition is implied by *inv*1':

cond(*j*, *k*) : *pat*[1...*k*] is a proper suffix of *pat*[1...*j*].

Therefore the shift is safe iff *k* is the smallest number satisfying *cond*(*j*, *k*). Denote this number by *Bord*[*j*]. The function *Bord* is called a failure function because it helps us at the time of a failure (mismatch). It is the crucial function. It is stored in a table with the same name. The failure function allows us to compute the length of the smallest safe shift, *s*=*j*-*Bord*[*j*]. Note that *Bord*[*j*] is precisely the length of the border of *pat*[1...*j*], and that the length of the smallest safe shift is the smallest period of *pat*[1...*j*] (see Section 2.4). In the case *j*=0, that is, when *pat*[1...*j*] is the empty word, we have a special situation. Since, in this situation, the length of the shift must be 1, we then define *Bord*[0]=–1. Now we have an improved version of algorithm *brute_force1*.

```
function MP : boolean;   { algorithm of Morris and Pratt }
    var : i, j integer;
```

```
begin
    i := 0; j := 0;
    while i ≤ n-m do begin
        while j<m and pat [j+1]=text [i+j+1] do j = j+1;
        if j=m then return(true);
        i := i+j-Bord [j]; j := max(0, Bord [j]);
    end;
    return(false);
end;
```

Lemma 3.2 The time complexity of the algorithm MP is linear in the length of the text. The maximal number of character comparisons executed is $2.n-m$.

Proof Let $T(n)$ be the maximal number of symbol comparisons "$pat[j+1]=text[i+j+1]$?" executed by the algorithm MP. There are at most n-m+1 unsuccessful comparisons (at most one for any given i). Consider the sum $i+j$. Its maximal value is n and minimal value is 0. Each time a successful comparison is made the value of $i+j$ increases by one unit. This value, observed at the time of comparing symbols, never decreases. Hence, there are at most n successful comparisons. Also observe that if the first comparison is successful then we have no unsuccessful comparison for position i=0. Finally, we conclude:

$$T(n) \leq n+n-m=2.n-m.$$

For $pat=ab$ and $text=aaaa\ldots a$ we have $T(n)=2.n-m$ (in this case m=2). ♦

```
procedure compute_borders_1;
{ compute the failure table Bord for pattern pat }
{ a version of algorithm MP with text=pat }
    var i, j : integer;
begin
    i := 1; j := 0;
    while i ≤ m–1 do begin
        while i+j<m and pat [j+1]=pat [i+j+1] do begin
            j := j+1; if Bord [i+j]=–1 then Bord [i+j] := j;
        end;
        i := i+j-Bord [j]; j := max(0, Bord [j]);
    end;
end;
```

The problem of computing the table *Bord* remains, and our aim is to derive a linear-time algorithm. We present two solutions. The first is to use algorithm MP to compute *Bord*. This, at first glance, can appear contradictory because *Bord* is needed inside the algorithm. However, we compute *Bord* in parts; whenever a value *Bord*[j] is needed in the computation of *Bord*[i] for i>j then *Bord*[j] is already computed. This is a kind of dynamic programming.

Suppose that *text* =*pat*. We apply algorithm MP starting with i=1 (for i=0 nothing interesting happens) and continue with i=2, 3, …., m–1. Then *Bord*[r]=j>0 whenever i+j=r in a successful comparison for the first time. If *Bord*[r]>0 then such a comparison will take place. Assume that initially *Bord*[j]=–1 for all j≥0.

Of course, we can delete the statement "**if** j=m **then return**(true);" because we are not solving the string-matching problem at this point. In fact, our interest here is only a side effect of the algorithm MP.

The complexity of algorithm *compute_borders_1* is linear. The argument is the same as for algorithm MP. Next, we present another linear-time algorithm computing table *Bord*.

```
procedure compute_borders;
{ compute the failure table Bord for pat, second version }
    var t, j : integer;
begin
    Bord [0] := -1; t := -1;
    for j := 1 to m do begin
        while t≥0 and pat [t+1]≠pat [j] do t := Bord [t];
        t := t+1; Bord [j] := t;
    end;
end;
```

The correctness of the algorithm *compute_borders* essentially derives from Proposition 2.2, or from the fact that if *Bord*[j]>0 then *Bord*[j]=t+1, where t=$Bord^k$[j] and k is the smallest positive integer such that *pat*[$Bord^k$[j]+1]=*pat*[j].

Lemma 3.3 The maximum number of character comparisons executed by algorithm *compute_borders* is 2.m–3.

Proof The complexity can be analyzed using a so-called "store principle." Interpret t as the number of items in a store. Note that when t<0, no comparison is done, and t becomes null. Therefore, we can consider that the store is initially empty. For each j running from 2 to m, we add at most one item (at statement t := t+1). However, whenever we execute the statement "t := *Bord*[t]," the value of t strictly decreases, which can be interpreted as deleting a nonzero number of items from the store. The total number of items inserted does not exceed m–2. Hence, the total number of deletions and executions of statement "t := *Bord*[t]" for unsuccessful comparisons "*pat*[t+1]*pat*[j]" does not exceed m–2.

For each j running from 2 to m, there is at most one successful comparison. The total number of successful comparisons then does not exceed m–1. Hence, the total number of comparisons does not exceed 2.m–3. ♦

The notion of failure function can be applied in a different way to the string-matching problem, as follows. Let w = *pat*&*text*, where & is a special new symbol. Compute the table *Bord* for this string. Then the pattern *pat* ends in *text* at all positions j with $Bord[j]=m$, where m is the size of *pat*. Such an abstract algorithm was given in [FP 74].

We show another application of the failure function computation through algorithm *compute_borders*, the computation of *Maxsuf*(*pat*), lexicographically maximal suffix of *pat*. When computing the value of $Bord[j]$ we examine consecutive values of the integer variable t. These values correspond to all consecutive proper suffixes $pat[1...t]$ of the word $pat[1...j-1]$, which are also prefixes of this word. Such examination could be useful in the computation of maximal suffixes because of the following fact:

If $pat[q+1...j-1]$ is the maximal suffix of $pat[1...j-1]$, and $pat[t+1...j]$ is the maximal suffix of $pat[1...j]$, then $pat[t+1...j-1]$ is a prefix and suffix of $pat[q+1...j-1]$. Hence, to find the next maximum suffix it is adequate to examine extensions of suffixes of the last maximal suffix (such suffixes can be presented using failure table *Bord*). However, the following unexpected feature of such an approach is that we have to compute borders of $pat[q+1...j-1]$, the current maximal suffix, and simultaneously its failure function. Therefore we have to compute the table *Bord* for the dynamically changing pattern $pat[q+1...j-1]$. The trick is that the failure table we need is essentially a prefix of the previous one.

```
function Maxsuf(pat): integer;
{ computes Maxsuf(pat) }
{ application of failure function computation }
    var ms, j, t: integer;
begin
    ms := 0; Bord [0] := -1;
    for j := 1 to m do begin
        t := Bord [j-1];
        while t≥0 and pat [ms+t+1]>pat [j] do begin
            ms := j-1-t; t := Bord [t];
        end;
        while t≥0 and pat [ms+t+1]>pat [j] do t := Bord [t];
        Bord [j-sm] := t+1;
    end;
    return(ms); { Maxsuf(pat) is pat [ms+1...m] }
end;
}
```

Denote $x(j)=Maxsuf(pat[1...j])$. If we have computed table *Bord* for $x(j-1)$, then $x(j)=x'pat[j]$, where x' is a prefix of $x(j-1)$. Hence, *Bord* is already computed for all positions corresponding to symbols of x'; only one entry of *Bord* should be updated (for the last position). We modify the algorithm for the failure function maintaining the following invariant:

$$pat[ms+1...j]=Maxsuf(pat[1...j]),$$

and the values of table *Bord* are correct values of the failure function for the word $pat[ms+1...j]$.

In algorithm *Maxsuf*, parameter t runs through lengths of suffixes of $pat[ms+1...j]$. The algorithm above computes *Maxsuf(pat)*, as well as the failure function for it, in linear time. Computation also requires linear extra space for the table *Bord*. In Chapter 13 we present a linear-time constant-space algorithm for the computation of maximal suffixes. It will be also seen why maximal suffixes are so interesting.

We have not yet taken into account the full invariant *inv*1 of algorithm *brute_force1*, but only its weaker version *inv*1'. We have left the mismatch property apart. We now develop a new version of algorithm MP that incorporates the mismatch property. The resulting algorithm, called KMP, improves the number of comparisons performed on a given letter of the text.

The clue for improvement is the following: assume that a mismatch in algorithm MP occurs on the letter $pat[j+1]$ of the pattern. The next comparison is between the same letter of the text and $pat[k+1]$ if $k=Bord[j]$. But if $pat[k+1]= pat[j+1]$, the same mismatch recurs. Therefore, we must avoid considering the border of $pat[1...j]$ of length k in this situation.

For $m>j\geq 0$ we consider a condition stronger than *cond*(j, k) by "one-comparison" information:

$$s\_cond(j, k) : (pat[1...k] \text{ is a proper suffix of } pat[1...j] \text{ and } pat[k+1] \neq pat[j+1]).$$

We then define $s\_Bord[j]$ as k, when k is the smallest integer satisfying $s\_cond(j, k)$, and as -1 otherwise. Moreover, we define $s\_Bord[m]$ as $Bord[m]$. We say that $s\_Bord[j]$ is the length of the longest *strict border* of $pat[1...j]$. This notion of strict-borderness is not intrinsic, but is defined only for prefixes of the pattern. Figure 3.2 illustrates the difference between functions *Bord* and *s_Bord* on pattern *abaab*.

The algorithm KMP is the algorithm MP in which table *Bord* is replaced by table *s_Bord*.

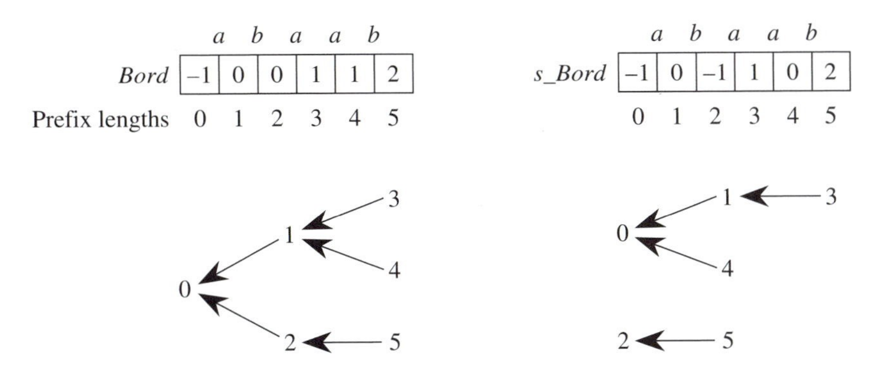

Figure 3.2: Functions *Bord* and *s_Bord* for pattern *abaab*.

```
function KMP : boolean; { algorithm of Knuth, Morris, and Pratt }
{ version of MP with table Bord replaced by s_Bord }
    var i, j : integer;
begin
    i := 0; j := 0;
    while i ≤ n-m do begin
        while j<m and pat [j+1]=text [i] do j := j+1;
        if j=m then return(true);
        i := i+j-s_Bord [j]; j := max(0, s_Bord [j]);
    end;
    return(false);
end;
```

The table *s_Bord* is more effective in the following on-line version of algorithm KMP. Assume that the text ends with the special end marker $. Each time we process the current input symbol, we then output 1 if the part of the text read so far ends with the pattern *pat*; otherwise we output 0.

```
Algorithm KMP;
{ on-line linear version of KMP search }
    var j : integer; symbol : char;
begin
    read(symbol); j := 0;
    while symbol ≠ end of text do begin
        while j<m and pat [j+1]=symbol do begin
            j := j+1; if j=m then write(1) else write(0);
            read(symbol);
        end;
        if s_Bord [j]=-1 then begin
            write(0); read(symbol); j := 0;
        end else
            j := s_Bord [j];
    end;
end;
```

Computation of strict borders of prefixes of the pattern *pat* relies on the following observation. Let $t=Bord[j]$. Then, $s\_Bord[j]=t$ if $pat[t+1]\neq pat[j+1]$. Otherwise, the value of $s\_Bord[j]$ is the same as the value of $s\_Bord[t]$ because $pat[t+1]= pat[j+1]$. The next algorithm applies this fact and is built on algorithm *compute_borders*. Its output is the global table *s_Bord*. Also note that the strict border of *pat* itself is its border, as if *pat* were followed by a marker.

Example Consider the pattern aba^{m-2}. Strict borders computed by the procedure *compute_s_borders* are given by the following table *s_Bord*:

$s\_Bord\,[0] = -1$, $s\_Bord\,[1] = 0$, $s\_Bord\,[2] = -1$, and $s\_Bord\,[j] = 1$,
for $3 \leq j \leq m$.

This is a worst case for which exactly $3m–5$ symbol comparisons are performed by the algorithm. ♦

```
procedure compute_s_borders;
{ compute table s_Bord for pattern pat }
    var t, j : integer;
begin
    s_Bord [0] := -1; t := -1;
    for j := 1 to m do begin
        { t equals Bord [j-1] }
        while t≥0 and pat [t+1]≠pat [j] do t := s_Bord [t];
        t := t+1;
        if j = m or pat [t+1]≠pat [j+1] then s_Bord [j] := t
        else s_Bord [j] := s_Bord [t];
    end;
end;
```

Denote by *delay*(m) the maximal time (measured as the number of statements "j :=*s_Bord*[j]") elapsed between two consecutive reads, for patterns of length m. By *delay_Bord*(m), we denote the corresponding time when *Bord* is used instead of *s_Bord*.

The large gap between *delay*(m) and *delay_Bord*(m) can be seen in the following example: *pat*=*aaaa…a* and *text*=$a^{m-1}ba$. In this case *delay*(m)=1 while *delay_Bord*(m) is linear in the length of *pat*. The value of *delay*(m) is generally small.

Lemma 3.4 The time *delay*(m) is $O(\log m)$, and the bound is tight.

Proof The proof of Lemma 3.4 is a consequence of the Periodicity Lemma of Section 2.4. The pattern g_k defined in Section 2.4 as a prefix of the k-th Fibonacci word (for $k \geq 3$) yields a sequence of exactly $k–3$ strict borders. The delay is then $k–3$, which is exactly of order $\log(m)$, if m=length(g_k). ♦

Observe that for texts on binary alphabets *delay*(m)≤1. This means that in this case, the algorithm is real-time. However, if the patterns are over the alphabet $\{a, b\}$ and texts over $\{a, b, c\}$, then the delays can be logarithmic, as shown by the above example.

It is an interesting exercise in program transformations to modify algorithm KMP to achieve a real-time computation independently of the size of the alphabet. This means that the time between two consecutive reads must be bounded by a constant. The crucial observation is that if we execute "j :=*s_Bord*[j]," then we know that for the next j–*s_Bord*[j] input symbols the output value will be 0 ("no match"). This allows for dispersal of output actions between input actions (reading symbol) in such a way that the time between consecutive writes-reads is bounded by a constant.

To do so, we can maintain up to m last symbols of the text *text* in a table. We leave details to the reader. It is interesting to observe that real-time condition can be achieved by using any of the tables *Bord*, *s_Bord*.

In this way, we sketched the proof of the following result (which becomes much more difficult if the model of the computation is a Turing machine).

Theorem 3.5 There is a real-time algorithm for string matching on a random access machine. The algorithm is a version of the KMP algorithm and uses $O(m)$ space. Complexities do not depend on the size of the alphabet.

3.2 The Simon algorithm

The Simon algorithm is a further improvement on algorithm *brute_force1*. The difference between this and previous algorithms is due to the way shifts are done. Recall that the main invariant of *brute_force1* is

$$inv1(i, j) : pat[1 \ldots j]=text[i+1 \ldots i+j] \text{ and } pat[j+1] \neq text[i+j+1],$$

occurring when a shift is to be done. In MP algorithm, the length of the shift is the period of $pat[1 \ldots j]$ corresponding to the border of this word. In KMP algorithm, a different notion of period is considered that may be perceived as compatible with the letter $text[i+j+1]$ that yields the mismatch. This corresponds to the notion of strict borders.

According to the information gathered so far on the text, the best length of the shift would be the period of $pat[1 \ldots j]text[i+j+1]$. Pre-computing all periods of ua (u prefix of pat, a possible letter of the text) takes time and space $O(|A|.|pat|)$, which depends on the size of the alphabet. This kind of solution is not desirable, especially for short patterns.

This solution is equivalent to considering the minimal deterministic automaton recognizing the language A^*x. Indeed, algorithms MP and KMP implicitly use a representation of this automaton. We also consider another implementation here that leads to a faster algorithm. The relationship of all these algorithms to automata is given in Chapter 7.

The idea exploited in the Simon algorithm is that we can only record nontrivial periods of words $pat[1 \ldots j]text[i+j+1]$, namely, those that are less $j+1$. This leads to consideration of what we call *tagged borders* of prefixes of the pattern *pat*. For each letter a of the alphabet A, distinct from letter $pat[j+1]$ if $j<m$, the border of $pat[1 \ldots j]$ tagged by a is its largest border $pat[1 \ldots k]$ such that $pat[k+1]=a$.

Implementation of the Simon algorithm requires lists of tagged borders (see Figure 3.3). They are organized as follows. A table *First* of size m gives pointers to another table, *t_Bord* of size $2.m$, that contains the lengths of tagged borders. Lists of tagged borders end with –1, and are in order of decreasing lengths.

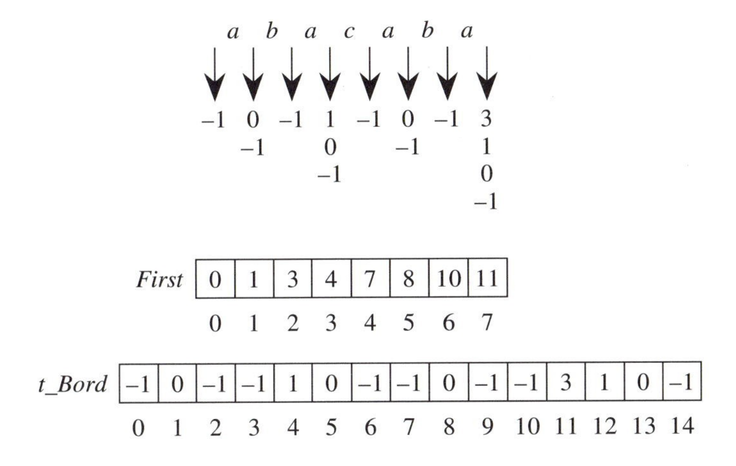

Figure 3.3: Lists of tagged borders for *pat=abacaba*, and their implementation.

Algorithm MPS below assumes that the lists associated with all prefixes of the pattern have been pre-computed. Figure 3.4 illustrates the difference between the three algorithms MP, KMP, and MPS.

```
function MPS : boolean;
{ algorithm of Simon, improvement on KMP }
   var i, j, p, k : integer;
begin
   i := 0; j := 0;
   while i ≤ n-m do begin
       while j<m and pat[j+1]=text[i+j+1] do j := j+1;
       if j=m return(true);
       p := First[j]; k := t_Bord[p];
       while k>-1 and pat[k+1]≠text[i+j+1] do begin
            p := p+1; k := t_Bord[p];
       end;
       i := i+j-k; j := k+1;
   end;
   return(false);
end;
```

Theorem 3.6 Algorithm MPS works in $O(n)$ time for a text of length n. The maximum number of character comparisons it executes is less than $2.n$. The delay between two inspections of a character of the text is $O(|A|)$, where A is the alphabet of the pattern.

Proof It is quite obvious that all comparisons executed by algorithm MPS are also executed during a run of algorithms KMP or MP. This yields less than $2.n$ compari-

```
text    .......abaabac..............
               abaabaa
                  abaabaa
                    abaabaa
                     abaabaa
                      abaabaa
```
(i) MP. 4 comparisons on letter c.

```
text    .......abaabac..............
               abaabaa
                    abaabaa
                     abaabaa
                      abaabaa
```
(ii) KMP. 3 comparisons on letter c.

```
text    .......abaabac..............
               abaabaa
                    abaabaa
                      abaabaa
```
(iii) MPS. 2 comparisons on letter c.

Figure 3.4: Behavior of MP, KMP, and MPS.

sons. All extra operations, including the work on lists of tagged borders, is proportional to this number. Hence, we get $O(n)$ time. Since the number of tagged borders of a given prefix of *pat* is at most $|A|$, no more than $|A|$ letter comparisons are executed on a given symbol of the text. This yields the delay $O(|A|)$. ♦

On a fixed alphabet (of patterns), algorithm MPS runs in real time. If the alphabet is potentially infinite, the algorithm can be transformed into a real-time algorithm using a method similar to the one used for algorithms MP and KMP (see Section 3.1).

Preprocessing the lists of tagged borders can be accomplished with a modified version of algorithm MPS. Computation is executed in (almost) on-line fashion: tagged borders related to a prefix u of *pat* are computed just after all tagged borders of its prefixes have been computed. We briefly describe how to compute tagged borders for u. Assume that $u=pat[1...j]$ and that $Border(u)=pat[1...k]$ (MPS is used to find it.) If $j<m$ and $pat[j+1]\neq pat[k+1]$, $pat[1...k]$ is the longest border tagged by $pat[k+1]$. Otherwise, $pat[1...k]$ is not a tagged border. All other tagged borders are those of $pat[1...k]$ which are not tagged by $pat[j+1]$. Thus, their computation is mainly a copy of list. The details are left to the reader.

The fact that preprocessing of lists can be realized in linear time and space is contingent upon the following property of tagged borders.

Proposition 3.7 For a pattern of length m, the number of tagged borders of its prefixes is at most m. The bound is tight.

Proof Let $pat[1...k]$ be a border of $pat[1...j]$ tagged by a. Associate with it the length (shift) $j-k$. Then, no other tagged border can be associated with the same value. Therefore, since $0<j-k \leq m$, the number of tagged borders is at most m. This number is exactly m for words of the form ab^{m-1}. ♦

3.3 String matching by duels

In this section we present a nonclassical string-matching algorithm in which the preprocessing phase is closely related to borders and to KMP algorithm. We also introduce an interesting new operation called the *duel*. A more essential use of this operation will be seen in the chapter on optimal parallel string matching and two-dimensional pattern matching. Hence, this section can be treated as a preparation for more advanced algorithms to be presented later.

We assume in this section that the pattern is non-periodic, which means that the smallest period of the pattern *pat* is larger than $|pat|/2$. This assumption implies that two consecutive occurrences of the pattern in the text (if any) are at distance greater than $|pat|/2$. However, it is not clear how to use this property for searching the pattern. We process as follows: after a suitable preprocessing phase, given two close positions in the text, we eliminate one of them as a candidate for a match. This leads to the idea of a duel.

The basic table which enables us to search for the pattern created by a duel-based algorithm can be computed either as a side effect of the Knuth-Morris-Pratt algorithm, or by use of the table *Bord*. Duels are performed at search phase. We define the following *witness table WIT* : for $0<i<|m|$,

$WIT[i]$ = **any** k such that $pat[i+k] \neq pat[k]$,
$WIT[i] = 0$, if there is no such k.

This definition is illustrated in Figure 3.5.

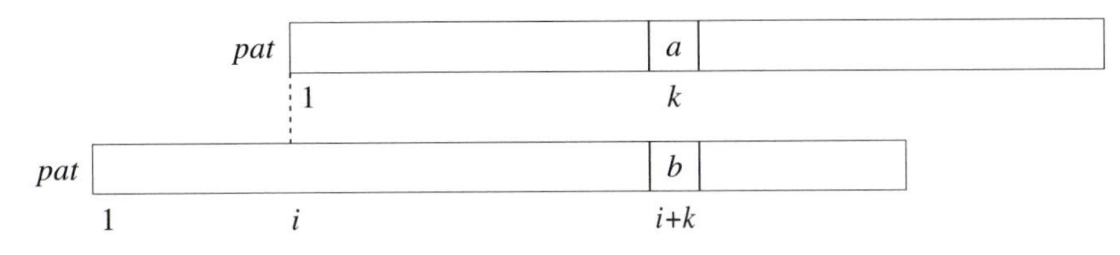

Figure 3.5: Witness of mismatch ($a \neq b$), $k = WIT[i]$.

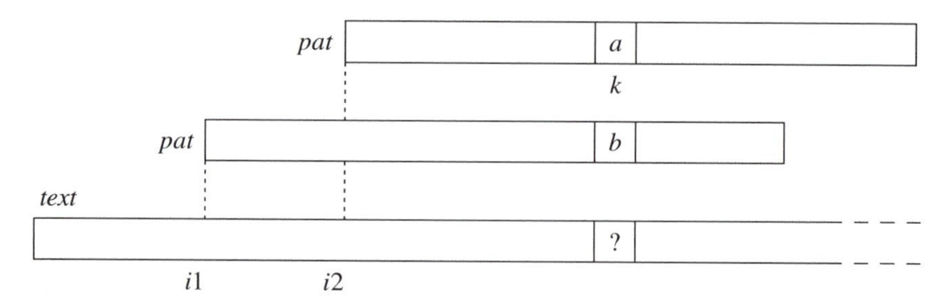

Figure 3.6: Duel between two inconsistent positions *i1* and *i2*. One of them is eliminated by comparing symbol "?" in the text with *a* and *b*.

A position *i1* is said to be *in the range of* a position *i2* iff $|i1-i2|<m$. We say that two positions $i1<i2$ in the text are *consistent* iff *i2* is not in the range of *i1*, or if $WIT[i2-i1]=0$. If the positions are not consistent, then we can remove one of them as a candidate for the starting position of the pattern just by considering position $i2+WIT[i2-i1]$ in the text. This is the operation called a duel (see Figure 3.6). Let $i = i2-i1$, and $k=WIT[i]$. Assume that we have $k>0$, that is, positions *i1*, *i2* are not consistent. Let $a=pat[k]$ and $b=pat[i+k]$, then $a \neq b$. Let *c* be the symbol in the text at position $i2+k$, that is indicated by "?" in Figure 3.6. We can eliminate at least one of the positions *i1*, *i2* as a candidate for a match by comparing *c* with *a* and *b*. In some situations, both positions could be eliminated. For simplicity, however, the algorithm below always removes exactly one position. Let us define (recall that $a=pat[WIT[i2-i1]]$):

duel(*i1*, *i2*) = (**if** $a=c$ **then** *i2* **else** *i1*).

The position that "survives" is the value of *duel*, the other position is eliminated.

Assume the witness table is computed. It is then possible to reduce the search for *pat* in *text* to the search for pattern 11...111 (repetition of *m* 1s) in a text of 0s and 1s. This last problem is obviously simpler than the general string-matching problem and can be solved in linear time (with essentially one counter).

The following property of consistent positions (transitivity) is crucial for the correctness of the algorithm. It is called the *consistency property*:

let $i1 < i2 < i3$,
if *i1*, *i2* are consistent and *i2*, *i3* are consistent then also *i1*, *i3* are consistent.

Using this property we are able to eliminate a set of candidate positions from the text in such a way that all remaining positions are pairwise consistent. This can be done using the mechanism of stack (pushdown-store). Assume we have a stack of positions satisfying the property: positions are pairwise consistent, and in increasing order (from the top of the stack). Then, if we push a position both smaller than the top position and consistent with it onto the stack, the stack retains the property.

A set of consistent positions is *complete* if an occurrence of the pattern cannot start at any other position (position not in this set). We say that a position x in the text agrees with a candidate position y, iff the symbol at position x agrees with the corresponding symbol of the pattern when it is placed at position y (that is, $text[x]=pat[x-y]$). Assume that S is a complete set of consistent positions, x is any position in the text, and y is any candidate position in S such that x is in the range of y with $x>y$. Then, as a consequence of the consistency property, we have the following:

x agrees with y iff x agrees with all positions in S.

Hence, it is sufficient to check the agreement of each position with any position from a set of consistent positions. In this checking, we flag x with the value 0 or 1 depending on the agreement. Using this feature, the string matching easily reduces to the matching of unary patterns (patterns consisting entirely of ones).

The duel-based algorithm uses an additional zero-one vector, called *text*1. The value of the vector *text*1 computed by the algorithm satisfies:

pattern 1^m occurs at position i in *text*1 iff the original pattern occurs at i in the text.

```
function String_searching_by_duels: boolean;
{ Let S be a stack of positions }
begin
   S := empty stack;
   for i := n downto 1 do begin
         push i on the stack S;
         while |S|≥2 and
                        the two top elements i1, i2 of S are inconsistent do
              replace them in S by the single element duel(i1, i2);
   end;
   mark in the text all position that are in S;
   { all marked positions are pairwise consistent }
   for i := 1 to n do begin
         k := first marked position to the left of i, including i;
         if k undefined or pat [i-k+1]≠text [i] then text1 [i]:=0
         else text1 [i]:=1;
   end;
   if text1 contains the pattern 1^m return(true)
   else return(false);
end;
```

The algorithm obviously has linear-time complexity. Moreover, this complexity does not depend on the size of the alphabet. The basic component that remains to be shown is the computation of the witness table *WIT* on the pattern.

Later we shall see, in the case of parallel computations, that the fact that *any* position k for a witness is possible has great importance. This is sufficient and makes it easier to compute in parallel. However, in the case of sequential computations, we can take the smallest such position as a witness. This leaves us to define *WIT*[*i*]=*PREF*[*i*], for each position *i*. Section 4.6 contains both the definition of *PREF*, and a linear-time algorithm to compute it. The complexity of this latter algorithm is independent of the size of the alphabet.

Theorem 3.8 String matching by duels takes linear time (search phase and preprocessing phase).

We believe that matching by duels is one of the basic algorithms, since the idea of duels is the key to the optimal parallel string matching of Vishkin presented later in Chapter 14. Historically, however, the first optimal (for fixed alphabets) parallel string-matching algorithm uses the idea of a sieve. This algorithm implicitly uses another type of duels that we call *expensive duels*. Its advantage is that no additional table, like the one of witnesses, is needed. Its drawback is that the resulting algorithm is not optimal. We only show the use of expensive duels for a special type of pattern. Assume that the size of the pattern is a power of two. Assume also that *pat*$[1...2^k]$ is non-periodic, for each $k \geq 1$. Here such patterns are called *special* patterns. An example of a special pattern is "wojtekrytter."

```
function expensive_duel(i, j) : integer;
begin
    k := log2(j-i)+1;
    if text[i+1..i+2^k]=pat[1..2^k] then kill j else kill i;
    return the survival;
end;
```

The defined function *expensive_duel* is similar to the duel function, however its computation is much more expensive. This is the reason for the name "expensive duel." In the algorithm, we partition the text into disjoint blocks of size 2^k. We call them *k*-blocks.

```
algorithm String_searching_by_expensive_duels;
{ assume n-m+1 and m are powers of two }
{ assume that the pattern is special }
begin
    initially all positions in [1..n-m] are survivals;
    for k := 1 to logn do begin
        let i and j be the only survivals in the k-blocks;
        make expensive_duel(i, j) a survival;
    end;
    { there are O(n/m) survivals }
    for each survival position i do
        check occurrence of pat at position i naively;
end;
```

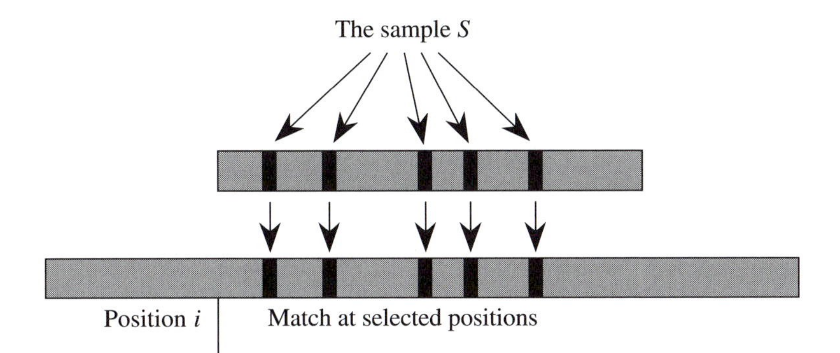

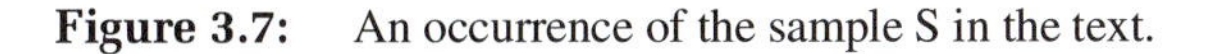

Figure 3.7: An occurrence of the sample S in the text.

Let us observe the very parallel nature of this algorithm at a given stage k, the actions on all k-blocks can be performed simultaneously.

Theorem 3.9 Assume that all prefixes $pat[1..2^k]$ are non-periodic. Then, the algorithm *String_searching_by_expensive_duels* takes $O(n \log m)$ time.

Proof At stage k we consider only $O(n/2^k)$ survivals. Each expensive duel at this stage takes $O(2^k)$ time. There are $m/2$ stages. Together this gives $O(n \log m)$ time. This completes the proof. ♦

The expensive duels are restricted in use. Historically, however, they appeared before the concept of the duel appeared. This is the only reason it is reported here.

3.4 String matching by sampling

Both Knuth-Morris-Pratt algorithm and Boyer-Moore algorithm (in the next chapter) scan symbols at consecutive positions on the pattern. The first one scans a prefix of the pattern, and the second one scans a suffix of the pattern. In this section, we show an algorithm that first scans a sequence of not necessarily consecutive positions on the pattern, and then, in case of success, completes the scan of the pattern. The first scanning sequence is called a **sample**.

A sample S for the pattern *pat* is a set of positions on *pat*. A sample S occurs at position i in the text iff $pat[j]=text[i+j]$ for each j in S (see Figure 3.7).

A sample S is called a **good sample** iff it satisfies the two conditions (see Figure 3.8):

1. S is small : $|S| = O(\log m)$;
2. there is an integer k such that if S occurs at position i in the text then no occurrence of the pattern starts in the segment $[i\text{-}k..i+m/2-k]$ (this segment is called the *desert*), except perhaps at i.

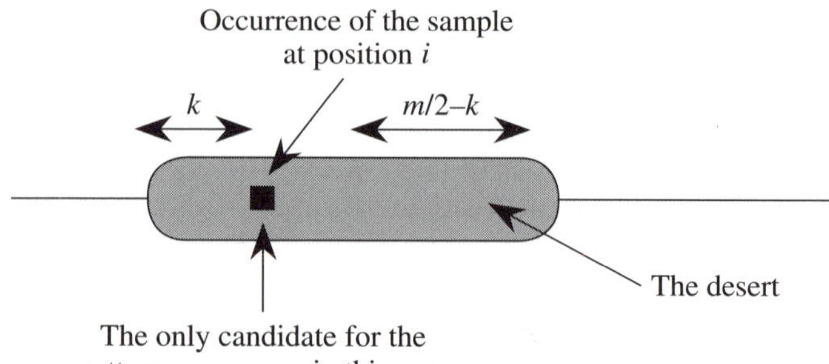

Figure 3.8: An occurrence of the sample, and the desert area.

If the pattern has period p, then $pat[1..k]$ is called the non-periodic part of the pattern, where $k=\min(m, 2.p-1)$.

Theorem 3.10 Assume we are given the period of the pattern *pat*, and a good sample of its non-periodic part, then, the search for *pat* can be done in $O(n \log n)$ time with only $O(\log m)$ additional memory space.

Proof Assume for a moment that the pattern itself is non-periodic. Let us partition the input text into *windows* of size $m/2$. We consider each window separately, and find the first and the last occurrences of the sample in the window (if there are at least two occurrences). These occurrences only are possible candidates for an occurrence of the pattern. Each of these occurrences is checked in a naive way (constant-size additional memory is sufficient for that). This proves that the non-periodic part of the pattern can be found in the text with the required complexity. The general case of periodic patterns is left as an exercise. One has to find sufficiently many consecutive occurrences of the non-periodic part of the pattern. An additional counter is needed to remember the number of consecutive occurrences. This completes the proof. ♦

Theorem 3.11 If the pattern is non-periodic then it has a good sample S. The sample can be constructed in linear time.

Proof Assume we have computed the witness table *WIT* (see Section 3.3). Let us consider potential occurrences of the pattern at positions 1, 2, ..., $m/2$ of some imaginary text. Let us identify these pattern occurrences with numbers 1, 2, ..., $m/2$. The occurrence corresponding to the i-th position is called the i-th row. If we draw a vertical line at a position j, then it can intersect a given i-th row or not. If it intersects, then there is a symbol at the point of intersection (see Figure 3.9). Let us denote this symbol by $symbol(i, j)$.

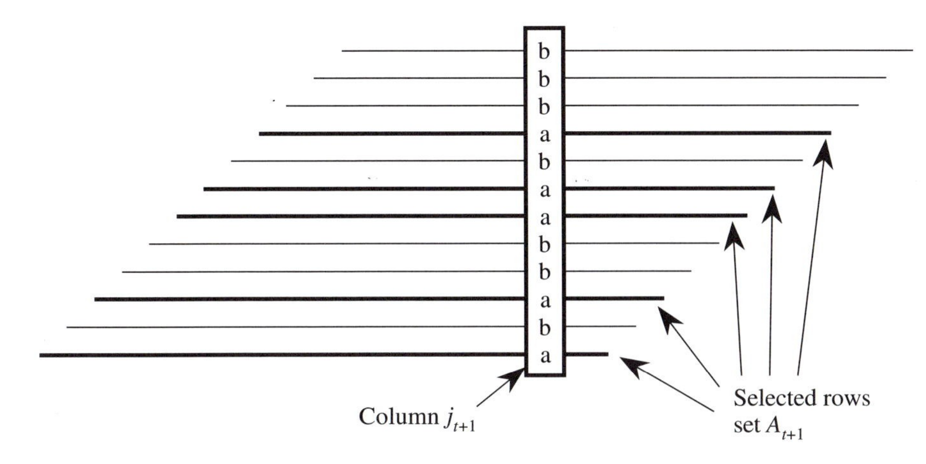

Figure 3.9: The set $A_{t+1} = \{ i \in A_t : symbol(i, j_{t+1})=a\}$ is the smaller one.

Claim 1 Let $i1$, $i2$ be two different elements of $[1...m/2]$. Then, there is an integer j such that the j-th column intersects both rows $i1$ and $i2$; moreover, $symbol(i1, j) \neq symbol(i2, j)$. The integer j can be found in constant time if the witness table of the pattern is precomputed.

The claim is a reformulation of the property of non-periodicity. Due to non-periodicity, for occurrences of the pattern placed at positions $i1$, $i2$ there is a mismatch position j given by the $j = i2+WIT[i2-i1]$. This means that if we look at the vertical column placed at position j, then this column intersects occurrences of the pattern with different symbols.

Claim 2 Let J be any set of rows. If a vertical column intersects the first and the last row of J, then it intersects all the rows of J.

We now prove an equivalent geometrical formulation of the thesis of the theorem.

Claim 3 There is a row i and a set J of $O(\log m)$ vertical columns placed at positions $j_1, j_2, ..., j_k$ such that:

1. all columns in J intersect the row i;
2. if $r \neq i$ (r in $[1..m/2]$), then there is a column j in J intersecting rows i and r such that $symbol(i, j) \neq symbol(r, j)$.

Proof (of the claim) We construct the set J and the row i by the algorithm below, which ends the proof of the theorem. ♦

```
Algorithm Find_good_sample;
begin
   J := empty set; A0 := [1..m/2]; t := 0;
   while |At| > 1 do begin
         find any column jt+1 which intersects all rows of At
         with two different symbols at intersection points;
         { use Claim 1 and Claim 2 }
         Let a, b be the two different symbols at intersections;
         At+1 := smaller of the two sets  {i ∈ At / symbol(i,jt+1)=a}
                                       and {i ∈ At / symbol(i,jt+1)=b};
         add jt+1 to J; t := t+1;
   end;
   let i be the unique element of At;
   return(J, i);
end;
```

Bibliographic notes

The first lower bound on the maximal number of symbol comparisons for string matching has been given by Rivest [Ri 77]. Recent improvements on this bound are by Colussi, Galil, and Giancarlo [CGG 90], and by Cole and Hariharan [CH 92].

MP algorithm is from Morris and Pratt [MP 70]. The fundamental algorithm considered in this chapter (KMP algorithm) has been designed by Knuth, Morris, and Pratt [KMP 77]. Our exposition is slightly different than in this paper.

That the computation of borders is equivalent to string matching has been noticed in [FP 74]. An application of the notion of failure table to the computation of maximal suffixes can be found in [Bo 80]. A more efficient algorithm computing maximal suffixes based on the ideas of Duval [Du 83] is found in Chapter 13 (see also Chapter 15).

Simon's Algorithm is reported from the author. It has recently been proven by Hancart [Ha 93] that Simon's algorithm can be transformed into a searching algorithm reaching the optimal bound of $(2-1/m).n$ comparisons among algorithms using a 1-letter window on the text. This bound has been found independently by Breslauer, Colussi, and Toniolo for a variant of string matching called the string prefix-matching problem [BCT 93].

A criterion that says whether an on-line algorithm can be transformed into a real-time algorithm has been shown by Galil in [Ga 81]. The principle applies to MP, KMP, and MPS algorithms.

The idea of duels can be attributed to Vishkin who applied it to the design of parallel algorithms [Vi 85] (see Chapter 14). Expensive duels are implicitly considered by Galil in [Ga 85]. The notion of sample is from Vishkin [Vi 90].

Selected references

[Ga 81] Galil, Z., "String matching in real time," *J. ACM* 28 (1981) 134–149.

[KMP 77] Knuth, D.E., Morris, Jr. J.H., & Pratt, V.R., "Fast pattern matching in strings," *SIAM J.Comput.* 6 (1977) 323–350.

4

The Boyer-Moore algorithm and its variations

In this chapter we describe another basic approach to string matching. It is introduced as an improvement on the second version of the naive algorithm of Chapter 3. We can get another naive string-matching algorithm, similar to *brute_force1* of the previous chapter, if the scan of the pattern is done right to left. This algorithm has quadratic worst-case behavior, but (similarly to algorithm *brute_force1*) its average-time complexity is linear.

In this section we discuss a derivative of this algorithm—the Boyer-Moore string-matching algorithm. The main feature of this algorithm is that it is efficient in the sense of worst-case (for most variants) as well as average-case complexity. For most texts and patterns the algorithm scans only a small part of the text because it performs "jumps" on the text. The algorithm applies to texts and patterns that reside in main memory.

```
function brute_force2 : boolean;
    var i, j : integer;
begin
    i := 0;
    while i ≤ n-m do begin
        { right-to-left scan of pat }
        j := m;
        while j>0 and pat [j]=text [i+j] do j := j-1;
        if j=0 then return(true);
```

```
        { inv2(i, j) }
        i := i+1; { length of shift = 1 }
    end;
    return(false);
end;
```

4.1 The Boyer-Moore algorithm

Algorithm BM can be viewed as an improvement on the initial naive algorithm *brute_force2*. We try again, as we did in the previous chapter for algorithm *brute_force1*, to analyze this algorithm, and to look carefully at information the algorithm wastes. This information is related to the invariant in the algorithm *brute_force2*

$inv2$: $pat[j+1\ldots m]=text[i+j+1\ldots i+m]$ and $pat[j]\neq text[i+j]$.

Denote by $inv2'$ the weaker invariant: $pat[j+1\ldots m]=text[i+j+1\in i+m]$. The information gathered by the algorithm is "stored" in the value of j. However, at the next iteration, this information is erased as j is set to a fixed value. Suppose that we want to make bigger shifts using the invariant.

The shift s is said to be *safe* iff we are certain that between i and $i+s$ there is no starting position for the pattern in the text. Suppose that the pattern appears at position $i+s$ (see Figure 4.1, where the case $s<j$ is presented). Then, the following conditions hold:

$cond1(j, s)$: for each k s.t. $j<k\leq m$, $s\geq k$ or $pat[k-s]=pat[k]$,
$cond2(j, s)$: if $s<j$ then $pat[j-s]\neq pat[j]$.

We define two kinds of shifts, each associated with a suffix of the pattern represented by position j ($<m$), and defined by its length:

$D1[j] = \min\{s>0: cond1(j, s) \text{ holds}\}$,
$D[j] = \min\{s>0: cond1(j, s) \text{ and } cond2(j, s) \text{ hold}\}$.

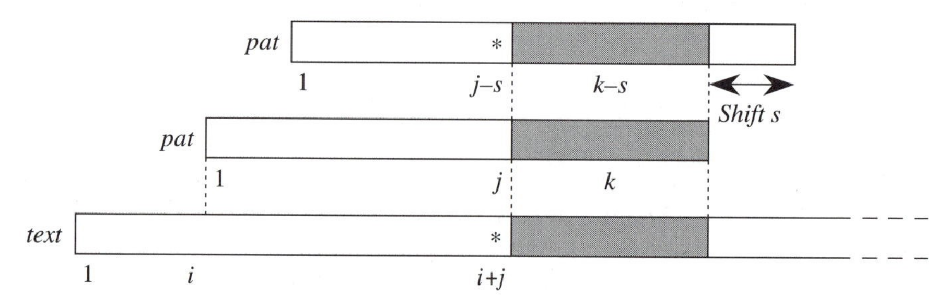

Figure 4.1: The case $s<j$.

```
                                cababababa
                              cababababa
                            cababababa
                          cababababa                          cabababa
                        cababababa                          cababababa
text = ............aaaaaaaabababababa ...... . aaaaaaaabababababa...
```

Figure 4.2: BM with $D1$-shifts makes 30 comparisons (left); BM with D-shifts makes only 12 comparisons (right).

We also define $D1[m]=D[m]=m-Bord[m]$. The BM algorithm is a version of *brute_force2* in which, in a mismatch situation, a shift of length $D[j]$ is executed instead of one position shift.

```
function BM: boolean;
{ improved version of brute_force2 }
begin
    i := 0;
    while i ≤ n-m do begin
        j := m;
        while j>0 and pat[j]=text[i+j] do j := j-1;
        if j=0 then return(true);
        { inv2(i, j) }
        i := i+D[j];
    end;
    return(false);
end;
```

Let us compare a run of this algorithm with a run of the similar algorithm which uses $D1$ instead of D. The history of the computations for *pat=cababababa* and *text=aaaaaaaaaabababababa* is presented in Figure 4.2.

We show that the time complexity of preprocessing the pattern (compute table D) is linear. We use a close correspondence between situations in relation to the definition of function D and configurations appearing in the algorithm that computes failure functions (see Figure 4.1 and Figure 4.3).

Let $x1$ be the reverse of *pat*. In the algorithm following, we update values of certain entries of table D according to what is shown in Figure 4.3. This figure is related only to the case $D[j1]<j1$, which is considered during the first stage of the algorithm. We treat other entries of the table D in the second stage separately (see Figure 4.4).

```
procedure compute_D;
begin
    { m is the maximal possible length of shifts }
    for j1 := 1 to m do D[j1] := m;
    { first stage is a partial computation of table D; }
```

```
    { correct values are computed for D[j1]<j1; }
    { we compute table Bord for the reverse of pat; }
    { D is computed as a side-effect }
    Bord[0] := 0; Bord[1] := 0; t := 0;
    for j := 2 to m do begin
        while x1[j]<>x1[t+1] and t>0 do begin { see Figure 4.3 }
            s:=j-t-1; j1:=m-t; D[j1]:=min(D[j1], s); t:=Bord[t];
        end;
        if x1[t+1]=x[j] then t := t+1
        else { see Figure 4.3 with t=0 } D[m] := min(D[m], j-1);
        Bord[j] := t;
    end;
    { second stage correct values for D[j1]≥j1, see Figure 4.4 }
    { consider the failure function Bord for the pattern pat }
    { not for the reverse of pat, as before }
    t := Bord[m]; q := 0;
    while t>0 do begin
        s := m-t;
        for j := q to s do D[j] := min(D[j], s);
        q := s+1; t := Bord[t];
    end;
end;
```

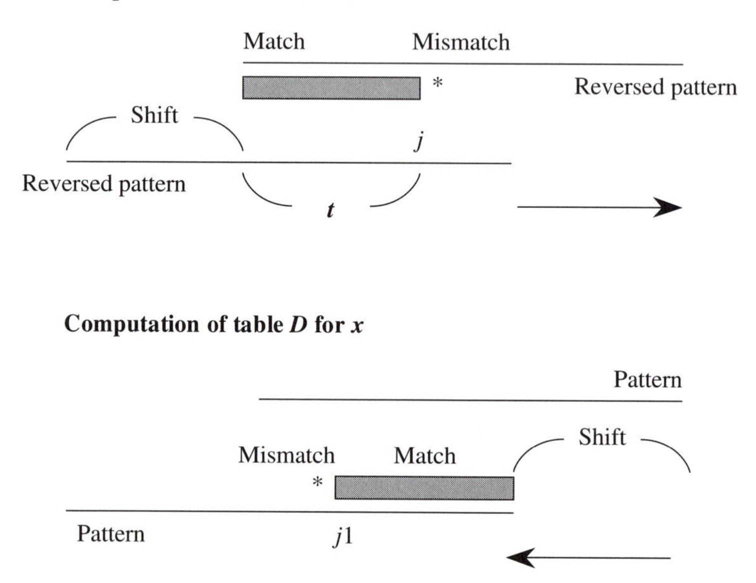

Figure 4.3: Situation just before executing $t = Bord[t]$ in the computation of failure function *Bord* for the reverse of pattern. We know that $D[j1] \geq s$, where $s=j-t-1$ and $j1=m-t$.

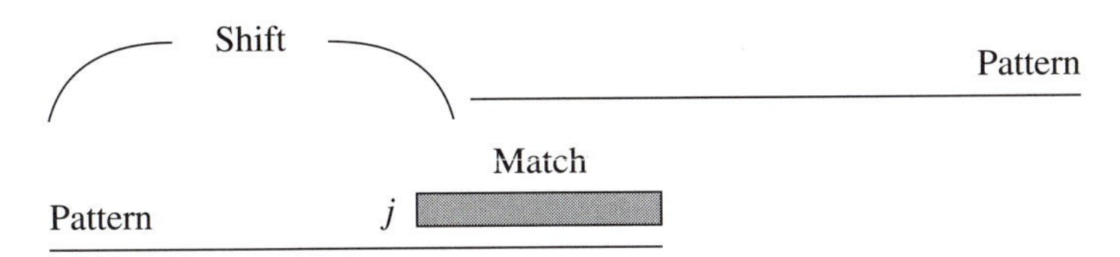

Figure 4.4: Case $D[j] \geq j$: $D[j]=s=m-t$, where $t=Bord^k[m]$ for some $k>0$.

Boyer and Moore introduced also another "heuristic" useful in increasing lengths of shifts. Suppose that we have a situation, where $symb=text[i+j]$ (for $j>0$), and *symb* does not occur at all in the pattern. Then, in the mismatch situation, we can make a shift of length $s=j$. For example, if $pat=a^{100}$ and $t=(a^{99}b)^{10}$, then we can always shift 100 positions, and eventually make only 10 symbol comparisons. For the same input words, algorithm BM makes 901 comparisons. But, if we take $pat=ba^{m-1}$ and $text=a^{2.m-1}$, the heuristic used alone (without using table *D*) leads to a quadratic-time complexity. Let *last*(*symb*) be the last position of an occurrence of symbol *symb* in *pat*. If there is no occurrence, *last*(*symb*) is set to zero. Then, we can define a new shift, replacing instruction "$i := i+D[j]$" of BM algorithm by

"$i := i+\max(D[j], j\text{-last}(text[i+j]))$".

Shift of length $j-last(text[i+j])$ is called an *occurrence shift*. In practice, it may improve the time of the search for some inputs, though theoretically it is not entirely analyzed. If the alphabet is binary, and more generally for small alphabets, the occurrence heuristics have little effect, therefore, it is almost useless.

4.2* Analysis of Boyer-Moore algorithm

The tight upper bound for the number of comparisons done by BM algorithm is approximately 3.*n*. The proof of this is rather difficult but it yields a fairly simple proof of a 4.*n* bound. The fact that the bound is linear is completely nontrivial, and surprising in view of the quadratic behavior of BM algorithm when modified to search for all occurrences of the pattern. The algorithm uses variable *j* to enlarge shifts, but afterwards "forgets" about the checked portion of the text. In fact, the same symbol in *text* can be checked a logarithmic number of times. If we replace *D* by *D*1, then the time complexity becomes quadratic (counterexample is given by text and patterns with the same structure as in the Figure 4.2: $pat=ca(ba)^k$ and $text=a^{2.k+2}(ba)^k$). Hence, one small piece of information (the "one bit" difference between invariants *inv*2, *inv*2') considerably reduces the time complexity in the worst case. This contrasts with improved versions of algorithm *brute_force1*, where *inv*1 and *inv*1' present similar complexities (see Chapter 3). The difference between the usefulness of information with one mismatched symbol gives evidence of the great importance of a (seemingly) technical difference in scanning the pattern left-to-right versus right-to-left.

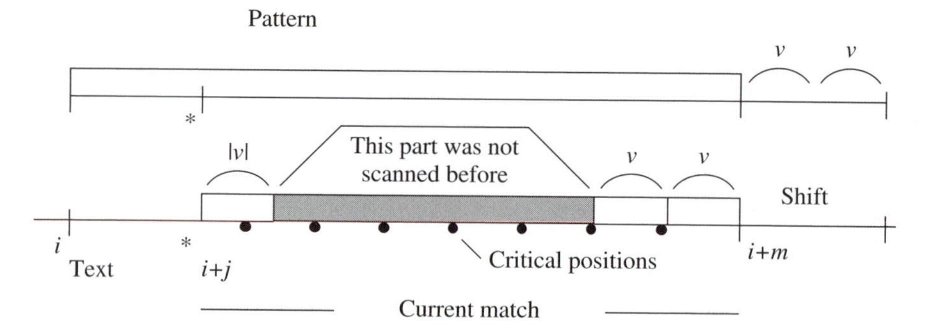

Figure 4.5: The part $text[i+j-1 \ldots i+m]$ of the text is the current match; v denotes the shortest full period of the suffix of the pattern, v is a period of the current match. Shaded area is the "forbidden" part of the text.

Assume that, in a given non-terminating iteration, BM algorithm scans the part $text[i+j-1 \ldots i+m]$ of the text and then makes the shift of length $s=D[j]$, where $j>0$ and $s>(m-j)/3$. By *current match*, we mean the scanned part of text without the mismatched letter (see Figure 4.5).

Lemma 4.1 Let s be the value of the shift made in a given non-terminating iteration of BM algorithm. Then, at most $3.s$ positions of the text scanned at this iteration have been scanned in previous iterations.

Proof It is easier to prove a stronger claim:

> /*/positions in the segment $text[i+j+k \ldots i+m-2.k]$ are scanned in this iteration for the first time, where k is the size of the shortest full period v of $pat[m-s \ldots m]$.

In other words: only the first k and the last $2.k$ positions of the current match could have been read previously. Denote by v the shortest full period v of $pat[m-s+1 \ldots m]$. The following property of the current match results from the definition of the shift:

(Basic property) v is a period of the current match, and v is a suffix of the pattern.

We introduce the notion of *critical position* in the current match. This is an internal position in this match in which the distance from the end of the match is a nonzero multiple of k, (see Figure 4.5). We say that a previous match ends at a position q in the text, if, in some previous iteration, the end of the pattern was positioned at q.

Claim 1 No previous match ends at a critical position of the current match.

Proof (of the claim) The position i+m is the end position of the current match. It is easy to see that if a critical point of the current match is the end of the match in a previous iteration i, then, in the iteration immediate after i, the end of the pattern is at position i+m+$shift$. Hence, the current match under consideration would not exist—a contradiction. This proves the claim. ♦

Claim 2 The length of the overlap of the current match and the previous match is smaller than k.

Proof (of the claim) Recall that by a match we mean a scanned part of the text without the mismatch position. The period v is a suffix of the pattern. We already know, from Claim 1, that the end of the previous match cannot end at a critical position. Hence, if the overlap is at least k long, then v occurs inside the current match with the end position not placed at a critical position. The primitive word v then properly overlaps itself in a text in which the periodicity is v. But, this is impossible for primitive words (due to periodicity lemma). This proves the claim. ♦

Claim 3 Assume that a previous match ends at position q inside a forbidden one, and is completely contained in the current match. Then, there is no critical point (in the current match) to the right of q.

Proof (of the claim) Suppose there is a critical position r to the right of q. It is then easy to see that r–q is a good candidate for the shift in the BM algorithm. The algorithm takes the smallest such candidate as an actual shift. If the shift is smaller than r–q, we have a new position $q1<r$. Then, we will have a sequence of previous matches with end positions $q1$, $q2$, $q3$,… . This sequence terminates in r, otherwise we would have an infinite increasing sequence of natural numbers smaller than r, which is impossible. This contradicts Claim 1, since we have a previous match ending at a critical position in the current match. This completes the proof of the claim. ♦

Proof of lemma Now we are ready to show that /*/ holds. The proof is by contradiction. Assume that in some earlier iteration we scan the "forbidden" part of the text (shaded in Figure 4.5). Let q be the end position of the match in this iteration. Then q is not a critical position, and this match is contained completely in the current match (its overlap with the current match is shorter than k and q lies too far from the beginning of the current match). By the same argument, the rightmost critical position in the current match is to the right of q. Hence, we have found a previous match that is completely contained in the current match and in which the end position lies to the left of some critical position. This is impossible, however, due to Claim 3. This completes the proof of the lemma. ♦

Theorem 4.2 The Boyer-Moore algorithm makes at most $4n$ symbol comparisons to find the first occurrence of the pattern (or to report no matches). The linear-time complexity of the algorithm does not depend on the size of the alphabet.

Proof The cost of each non-terminating iteration can be split into two parts:

1. the cost of scanning some positions of the text for the first time,
2. three times the length of the shift.

The total cost of all non-terminating iterations can be estimated by separately totaling all costs of type (1), this gives at most n, and all costs of type (2), which gives at most $3.(n-m)$. The cost of a terminating iteration is at most m. Hence, the total cost of all iterations is upper bounded by:

$$n + 3(n-m) + m \leq 4.n.$$

This completes the proof. ♦

4.3 Galil's improvement

If we wish to find all occurrences of *pat* in *text* with a modified BM algorithm, then the complexity can become quadratic. The simplest counterexample is given by a text and a pattern over a one-letter alphabet. Observe a characteristic feature of this counterexample: high periodicity of the pattern. Let p be the period of the pattern. If we discover an occurrence of *pat* at some position in *text*, then the next shift must naturally be equal to p. Afterward, we have only to check the last p symbols of *pat*. If they match with the text, then we can report a complete match without inspecting all other $m-p$ symbols of *pat*. This simple idea is embodied in the algorithm below. The variable named *memory* "remembers" the number of symbols that we have not inspected (*memory*=0 or *memory*=$m-p$). In fact, it remembers the prefix of the pattern that matches the text at the current position. This technique is called *prefix memorization*. The correctness of the following algorithm is straightforward. The period of *pat* can be precomputed from algorithms of Chapter 3 (see MP algorithm).

```
procedure BMG;
{ BM algorithm with prefix memorization; p = period(pat) = D[0]}
begin
   i := 0; memory := 0;
   while i ≤ n-m do begin
      j := m;
      while j>memory and pat [j]=text [i+j] do j := j-1;
```

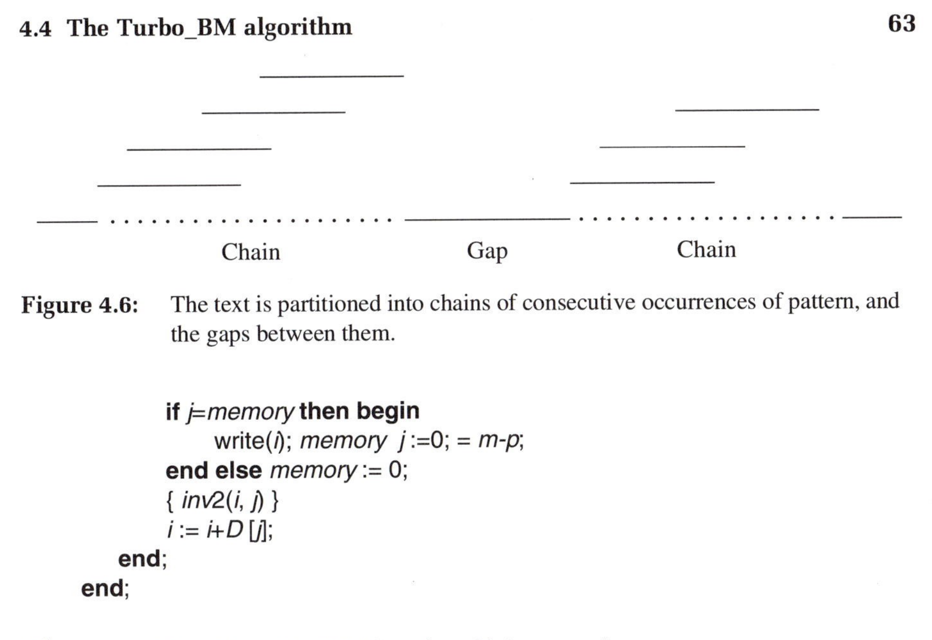

Figure 4.6: The text is partitioned into chains of consecutive occurrences of pattern, and the gaps between them.

```
            if j=memory then begin
                write(i); memory j:=0; = m-p;
            end else memory:= 0;
            { inv2(i, j) }
            i := i+D [j];
        end;
    end;
```

Theorem 4.3 Algorithm BMG makes $O(n)$ comparisons.

Proof It is a natural consequence of the previous theorem that the complexity is $O(n+r.m)$, where r is the number of occurrences of *pat* in *text*. This is because between any two consecutive occurrences of *pat* in *text*, BMG does not make more comparisons than the original BM algorithm. Hence, if $p \geq m/2$, since $r \leq n/p$, $n+r.m$ is $O(n)$.

We have yet to consider the case $p<m/2$. In this case, we can group occurrences of the pattern into chains of positions differing only by p (for two consecutive positions in a group) (see Figure 4.6).

Within each such chain every text symbol is inspected at most only once. The gaps between chains are larger than $m/2$, and, inside each such gap, BMG works no slower than BM algorithm. An argument similar to that used in the case of large periods can now be applied. This completes the proof. ♦

4.4 The Turbo_BM algorithm

In this section we present another efficient version of the Boyer-Moore algorithm. The modification looks superficial, but it actually improves the worst-case complexity. The number of letter comparisons executed during the search becomes less than $2.n$. The BMG algorithm of Section 4.3 implements a prefix memorization after the discovery of an occurrence of the pattern. In the present version, we implement a factor memorization. This does not occur only after occurrences of the pattern.

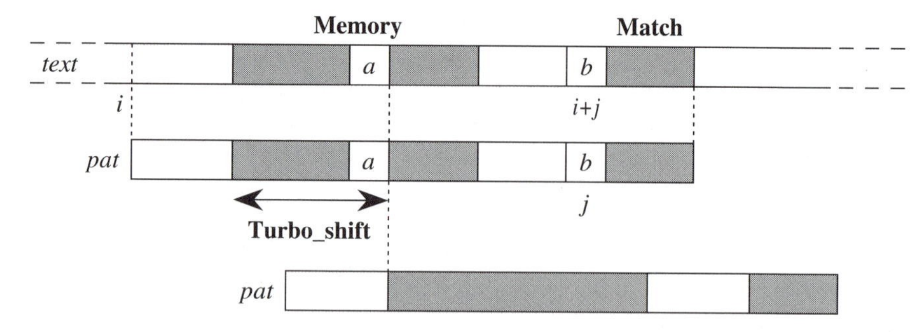

Figure 4.7: Turbo-shift equals *memory-match*. Distinct letters a and b in text are at distance h, and h is a period of the right part of the pattern.

In the new approach no extra preprocessing is needed. The only table to keep from BM algorithm is the original table of shifts, D. All extra memory is of a constant size (two integers). The resulting algorithm, Turbo_BM, forgets all its history except the most recent one, and its behavior again has a "mysterious" character. Despite that, the complexity is improved and the analysis is simple.

The main feature of Turbo_BM algorithm is that, during the search of the pattern, one factor of the pattern matching the text at the current position is memorized (this factor can be empty). This has two advantages:

- it can lead to a jump over the memorized factor during the scanning phase,
- it allows us to perform what we call a *Turbo-shift.*

In Turbo_BM algorithm, two kinds of shifts are considered: ordinary shifts of BM algorithm (called a D-shift), and Turbo_shifts. We now explain what a *Turbo_shift* is. Let x (**match**) be the longest suffix of *pat* that matches the text at a given position. Let also y (**memory**) be the memorized factor that matches the text at the same position. We assume that x and y do not overlap (i.e., for some non-empty word z, yzx is a suffix of *pat*). For different letters a and b, ax is a suffix of *pat* aligned with bx in the text (see Figure 4.7). The only interesting situation is when y is non-empty, which only occurs immediately after a D-shift. Let *shift* be its length. A *Turbo_shift* can occur when x is shorter than y. In this situation, ax is a suffix of y. Thus, letters a and b occur at distance *shift* in the text. But, suffix yzx of *pat* has period *shift* (by definition of the D-shift), and thus, this word cannot overlap both occurrences of letters a and b in the text. As a consequence, the smallest safe shift of the pattern is $|y|-|x|$, which we call the *Turbo_shift.*

In a first approximation, the length of the current shift in Turbo_BM algorithm is max($D[j]$, *Turbo_shift*). In fact, in a case where the D-shift does not apply ($D[j]$ is smaller than *Turbo_shift*), the length of the actual shift is made greater than the length of the matched suffix of *pat*. The proof of correctness of this second feature is similar to the above argument. It is explained in Figure 4.8.

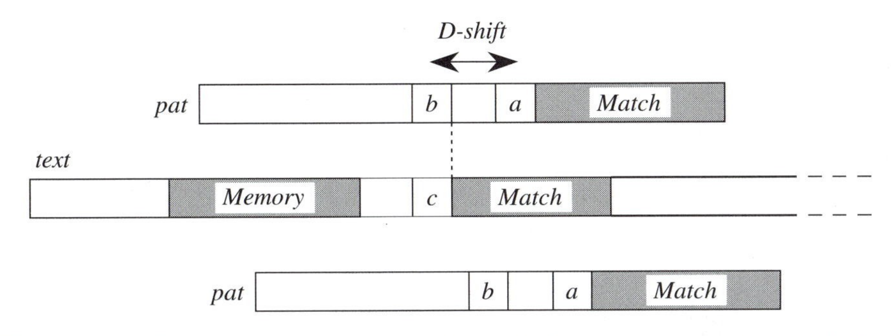

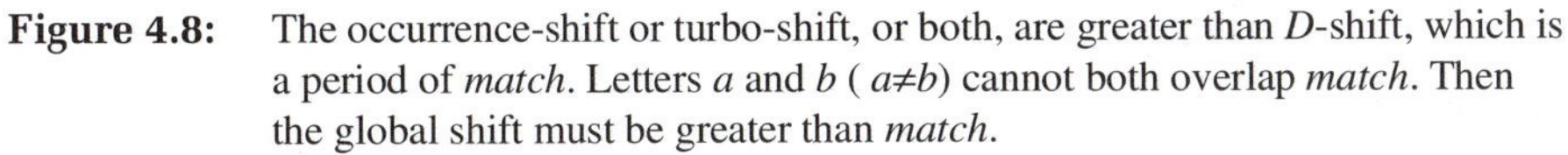

Figure 4.8: The occurrence-shift or turbo-shift, or both, are greater than *D*-shift, which is a period of *match*. Letters *a* and *b* ($a \neq b$) cannot both overlap *match*. Then the global shift must be greater than *match*.

From the above discussion, and the analysis of BM algorithm itself, it is logical to derive a correctness proof of Turbo_BM algorithm. The algorithm, given as follows, finds all occurrences of the pattern, not simply the first one, as BM algorithm does.

```
procedure Turbo_BM; { BM algorithm with factor memorization }
begin
   i:=0; memory:=0;
   while i ≤ n-m do begin
      j := m;
      while j > 0 and pat [j] = text [i+j] do
         if memory ≠ 0 and j = m-shift then j := j-memory { jump }
         else j := j-1;
      if j = 0 then report match at position i;
      match := m-j; Turbo_shift := memory-match;
      shift := max(D [j], Turbo_shift);
      if shift > D [j] then begin
         i := i+max(shift, match+1); memory := 0;
      end else begin
         i := i+shift; memory := min(m-shift, match);
      end;
   end;
end;
```

In the above Turbo_BM algorithm, we deal only with ordinary shifts of BM algorithm. We have already discussed, at the end of Section 4.1, how to incorporate occurrence shifts into BM algorithm. The version of Turbo_BM including occurrence shifts is obtained by a simple modification of the instruction computing variable *shift*. It becomes

```
shift := max(D [j], j-last(t [i+j]), Turbo_shift).
```

In the case where an occurrence shift is possible, the length of the shift is made greater than *match*. This is similar to the case where a turbo-shift applies. The proof of correctness is again similar and explained by Figure 4.8. The length of the D-shift of BM algorithm is a period of the segment *match* of the text (see Figure 4.8). At the same time, we have two distinct symbols a,b in which the distance is the period of the segment. Hence, the shift has length at least $match+1$. This shows the correctness of Turbo_BM algorithm with occurrence shifts.

4.5* Analysis of Turbo_BM algorithm

The analysis of the time complexity of Turbo-BM is far simpler than that of BM algorithm. Moreover, the maximum number of letter comparisons executed during the search (of all occurrences of the pattern) is less the "canonical" $2.n$. (Recall that it is approximately $3.n$ for BM algorithm to find the first occurrence of the pattern.) Therefore, Turbo_BM algorithm is superior to BM algorithm in two aspects:

- the time complexity is improved,
- the analysis is simpler.

Theorem 4.4 The Turbo_BM algorithm (with or without the occurrence heuristics) is linear. It makes less than $2.n$ comparisons.

Proof We break the search down into stages. Each stage is also divided into the two operations: scan and shift. At stage k, we call Suf_k the suffix of the pattern that matches the text and suf_k its length. It is preceded by a letter that does not match the aligned letter in the text (in the case Suf_k is not the p itself). We also call $shift_k$ the length of the shift done at stage k.

Consider 3 types of stages according to the nature of the scan, and of the shift:

i. stage followed by a stage with jump,
ii. no type (i) stage with long shift.
iii. no type (i) stage with short shift,

We say that the shift at stage k is short if $2.shift_k < suf_k+1$. The idea of the proof is to amortize comparisons with shifts. We define $cost_k$ as follows:

- if stage k is of type (i), $cost_k = 1$;
- if stage k is of type (ii) or (iii), $cost_k = suf_k+1$.

In the case of type (i) stage, the cost corresponds to the mismatch comparison. Other comparisons executed during the same stage are reported to the cost of the next stage. Therefore, the total number of comparisons executed by the algorithm is the sum of costs. We want to prove $\Sigma costs < 2.\Sigma shifts$. In the second Σ, the length of the last shift is replaced by m. Even with this assumption, we have $\Sigma shifts \leq |t|$, and if the above inequality holds, the result $\Sigma costs < 2.|t|$.

For stage k of type (i), $cost_k$ (=1) is obviously less than $2.shift_k$, because $shift_k > 0$. For stage k of type (ii), $cost_k = suf_k+1 \leq 2.shift_k$, as per the definition of long shifts.

We have yet to consider stages of type (iii). Since, in this situation, we have $shift_k<suf_k$, the only possibility is for a D-shift to be applied at stage k. The memory is then set up. At the next stage $k+1$, the memory is not empty, which leads to a potential turbo-shift. The situation at stage $k+1$ is the general situation in which a turbo-shift is possible (see Figure 4.9). Before continuing the proof, we must first consider the two cases and establish inequalities (on the cost of stage k) that are used later.

Case (a): $suf_k+shift_k \leq |p|$ By definition of the turbo-shift, we have $suf_k - suf_{k+1} \leq shift_{k+1}$. Thus,

$$cost_k = suf_k+1 \leq suf_{k+1}+shift_{k+1}+1 \leq shift_k+shift_{k+1}.$$

Case (b): $suf_k+shift_k > |p|$ By definition of the turbo-shift, we have $suf_{k+1}+shift_k+shift_{k+1} \geq m$. Then

$$cost_k \leq m \leq 2.shift_k-1+shift_{k+1}.$$

We can speculate that at stage $k+1$, case (b) occurs, because this gives the higher bound on $cost_k$ (this is true if $shift_k \geq 2$; the case $shift_k=1$ can be treated directly).

If stage $k+1$ is of type (i), then $cost_{k+1} = 1$, and then $cost_k+cost_{k+1} \leq 2.shift_k+shift_{k+1}$, an even better bound than expected.

If stage $k+1$ is of type (ii), or even if $suf_{k+1} \leq shift_{k+1}$, we get what we expected, $cost_k+cost_{k+1} \leq 2.shift_k+2.shift_{k+1}$.

The last situation to consider is one in which stage $k+1$ is of type (iii) with $suf_{k+1} > shift_{k+1}$. This means, as previously mentioned, that a D-shift is applied at stage $k+1$. Thus, the above analysis also applies at stage $k+1$, and, since only case (a) can occur then, we get $cost_{k+1} \leq shift_{k+1}+shift_{k+2}$. We finally get $cost_k+cost_{k+1} \leq 2.shift_k+2.shift_{k+1}+shift_{k+2}$.

The last argument proves the first step of an induction: if all stages k to $k+j$ are of type (iii) with $suf_k > shift_k$, ..., $suf_{k+j} > shift_{k+j}$, then

$$cost_k+\ldots+cost_{k+j} \leq 2.shift_k+\ldots+2.shift_{k+j}+shift_{k+j+1}.$$

Let k' be the first stage after stage k such that $suf_k' \leq shift_k'$. Integer k' exists because the contrary would produce an infinite sequence of shifts with decreasing lengths. We then get

$$cost_k+\ldots+cost_k' \leq 2.shift_k+\ldots+2.shift_k',$$

which shows that $\Sigma cost_k \leq 2.\Sigma shift_k$, as expected. ♦

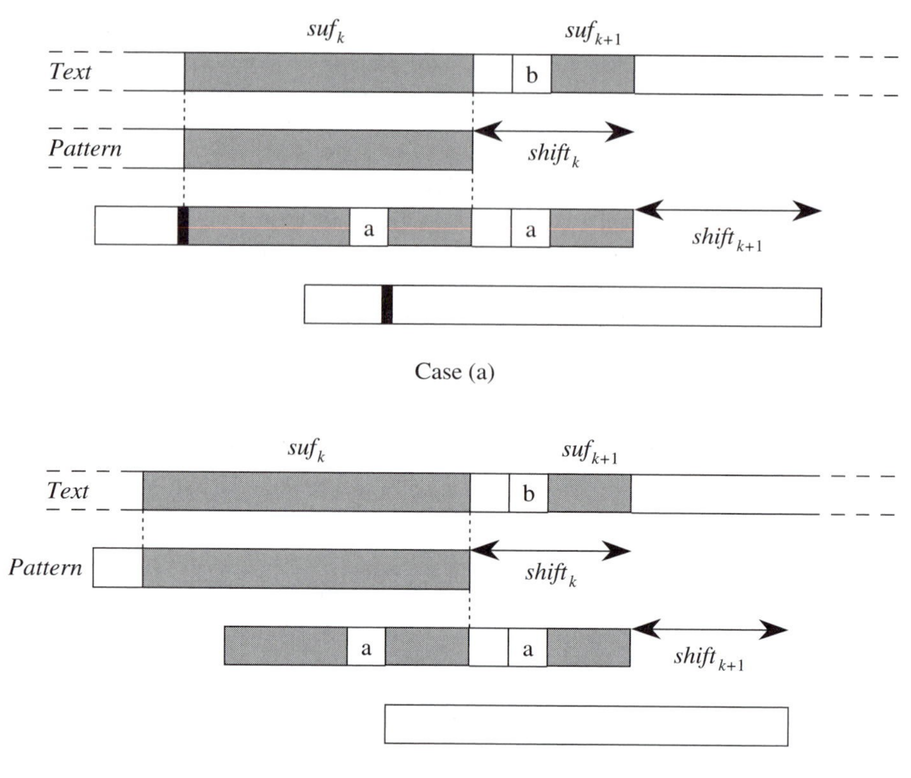

Figure 4.9: Costs of stages k and k+1 correspond to shaded areas plus mismatches. If $shift_k$ is small, then $shift_k + shift_{k+1}$ is large enough to partially amortize the costs.

4.6 Other variations of Boyer-Moore algorithm

In this section we describe two other possible improvements on BM algorithm. The first one is called AG algorithm. Its basic idea is to avoid checking the segments of the text that have already been examined. These segments have a very simple structure. They are all suffixes of pattern *pat*. Call these segments *partial matches*. Their length is the number of symbols that matched in one stage of BM algorithm.

It is convenient to introduce a table $S[j]$, analogous to $Bord[i]$ in the sense that *Bord* is defined in terms of prefixes, while S is defined in terms of suffixes. Value $S[j]$ is the length of the longest suffix of $pat[1\ldots j]$ that is also a suffix of $pat[1\ldots n]$.

Suppose that we scan text *text* right-to-left and compare it with pattern *pat* as in BM algorithm. Let us consider the situation in which segment $pat[j+1\ldots n]$ of *pat* has just been scanned. At this moment, the algorithm attempts to compare the j-th symbol of *pat* with the $(i+j)$-th of *text*. We further assume that attached to the current

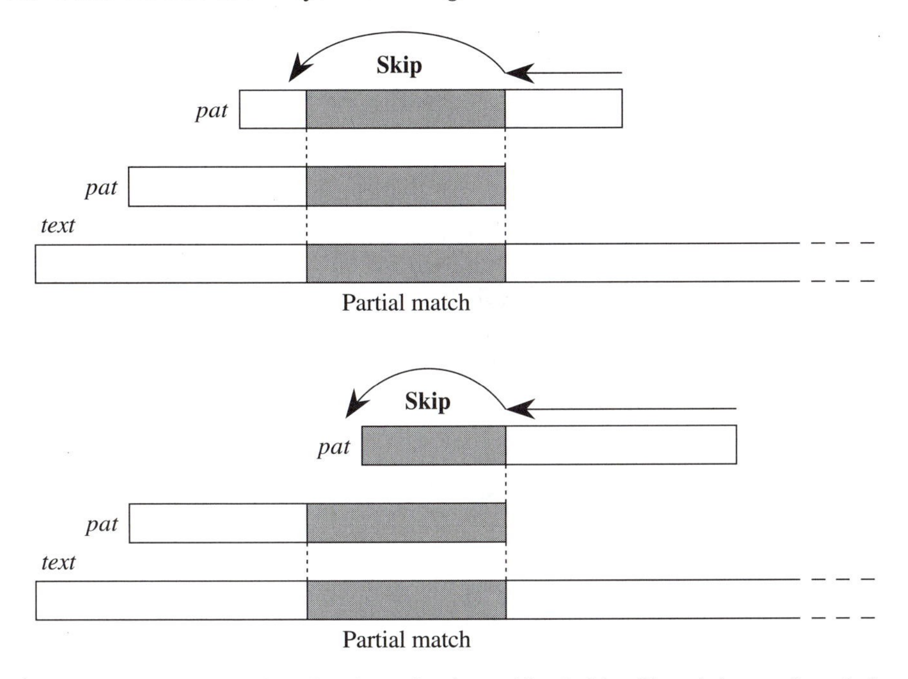

Figure 4.10: Two possible situations allowing a skip. A skip of length k saves k symbol comparisons. The skipped part is always a suffix of the pattern.

position of the text is the fact that a partial match of length k has previously ended here. Hence, we know that the next k positions to the left in the text give a suffix of the pattern. If $k \leq S[j]$, then we know that we can skip these k symbols (because they match), and continue comparing *pat* with *text* further to the left. If not, we know that our actual match is a failure because no suffix of length k ends at the current position of the pattern (except when $S[j]=j$). In AG algorithm, a variable *skip* remembers how many symbols can be skipped. More precisely, the condition that enables us to make a skip is:

$cond(k, j)$: $k \leq S[j]$ or ($S[j]=j$ and $j \leq k$].

```
function AG : boolean;
{ another improved version of brute_force2 }
begin
    i := 0; { PM is the table of partial matches,
              all its entries are initially set to zero }
    while i ≤ n-m do begin
        j := m;
        while j>0 and ((PM [i+j]≠0 and cond(PM [i+j]) or
                        (PM [i+j]=0 and pat [j]=text [i+j])) do
            j := j-max(1, PM [i+j]);
```

```
            if j=0 then return(true);
            PM [i+m] := m-j;
            { inv2(i, j) }
            i := i+D [j];
      end;
      return(false);
   end;
```

Recall that we have assumed a lazy evaluation from left to right of Boolean expression in algorithms. It is then easily seen that each symbol of the text is examined at most twice during a run of AG algorithm. We get a bound $2.n$ on the number of all symbol comparisons. Also, the number of times table *PM* is accessed is easily seen to be linear. Hence, AG algorithm has time complexity generally much better than that of BM algorithm. Moreover, the number of symbol comparisons made by AG algorithm never exceeds the number of comparisons made by BM algorithm. The main advantage of the latter algorithm is that it is extremely fast for many inputs (reading only small parts of text). This advantage is preserved by AG algorithm. Note that AG algorithm requires $O(m)$ extra space due to table *PM*.

We go to the computation of table *S* used in AG algorithm. Table *S* can be computed in linear time, in a similar manner as table *Bord*; one possibility is to compute it as a by-product of MP algorithm (when text=pattern). Indeed, rather than computing *S*, one can compute a more convenient table *PREF*. It is defined by

$PREF[i]=\max\{ j: pat[i\ldots i+j-1]$ is a prefix of $pat\}$.

The computation of table *S* is directly reducible to the computation of *PREF* for the reversed pattern. One of many alternative algorithms is based on the following observation:

if $Bord[i]=j$ then $pat[i-j+1\ldots i]$ is a prefix of *pat*.

This leads to the following algorithm.

```
procedure compute_PREF;
begin
   for i := 1 to n do PREF [i] :=0;
   for i := 1 to n do begin
      j := Bord [i]; PREF [i-j+1] := max(PREF [i-j+1], j);
   end;
end;
```

Though useful, this algorithm is not entirely correct. In the case of one-letter alphabets only one entry of table *PREF* will be accurately computed. But its incorrectness is quite "weak." If $PREF[k]>0$ after the execution of the algorithm then it is accurately computed. Moreover, *PREF* is computed for "essential" entries. The entry i is essential iff $PREF[i]>0$ after applying this algorithm. These entries partition interval $[1\ldots m]$ into subintervals for which further computation is relatively simple.

The computation of the table for nonessential entries is executed as follows: traverse the whole interval left-to-right and update $PREF[i]$ for each entry i. Take the first (to the left of i, including i) essential entry k with $PREF[k] \geq i-k+1$, then we execute $PREF[i]=\min(PREF[i-k+1], PREF[k]-(i-k))$. If there is no such essential entry, then $PREF[i]=0$. Apply this for an example pattern over a one-letter alphabet to see how it works.

BM algorithm is particularly fast for large alphabets, because shifts are likely to be long. For small alphabets, the average number of symbol comparisons is linear. We now design an algorithm making $O(n \log(m)/m)$ comparisons on the average. Hence, if m is of the same order as n, the algorithm makes only $O(\log n)$ comparisons. It is essentially based on the same strategy as BM algorithm, and can be treated as another variation of it. For simplicity, we assume that the alphabet has only two elements, and that each symbol of the text is chosen independently with the same probability. Let $r=2.\log m$.

```
Algorithm fast_on_average;
begin
    i := m;
    while i ≤ n do begin
        if text[i-r...i] is a factor of pat then
            compute occurrences of pat starting in [i-m...i-r]
            applying KMP algorithm
        else /* pattern does not start in [i-m...i-r] */
            i := i+m-r;
    end;
end;
```

Theorem 4.5 The *fast_on_average* algorithm works in $O(n \log m/m)$ expected time and (simultaneously) in $O(n)$ worst-case time if the pattern is preprocessed. The preprocessing of the pattern takes $O(m)$ time.

Proof A preprocessing phase is needed to efficiently check if $text[i-r...i]$ is a factor of pat in $O(r)$ time. Any of the data structures developed in Chapters 5 and 6 (a suffix tree or a dawg) can be used. Assume that text *text* is a random string. There are $2^r \geq m^2$ possible suffixes of *text*, and less than m factors of *pat* of length r. Hence, the probability that the suffix of length r of *text* is a factor of *pat* is no greater than $1/m$. The expected time spent in each of the subintervals $[1...m]$, $[m-r...2.m-r]$, ... is $O(m.1/m+r)$. There are $O(n/m)$ such intervals. Thus, the total expected time of the algorithm is of order $(1+r)n/m = n(\log(m)+1)/m$. This completes the proof. ♦

Bibliographic notes

The tight upper bound of approximately $3.n$ for the number of comparisons in BM algorithm has recently been discovered by Cole [Co 90]. The proof of the $4.n$ bound reported in Section 4.2 is from the same paper. A previous proof of the $4.n$ bound was presented earlier by Guibas and Odlyzko [GO 80], but with more complicated arguments. These authors also conjectured a $2.n$ bound which has since been proven false. Another combinatorial estimation of the upper bound on the number of comparisons needed by Boyer-Moore algorithm may be found in [KMP 77]. We recommend reading these proofs for readers interested in combinatorics on words.

The computation of the table of shifts, D, in the Boyer-Moore algorithm is from [Ry 80]. The algorithm presented in [KMP 77] contains a small flaw. The improved versions of BM algorithm, called BMG and AG algorithms, are from Galil [Ga 79], and Apostolico and Giancarlo [AG 86], respectively. The Boyer-Moore algorithm is still a theoretically fascinating algorithm: an open problem related to the algorithm questions the number of states of a finite-automaton version of the strategy "Boyer-Moore." It is not yet known whether this number is exponential or not. The problem appears in [KMP 77]. Baeza-Yates, Choffrut, and Gonnet discuss the question in [BCG 93] (see also [Ch 90]), and they design an optimal algorithm for building the automaton.

The Turbo_BM algorithm was recently discovered while working on another algorithm based on automata presented in Chapter 6. It is from Crochemore et al. [C-R 92].

Selected references

[BM 77] Boyer, R.S., & Moore J.S., "A fast string searching algorithm," *Comm. ACM* 20 (1977): 762–772.

[Co 91] Cole, R., "Tight bounds on the complexity of the Boyer-Moore pattern matching algorithm," in (*2nd annual ACM Symp. on Discrete Algorithms*, 1991): 224–233

[C-R 92] Crochemore, M., Czumaj, A., Gasieniec, L., Jarominek, S., Lecroq, T., Plandowski, W., & Rytter, W., "Speeding up two string matching algorithms," in (A. Finkel and M. Jantzen editors, *9th Annual Symposium on Theoretical Aspects of Computer Science*, Springer-Verlag, Berlin, 1992): 589–600.

[KMP 77] Knuth, D.E., Morris Jr., J.H., & Pratt, V.R., "Fast pattern matching in strings," *SIAM J.Comput.* 6 (1977): 323–350.

5

Suffix trees

The basic data structures in this book are those representing all factors of a given word. Their importance derives from a multitude of applications. We shall see some of these applications at the end of the present chapter, and in the following one.

This chapter is mainly devoted to the first of such basic data structures—the suffix tree. A related concept of subword graphs is also introduced in this chapter, but its full treatment is postponed until Chapter 6.

The basic item to be represented is the set *Fac*(*text*) of all factors of the text *text* of length n. It should be represented efficiently: this means that basic operations related to the set *Fac*(*text*) should be performed quickly. In this chapter, we closely examine the structure of the set *Fac*(*text*). Its size is usually quadratic, but it has succinct representations that are of linear size only. We introduce three such representations in this chapter, namely, suffix trees, subword graphs, and suffix arrays. We shall see that the subword graph of a text has the same origin as its suffix tree: an uncompressed trie of suffixes of the text. This is the reason why we introduce them together with suffix trees.

We need a representation efficiently addressing a class of problems. Since *Fac*(*text*) is a set, the most typical problem is the membership problem: check if a given word x is in *Fac*(*text*). Denote the corresponding predicate by *FACTORIN*(x, *text*). The utility of the data structure for this problem means that *FACTORIN*(x, *text*) can be computed in time $O(|x|)$, even if x is much shorter than *text*. The complexity here is computed according to the basic operation of the data structure, *branching*, which takes a unit of time. Under the comparison model, branching can be implemented in order to take $O(\log|A|)$ time (A is the alphabet). Since it is a common practice to work on a fixed alphabet such as the ASCII alphabet, we can also consider that log|*A*| is a constant.

Note that most algorithms in Chapters 3 and 4 compute the value of *FACTORIN*(x, *text*) in time $O(|text|)$ once the word x has been preprocessed. The present data structure is thus well suited when the text is fixed. Such a data structure is also necessary for examples in Chapter 4 (see Section 4.6). We want the preprocessing time, used to build the representation of the text, to be realized in linear time with respect to the size of the text. Let us call representations satisfying these requirements "good" representations of *Fac*(*text*). We can summarize this in the following definition. The data structure D representing the set *Fac*(*text*) is *good* iff:

1. D has linear size;
2. D can be constructed in linear time;
3. D allows computing *FACTORIN*(x, *text*) in $O(|x|)$ time.

The main goal of this chapter is to show that suffix trees are *good* data structures. We also introduce a third data structure, a suffix array, which is "*almost*" *good*, but simpler. The "goodness" of subword graphs is the subject of the next chapter. The important algorithm of this chapter is McCreight's algorithm (Section 5.3). It processes the text from left to right, while the equivalent Weiner's algorithm traverses the text in the opposite direction.

5.1 Prelude to McCreight and Weiner algorithms

In this section we deal only with simple data structures for *Fac*(*text*), namely trees called *tries*. Algorithms presented here are intermediate steps in the understanding and design of McCreight and Weiner suffix tree constructions (Sections 5.2 and 5.3, respectively). The goal is to construct such trees in time proportional to their size.

Figure 5.1 shows the trie associated with text *aabbabbc*. In these trees, the links from a node to its sons are labeled by letters. In the tree associated to *text*, a path down the tree spells a factor of *text*. All paths from the root to leaves spell suffixes of *text*. And all suffixes of *text* are labels of paths from the root. In general, these paths do not necessarily end in a leaf. But for ease of the description, we assume that the last letter of *text* occurs only once and serves as a **marker**. With this assumption, no suffix of *text* is a proper prefix of another suffix, and all suffixes of *text* label paths from the root of the trie to its leaves.

Both Weiner and McCreight algorithms are incremental algorithms. They compute the tree considering suffixes of the text in a particular order. The tree is computed for a subset of the suffixes. A new suffix is then inserted into the tree. This continues until all suffixes are included in the tree. Let p be the next suffix to be inserted in the current tree T. We define the *head of* p, *head*(p, T), as the longest prefix of p

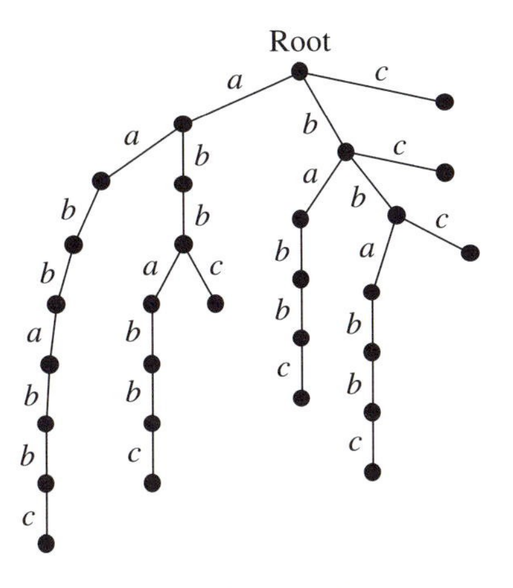

Figure 5.1: Trie of factors of *aabbabbc*. It has eight leaves corresponding to the eight non-empty suffixes.

occurring in T (as the label of a path from the root). We identify $head(p, T)$ with its corresponding node in the tree. Having this node, only the remaining part of p, say π, needs to be grafted onto the tree. After adding the path, below the head to a new leaf, the tree contains a new branch labeled by p, and this produces the next tree of the construction (see Figure 5.2). Below is the general scheme of both McCreight's and Weiner's algorithms. The method used to find the next head, and the grafting operation, are then explained.

```
Algorithm general-scheme;
begin
    compute initial tree T for the first suffix;
    leaf:= leaf of T;
    for i := 2 to n do begin
        { insert next suffix }
        localize next head as head(current suffix, T);
        let π be the string corresponding to path from head to leaf;
        create new path starting at head corresponding to π;
        leaf:= lastly created leaf;
    end;
end.
```

The main technique used by subsequent algorithms is called *Up-Link-Down*. It is the key to an improvement on a straightforward construction. From the most recently created leaf of the tree, it is used to find the node (the head) where a new path must be grafted. The data structure incorporates links that provide shortcuts in order to hasten the search for heads. The strategy Up_Link_Down works as follows:

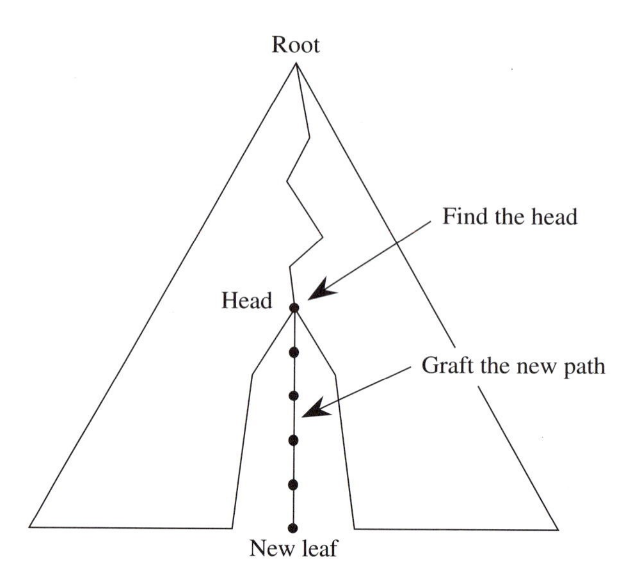

Figure 5.2: Insertion of the next suffix.

it goes up the tree from the last leaf until a shortcut through an existing link is possible; it then goes through the link and starts going down the tree to find the new head (see Figure 5.3). This is done using the procedure Up_Link_Down that works on the current tree and modifies it. This is somehow an abstract procedure, that admits two instances. The link mentioned in the procedure should be understood to be conceptual. The actual links used afterward depends on which algorithm is used to build the suffix tree.

```
function Up_Link_Down(link, q) : node;
{ finds new head, from leaf q }
begin
{ UP, going up from leaf q }
    v := first node v on path from q to root s.t. link[v]≠nil;
    if no such node then return nil;
    let π=a_j a_{j+1}...a_n be the string, label of path from v to q;
{ LINK, going through suffix link }
    head := link[v];
{ DOWN, going down to new head, making new links }
    while son(head, a_j) exists do begin
        v := son(v,a_j); head := son(head,a_j);
        link[v] := head; j := j+1;
    end;
    return (v, head);
end
```

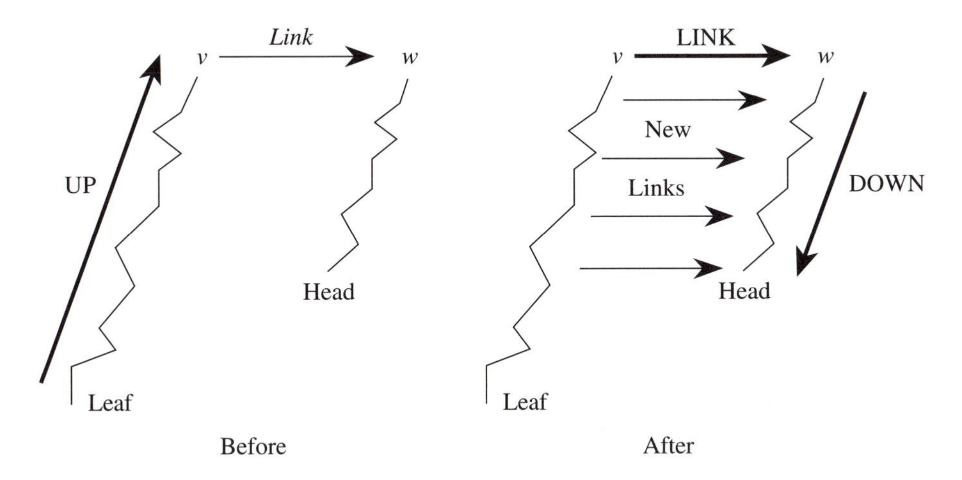

Figure 5.3: Strategy for finding the next head.

We also define the procedure Graft. Its aim is to construct a path of new nodes from the current head to a newly created leaf. It also updates links from the branch containing the previous leaf, for nodes that should point to newly created nodes.

```
procedure Graft(link, v, head, a_j a_{j+1}...a_n);
begin  w := head;
   for k := j to n do begin
         v := son(v,a_k); w := createson(w,a_k); link [v] := w;
   end;
{ w is the last leaf }
end
```

Function Up_Link_Down (or a variant in case of a compressed representation of a tree) is the basic tool used by algorithms that build suffix trees, and also those that build dawgs. The time of a single call to Up_Link_Down can be proportional to the length of the whole pattern. But the sum of all costs is indeed linear. When used to create a compressed suffix tree, a careful implementation of the function also yields an overall linear-time complexity.

For the pattern *text* = $a_1a_2\ldots a_n$, we define p_i as the suffix $a_ia_{i+1}\ldots a_n$. *Trie*(p_1, p_2 ..., p_i) is the tree in which the branches are labeled with suffixes p_1, p_2 ..., p_i. In this tree, $leaf_i$ is the leaf corresponding to suffix p_i. And $head_i$ is the first (bottom-up) ancestor of $leaf_i$ having at least two sons; it is the root if there is no such node.

For a node corresponding to a non-empty factor *aw*, we define the *suffix link* from *aw* to *w*, by *suf*[*aw*] = *w* (see Figure 5.7). Table *suf* will serve to create shortcuts during the construction of *Trie*(p_1, p_2 ..., p_n). It is the link mentioned in the procedure Up_Link_Down. In *Trie*(p_1, p_2 ..., p_i) it can be possible for *suf*[*v*] to be undefined, for some node v. This situation does not occur for (compressed) suffix trees. Below is the algorithm Left_to_Right that builds the suffix tree of *text*. It processes suffixes from p_1 to p_n.

Basic property of *Trie*($p_1, p_2 ..., p_i$): Let v be the first node on the path from $leaf_{i-1}$ to root such that $suf[v]$≠nil in $Trie(p_1, p_2 ..., p_{i-1})$. Then $head_i$ (in $Trie(p_1, p_2 ..., p_i)$) is a descendant of $suf[v]$.

```
Algorithm Left_to_Right(a1a2...an, n>0);
begin
    T := Trie(p1) with suffix link (from son of root to root);
    for i:=2 to n do begin
    { insert next suffix pi = aiai+1...an into T }
        { FIND new head }
        (v, head) := Up_Link_Down(suf, leafi-1); { head = headi }
        if head = nil then begin
        { root situation }
            let v be the son of root on the branch to leafi-1;
            suf [v] := root; head := root;
        end;
        { going down from headi to leafi creating a new path }
        let aj...an be the label of the path from v to leafi-1;
        Graft(suf, v, head, ajaj+1...an);
    end;
end.
```

Theorem 5.1 The algorithm Left_to_Right constructs the T=*Trie*(*text*) in time O(|T|).

Proof The time is proportional to the number of created links (*suf*), which is obviously proportional to the number of internal nodes, and then to the total size of the tree. ♦

The next algorithm, Right_to_Left, builds the suffix tree of *text*, as does the algorithm Left_to_Right, but processes the suffixes in reverse order, from p_n to p_1. The notion of link is modified. The node corresponding to a (non-empty) factor aw (a a letter) is linked to w. The link is called a a-link, and we note $link_a[w] = aw$ (see Figure 5.10). These links have the same purpose as the suffix links of the previous algorithm. The procedure Up_Link_Down will now use the a-links.

```
Algorithm Right_to_Left;
begin
    T := Trie(pn) with linkan from root to its son;
    for i := n-1 downto 1 do begin
    { insert next suffix pi = aiai+1...an into T }
        { FIND new head }
        head := Up_Link_Down(linkai, leafi+1); { head = headi }
```

```
        if head = nil then begin
        { root situation }
            v := root; head := createson(v,a_i); link_ai[v] := head;
        end;
        { GRAFT, going down from head_i to leaf_i creating a new path }
        let a_j a_j+1 ...a_n be the label of the path from v to leaf_i-1;
        Graft(link_ai, v, head, a_j a_j+1 ...a_n);
    end;
end.
```

Basic property of *Trie*($p_i, p_{i+1} ..., p_n$): Let v be the first node on the path from $leaf_{i+1}$ with $link_{ai}[v]\neq$nil in *Trie*($p_{i+1}, p_{i+2} ..., p_n$). Then $head_i$ (in *Trie*($p_i, p_{i+1} ..., p_n$)) is a descendant of $link_{ai}[v]$.

Theorem 5.2 The algorithm Right_to_Left constructs the trie *Trie*($p_1, p_2 ..., p_n$) in time O(|*T*|).

Proof The time is proportional to the number of links in the tree *Trie*($p_1, p_2 ..., p_n$). Note that the reversed links are exactly the suffix links (considered in McCreight algorithm). Since there is one link out to each internal node, we achieve the result. ♦

Sections 5.3 and 5.4 present versions of algorithms Left_to_Right and Right_to_Left respectively, adapted to the construction of suffix trees, that is, compacted tries.

5.2 Two compressed versions of the naive representation

Our approach to compact representations of the set of factors of a text is graph-theoretical. Let G be an acyclic rooted directed graph in which the edges are labeled with symbols or with words: *label*(e) denotes the *label* of edge e. The *label* of a path π (denoted by *label*(π)) is the composition of labels of its consecutive edges. The edge-labeled graph G represents the set:

$$words(G) = \{ \ label(\pi): \pi \text{ is a directed path in } G \text{ starting at the root } \}.$$

And we say that the labeled graph G represents the set of factors of *text* iff

$$words(G)=Fac(text).$$

The first approach to representing *Fac*(*text*) in a compact way is to consider graphs that are trees. The simplest type of labeled graphs G are trees in which their edges are labeled by single symbols. These trees are the *subword tries* considered in the previous section. Two examples of tries are shown in Figures 5.1 and 5.4. How-

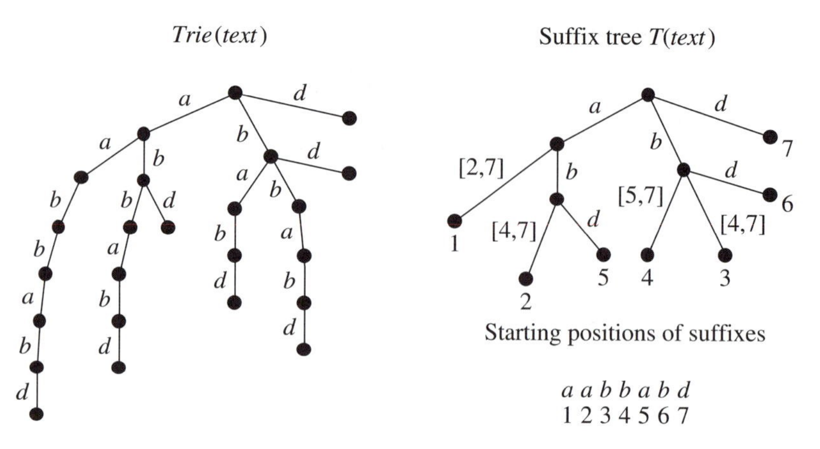

Figure 5.4: The tree *Trie*(*text*) and its compacted version, the suffix tree *T*(*text*), for *text* = *aabbabd*. *Trie*(*text*) has 24 nodes. *T*(*text*) has only 11 nodes.

ever, tries are not "good" representations of *Fac*(*text*), because they can be too large. If $text{=}a^n b^n a^n b^n d$, for example, then *Trie*(*text*) has a quadratic number of nodes.

We consider two kinds of succinct representations of the set *Fac*(*text*). They both result from compressing the tree *Trie*(*text*). Two types of compression can be applied, separately or simultaneously:

1. compressing chains (paths consisting of nodes of out-degree one), which produces the *suffix tree* of the text,
2. merging isomorphic subtrees (e.g., all leaves are to be identified), which leads to the *directed acyclic word graph* (dawg) of the text.

Each of these methods has its advantages and disadvantages. The first produces a tree, and this is an advantage because in many cases it is easier to deal with trees than with other types of graphs (particularly in parallel computations). Moreover, in the tree structure, leaves can be assigned specific values. The disadvantage of this method is the varying length of the labels of edges. An advantage of the second method (use of dawgs) is that each edge is labeled by a single symbol. This allows us to attach information to symbols on edges. And the main disadvantage of dawgs is that they are not trees!

The linear size of suffix trees is a trivial fact, but the linear size of dawgs is much less trivial. It is probably one of the most surprising facts related to representations of *Fac*(*text*).

Since we assume that no suffix of *text* is a proper prefix of another suffix (because of the right end marker), the leaves of *Trie*(*text*) are in one-to-one correspondence with (non-empty) suffixes of *text*. Let *T*(*text*) be the compacted version of *Trie*(*text*) (see Figure 5.4). Each chain π (path consisting of nodes of out-degree one) is compressed into a single edge e with $label(e)=[i, j]$, where $text[i...j]=label(\pi)$ (observe that a compact representation of labels is also used). Note that there is a certain nondeterminism here, because there can be several possibilities of choosing i and j representing the same factor $text[i...j]$ of *text*. Any such choice is acceptable. In this context we identify the label $[i, j]$ with the word $text[i...j]$. The tree $T(text)$ is called the suffix tree of the word *text*. Figure 5.4 presents the suffix tree $T(text)$ for $text=aabbabd$.

For a node v of $T(text)$, let $val(v)$ be $label(\pi)$, where π is the path from the root to v. Whenever it is unambiguous we identify nodes with their values, and paths with their labels. In this sense, the set of leaves of a suffix tree is the set of suffixes of *text* (since no suffix is a prefix of another suffix). This is the reason for the name: suffix tree. Such trees are also called subword (factor) trees.

Note that the suffix tree obtained by compressing chains has the following property: the labels of edges starting at a given node are words having different first letters. Therefore, the branching operation performed to visit the tree is reduced to comparisons on the first letters of the labels of outgoing edges.

In the following we consider the notion of *implicit nodes* that is now defined. The aim is to restore the nodes of $Trie(text)$ inside the suffix tree $T(text)$. We say that a pair (w, α) is an implicit node in T iff w is a node of T and α is a proper prefix of the label of an edge from w to a son of w. If $val(w) = x$, then $val((w, \alpha)) = x\alpha$. For example, in Figure 5.6, (w, ab) is an implicit node. It corresponds to a place where a real node is later created. The implicit node (w, α) is said to be a "real" node if α is the empty word. In this case (w, ε) is identified with the node w itself.

Lemma 5.3 The size of the suffix tree $T(text)$ is linear ($O(|text|)$).

Proof Let $n=|text|$. The tree $Trie(text)$ has at most n leaves, hence $T(text)$ also has at most n leaves. Thus, $T(text)$ has at most $n-1$ internal nodes because each internal node has at least two sons. Hence, $|T(text)| \leq 2n-1$. This completes the proof. ♦

There is a result similar to Lemma 5.3 concerning the size of dawgs. This question is the subject of Section 6.1.

5.3 McCreight's algorithm

A straightforward approach to the construction of $T(text)$ could be: first build $Trie(text)$, next, compress all its chains. The main drawback to this scheme is that the size of $Trie(text)$ can be quadratic, resulting in a quadratic time and space algorithm. Two other approaches are presented in this section (McCreight's algorithm) and the following one (Weiner's algorithm).

Both Weiner's algorithm and McCreight's algorithm are incremental algorithms. The tree is computed for a subset of consecutive suffixes. The next suffix is

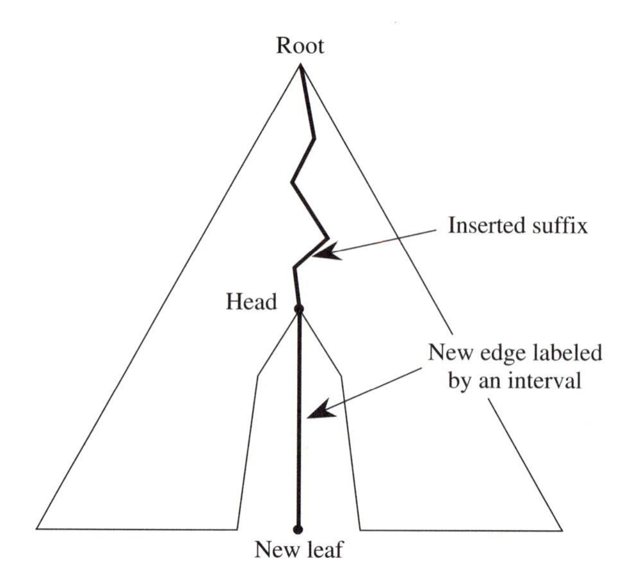

Figure 5.5: Insertion of a suffix into the tree.

then inserted into the tree, and this continues until all (non-empty) suffixes are included in the tree.

Consider the structure of the path corresponding to a new suffix p inserted into the tree T. Such a path is indicated by the bold line in Figure 5.5. Denote by *insert*(p, T) the tree obtained from T after insertion of the string p. The path corresponding to p in *insert*(p, T) ends in the most recently created leaf of the tree. Denote the father of this leaf by *head*. It may be that the node *head* does not already exist in the initial tree T (it is only an implicit node at the time of construction), and has to be created during the insert operation. For example, this occurs if we try to insert the path p = *abcdeababba* starting at v in Figure 5.6. In this case, one edge of T (labeled by *abadc*) has to be split. This is why we consider the operation *break* defined below.

The notion of implicit nodes introduced in Section 5.2 is considered conceptually in the construction. Let (w, α) be an implicit node of the tree T (w is a node of T, α is a word). The operation *break*(w, α) on the tree T is defined only if there is an edge outgoing from the node w in which the label δ has α as a prefix. Let β be such that $\delta = \alpha\beta$. The (side) effect of the operation *break*(w, α) is to break the corresponding edge, a new node is inserted at the breaking point, and the edge is split into two edges of respective labels α and β. The value of *break*(w, α) is the node created at the breaking point.

Let v be a node of the tree T, and let p be a subword of the input word *text* represented by a pair of integers l, r, where $p = text[l...r]$. The basic function used in the suffix tree construction is the function *find*. The value *find*(v, p) is the last implicit node along the path starting in v and labeled by p. If this implicit node is not real, it is

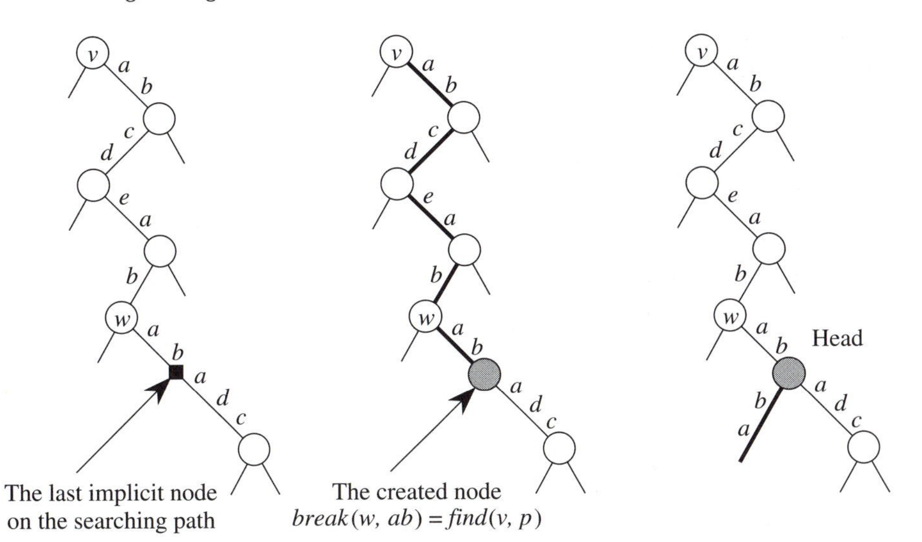

Figure 5.6: We attempt to insert the string $p = abcdeababba$. The result of *fastfind* and *slowfind* is the newly created node *find*(v, p). *fastfind* takes only 5 steps, while *slowfind* would take $O(|p|)$ steps.

(w, α) for some non-empty α, and the function *find* converts it into the "real" node *break*(w, α) (see Figure 5.6).

The important aspect of the algorithm is the use of two different implementations of the function *find*. The first one, called *fastfind*, deals with the situation when we know in advance that the searching path labeled by p is fully contained in some path starting at v. This knowledge allows us to find the searched node much faster using the compressed edges of the tree as shortcuts. If we are at a given node u, and if the next letter of the path is a, then we look for the edge outgoing from u for which the label starts with a. Only one such edge exists due to the definition of suffix trees. This edge leads to another node u'. We jump in our searching path at a distance equal to the length of the label of edge (u, u'). The second implementation of *find* is the function *slowfind* that follows its path letter by letter. The application of *fastfind* is a main feature of McCreight's algorithm, and plays a central part in its performance (together with links).

```
function fastfind (v : node; p : string) : node;
{ p is fully contained in some path starting at v }
begin
    from node v, follow path labeled by p in the tree using
    labels of edges as shortcuts; only first symbols on each
    edge are checked;
    let (ω, α) be the last implicit node;
```

```
        if α is empty then return w
        else return break (w, α);
    end

    function slowfind (v : node; p : string) : node;
    begin
        from node v, follow the path labeled by the longest possible
        prefix of p, letter by letter;
        let (w, α) be the last implicit node;
        if α is empty then return w
        else return break (w, α);
    end
```

McCreight's algorithm builds a sequence of compacted trees T_i in the order i=1, 2 …, n. The tree T_i contains the i-th longest suffixes of *text*. Note that T_n is the suffix tree $T(text)$, but that intermediate trees are not strictly suffix trees. An on-line construction is presented in Section 5.5. At a given stage of McCreight's algorithm, we have $T = T_{k-1}$, and we attempt to build T_k. The table of suffix links, called *suf*, plays a crucial role in the reduction of the complexity. In some sense, it is similar to the role of failure functions used in Chapter 3. If the path from the root to the node v is spelled by $a\pi$, then *suf*[v] is defined as the node corresponding to the path π (see Figure 5.7). In the algorithm, the table *suf* is computed at a given stage for all nodes, except for leaves and maybe for the present head.

McCreight's algorithm is a rather straightforward transformation of the algorithm Left_to_Right in Section 5.2. Here most of the nodes become implicit nodes.

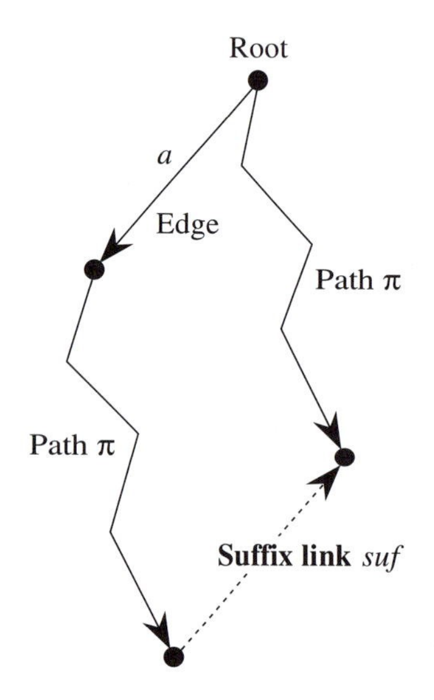

Figure 5.7: A suffix link *suf*.

But the algorithm functions similarly. The main difference appears in the procedure Up_Link_Down where *fastfind* is used whenever possible.

```
Algorithm scheme of McCreight's algorithm;
{ left-to-right suffix tree construction }
begin
    compute the two-node tree T with one edge labeled p1=text;
    for i:=2 to n do begin
        { insert next suffix pi = text [i..n] }
        localize headi as head(pi, T),
        starting the search from suf [father(headi-1)],
        using fastfind whenever possible;
        T := insert(pi, T);
    end
end.
```

The algorithm is based on the following two obvious properties:

1. $head_i$ is a descendant of the node $suf[head_{i-1}]$,
2. $suf[v]$ is a descendant of $suf[\mathrm{father}(v)]$ for any v.

The basic work performed by McCreight's algorithm involves localizing heads. If it is executed in a rough way (top-down search from the root), then the time is quadratic. The key to the improvement is the relation between $head_i$ and $head_{i-1}$ (Property 1). Hence, the search for the next head can start from some node deep in the tree, instead of from the root. This saves some work and the amortized complexity is linear. The behavior of McCreight's algorithm is illustrated in Figures 5.8 and 5.9.

```
Algorithm McCreight;
begin
    T := two-node tree with one edge labeled by p1=text;
    for i:=2 to n do begin
        { insert next suffix pi = text [i..n] }
        let β be the label of the edge (father[headi-1], headi-1);
        let γ be the label of the edge (headi-1, leafi-1);
        u := suf [father[headi-1]];
        v := fastfind(u, β);
        if v has only one son then
            { v is a newly inserted node } headi := v
        else headi := slowfind(v, γ);
        suf [headi-1] := v;
        create a new leaf leafi; make leafi a son of headi;
        label the edge (headi, leafi) accordingly;
    end
end.
```

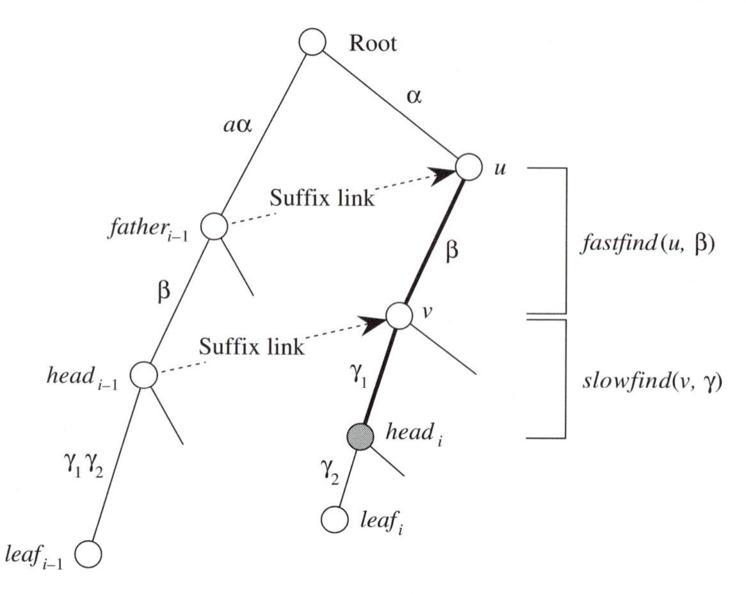

Figure 5.8: McCreight's algorithm: the case when v is an existing node.

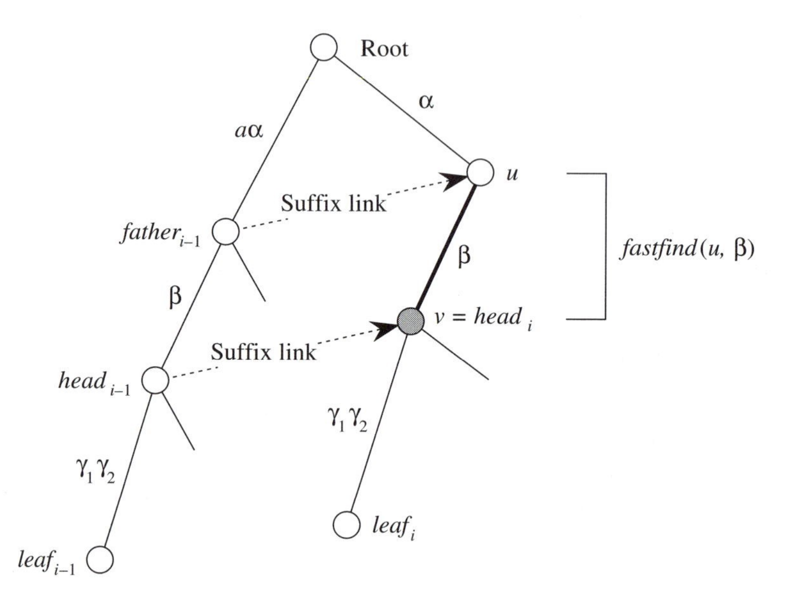

Figure 5.9: McCreight's algorithm: the case when $v = head_i$ is a newly created node.

Theorem 5.4 McCreight's algorithm has $O(n \log|A|)$ time complexity, where A is the underlying alphabet of the text of length n.

Proof Assume for a moment that the alphabet is of a constant size. The total complexity of all executions of *fastfind* and *slowfind* is estimated separately. Let $father_i$ = father($head_i$). The complexity of one run of *slowfind* at stage i is proportional to the difference $|father_i|$ - $|father_{i-1}|$, plus some constant. Therefore, the total time complexity of all runs of *slowfind* is bounded by $\Sigma(|father_i| - |father_{i-1}|) + O(n)$. This is obviously linear.

Similarly, the time complexity of one call to *fastfind* at stage i is proportional to the difference $|head_i| - |head_{i-1}|$, plus some constant. Therefore, the total complexity of all runs of *fastfind* is also linear.

If the alphabet is not fixed, we can then look for the edge starting with a given single symbol via binary search in $O(\log|A|)$ time. This is to be added as a coefficient and gives the total time complexity $O(n \log|A|)$. This completes the proof. ♦

Remarks There is a subtle point in the algorithm: the worst-case complexity is not linear if the function *slowfind* is used in the place of *fastfind*. The text $text = a^n$ is a counter-example to linear time. In fact, we could perform *slowfind* in all situations except in the following: when we are at the father of $head_i$ and we have to go down by one edge. In this situation, the work on this edge is not charged to the difference $|father_i| - |father_{i-1}|$ and it should be constant on this single edge. We can call it a *crucial* edge. The function *fastfind* guarantees that traversing the single edge takes constant time (only the first symbol on the edge is inspected and a length computed), while the function *slowfind* traverses the label of the single edge letter by letter, possibly spending a linear time on just one edge. However, the strategy (always use *slowfind* except for crucial edges) cannot be used because we do not know in advance whether an edge is crucial or not.

Another remark concerns a certain redundancy of the algorithm. In the case in which $suf[head_{i-1}]$ already exists, we could go directly to $suf[head_{i-1}]$ without making an "Up-Link-Down" tour. Such a situation can happen frequently in practice. Despite this redundancy, the algorithm is linear time. We suggest the reader may improve the algorithm by inserting a statement omitting the redundant computations. For clarity, this practical improvement is not implemented in the above version.

5.4 Weiner's algorithm

Weiner's algorithm builds the sequence of suffix trees $T_i = T(p_i)$ in the order $i=n$, $n-1$, ..., 1 (recall that $p_i = text[i\ldots n]$). At a given stage we have $T = T_{k+1}$, and we attempt to construct T_k. T is the suffix tree of the current suffix of the text. If the current suffix is denoted by p, the current value of T is the suffix tree $T(p)$.

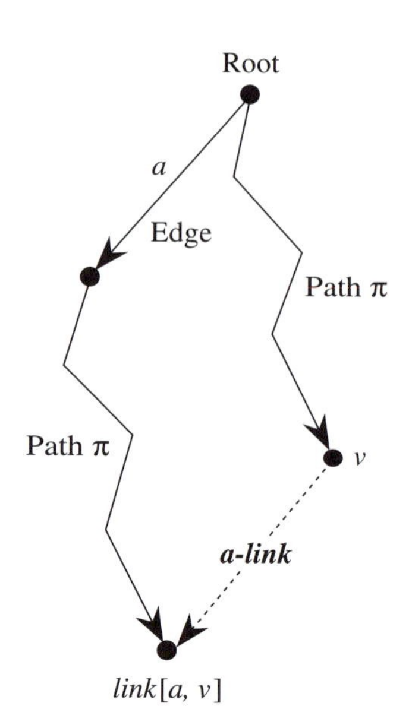

Figure 5.10: An a-link ($link_a$).

Weiner's algorithm is a rather straightforward transformation of the algorithm Right_to_Left in Section 5.2. Here again (as for McCreight's algorithm) most of the nodes become implicit nodes, and the algorithm works similarly. The notion of an a-link (see Figure 5.10) is used intensively to localize heads (see Section 5.2). Recall that $head_i$ is the node in T_i which is the father of the leaf corresponding to the most recently inserted suffix.

```
Algorithm scheme of Weiner's algorithm;
{ right-to-left suffix tree construction }
begin
   compute the suffix tree T with tables test and link for
   the one-letter word text[n]; { O(1) cost }
   for i := n-1 downto 1 do begin
       localize head as head(pi, T);     { denote it by headi }
       compute T = next(T, pi);
   end
end.
```

The basic work performed in the above algorithm involves localizing heads. If it is done in a rough way (top-down search from the root) then the total time is quadratic. The key to the improvement is the relation between $head_i$ and $head_{i+1}$. If we append the symbol $a = text[i]$ to $val(head_{i+1})$, then we obtain a prefix of $val(head_i)$ (recall that for a node v of the tree, $val(v)$ is the word label of the path from the root to v). Hence, in the search we can start from the node corresponding to this prefix, instead of from the root. This saves some work, and the amortized complexity becomes linear. However, the algorithm also becomes more complicated, as we need extra data structures for links.

To implement the idea described above, we maintain two additional tables *link* and *test* related to internal nodes of the current tree T. They satisfy invariants (p current suffix):

1. $test[a, v]$ = true iff $ax \in Fac(p)$, where $x = val(v)$;
2. $link[a, v] = w$ if $val(w) = ax$ for some node w of T,

   $link[a, v] = nil$ if there is no such node.

These tables inform us about possible left extensions of a given factor (label of node v).

Let us define the function *breaknode*($w1$, $w2$, $v1$, $v2$). This function is defined only for nodes $w1 \neq w2$ such that the label x of the path from $v1$ to $v2$ is a proper prefix of the label y of the edge ($w1$, $w2$). The value of *breaknode*($w1$, $w2$, $v1$, $v2$) is a (newly created) node w. As a side effect of the function, there are two created edges ($w1$, w) and (w, $w2$) with respective labels x and x' (x' is such that $y = xx'$). Moreover, $test[s, w]$ is set to $test[s, w2]$ for each symbol s, and $link[s, w]$ is set to nil. Hence, the edge ($w1$, $w2$) with label y is broken down into two edges with labels x and x' such that y=xx'. In the example of Figures 5.11 and 5.12, x=$text$[6...**7**], x'=$text$[8...14] and y=$text$[6...14]. It is clear that the cost to compute *breaknode*($w1$, $w2$, $v1$, $v2$) is proportional to the length of the path from $v1$ to $v2$.

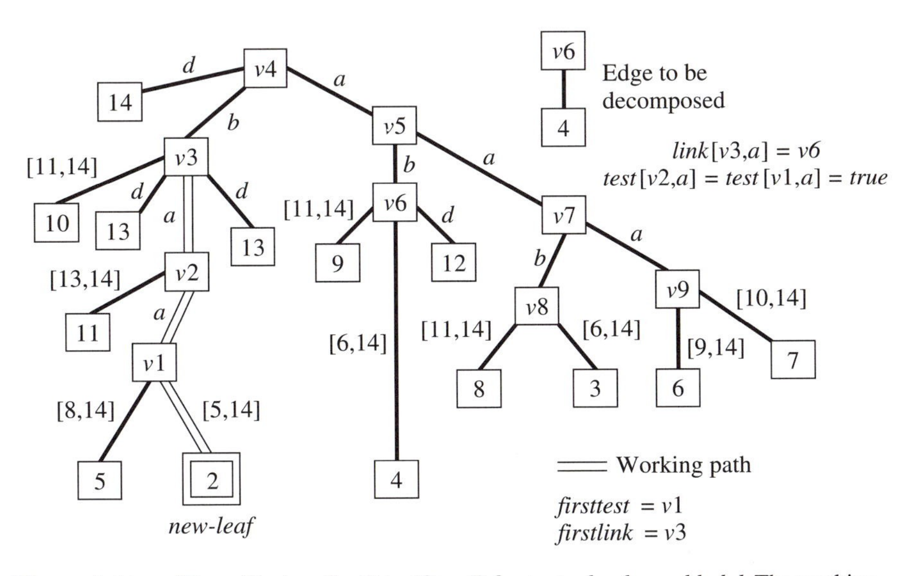

Figure 5.11: The suffix tree T_2=T($text$[2...n]) for $text$=$abaabaaaabbabd$. The working path starts in leaf 2 and goes up to the first node v such that $link[a, v] \neq nil$, here, v = $v3$. Let *firsttest* be the first node v' on this path with $test[a, v']$= true; $w1$=$link[a, v]$=$v6$, $depth(w1) \leq depth(v)+1$. The cost of one stage can be charged to the difference between the depths of leaves i+1 and i.

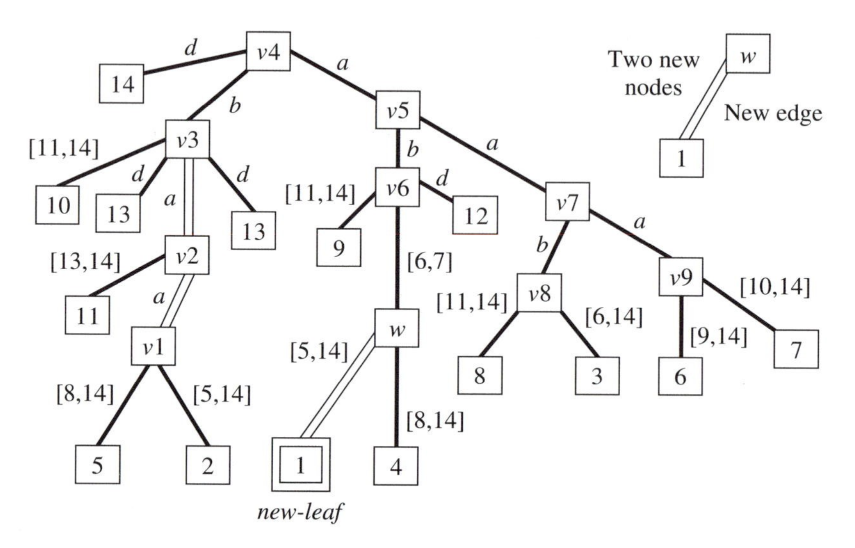

Figure 5.12: The suffix tree T_1 for *text=abaabaaaabbabd*.. The edge ($v6$, 4) of T_2 has been split into two edges. Generally, the tree T_i is a modification of tree T_{i+1} (in the example i=1). The edge ($w1$, $w2$) is split into ($w1$, w) and (w, $w2$). Here, we have w=*breaknode*($v6$, 4, $v3$, $v1$). The leaf 1 becomes the new active leaf.

We describe separately the simpler case of transforming T_{i+1} into T_i when a=*text*[i] does not occur in *text*[i+1...n]: a is a symbol not previously scanned. Define the procedure *newsymbol*(a). This function is defined only for the case described above. Let *active_leaf* be the most recently created leaf of T_{i+1}. The working path consists of all nodes on the path from *active_leaf* to root.

```
procedure newsymbol(a);
begin
   A new leaf i is created and connected directly to root;
   for all nodes v on the working path do test [a, v]:=true;
   link [a, active_leaf]:=i; active_leaf:=i;
end

Algorithm Weiner;
{ right-to-left suffix tree construction }
begin
   compute the suffix tree T with tables test and link for
   the one-letter word text [n]; { O(1) cost }
   active_leaf := the only leaf of T;
   for i := n-1 downto 1 do begin
   { T = T(p_{i+1}), active_leaf = i+1, construct T(p_i) }
      a := text [i];
      if a does not occur in p_{i+1} then newsymbol(a) else begin
```

```
            firsttest :=   the first node v on the path up from
                           active_leaf such that test [a, v]=true;
            firstlink :=   the first node v on the path up from
                           active_leaf such that link [a, v]≠nil;
            w1 := link [a, firstlink];
            if firstlink = firsttest then
            { no edge is to be decomposed } head := w1
            else begin
                w2 :=      son of w1 such that the label of path from
                           firstlink to firsttest starts with
                           the same symbol as the label of edge (w1, w2);
                head:=     breaknode(w1, w2, firstlink, firsttest);
            end;
            link [a, firsttest] := head; create new leaf with number i;
            link [a, active_leaf] := i;
            for each node v on the working path do test [a, v]:=true;
            active_leaf := i;
        end;
    end;
end.
```

The entire algorithm above is a version of Weiner's algorithm. One stage of the algorithm consists essentially of traversing a working path from *active_leaf* to the first node (*firstlink*) such that *link*[a, v]≠*nil*. In this way, the node *firsttest* is found. The tree is modified locally using the information about nodes *firstlink*, *firsttest* and w=*link*[a, *firstlink*]. One edge is added and (sometimes) one edge is split into two edges. The tables *link* and *test* are updated for nodes on the working path. The newly created leaf becomes the next active leaf (*active_leaf*).

Theorem 5.5 Weiner algorithm builds the suffix tree of a text in linear time (on fixed alphabet).

Proof One iteration (one stage) of the algorithm has a cost proportional to the length L of the working path (number of nodes on this path). However, it can be easily proven that

$$depth(link[a, firstlink]) \leq depth(firstlink)+1.$$

The cost of one stage can thus be charged to the difference between the depths of leaves i+1 and i (plus a constant c). The sum of all these differences ($depth(i)-depth(i+1)+c$) is obviously proportional to n. This completes the proof. ♦

The main inconvenience of Weiner's algorithm is that it requires tables indexed simultaneously by nodes of the tree and letters. From this point of view, McCreight's algorithm is simpler because it requires only one additional link independent of the alphabet.

5.5 Ukkonen's algorithm

In this section we give a sketch of an on-line construction of suffix trees. We drop the assumption that a suffix of the text cannot be a proper prefix of another suffix. Therefore, no marker at the end of the text is assumed. We denote the prefix of length i of the text by p^i. We add a constraint on the suffix tree construction: not only do we want to build $T(text)$, but we also want to build intermediate suffix trees $T(p^1)$, $T(p^2)$, ... $T(p^{n-1})$. However, we do not keep all these suffix trees in memory, because overall it would take a quadratic time. Rather, we transform the current tree, and its successive values are exactly $T(p^1)$, $T(p^2)$, ... $T(p^n)$. Doing so also requires that the construction take linear time (on fixed alphabet).

Goal1 (on-line suffix-tree construction): Compute the sequence of suffix trees $T(p^1)$, $T(p^2)$, ... $T(p^n)$ in linear time.

We shall first consider the uncompressed tree of suffixes of *text*, *Trie*(*text*). For simplicity, we assume that the alphabet is fixed. Our first goal is to solve an easier problem, an on-line construction of *Trie*(*text*) in reasonable time, i.e., proportional to the size of the output tree.

Goal2 (on-line uncompressed suffix-tree construction): Compute the sequence of suffix trees $Trie(p^1)$, $Trie(p^2)$, ... $Trie(p^n)$ in $O(Trie(p^n))$ time.

Let us examine closely how the sequence of uncompressed trees is constructed in Figure 5.13. The nodes corresponding to suffixes of the current text p^i are shaded. Let v_k be the suffix of length k of the current prefix p^i of the text. Identify v_k with its corresponding node in the tree. The nodes v_k are called *essential* nodes. In fact, additions to the tree (to make it grow up) are "essentially" created at such essential nodes. Consider the sequence v_i, v_{i-1}, ..., v_o of suffixes in decreasing order of their length. Compare such sequences in trees for p^3=*mam*, and for p^4=*mamm*, respectively (Figure 5.13). The basic property of the sequence of trees $Trie(p^i)$ is related to the way they grow. It is easily described with the sequence of essential nodes.

Basic growing property (*)Let v_j be the first node in the sequence v_{i-1}, v_{i-2}, ..., v_o of essential nodes, such that son(v_j, a_i) exists. Then, the tree $Trie(p^i)$ results from $Trie(p^{i-1})$ by adding a new outgoing edge labeled a_i, to each of the nodes v_{i-1}, v_{i-2}, ..., v_{j-1}, simultaneously creating a new son. If there is no such node v_j, then a new outgoing edge labeled a_i is added to each of the nodes v_{i-1}, v_{i-2}, ..., v_0.

Let *suf* be the same suffix link table as in Section 5.3. Recall that, if the node v corresponds to a factor ax (for a letter a), *suf*[v] is the node associated with x. We also assume that *suf*[*root*]=*root*. The sequence of essential nodes can be generated by iteratively taking suffix links from its first element, the leaf for which the value is the whole prefix read so far.

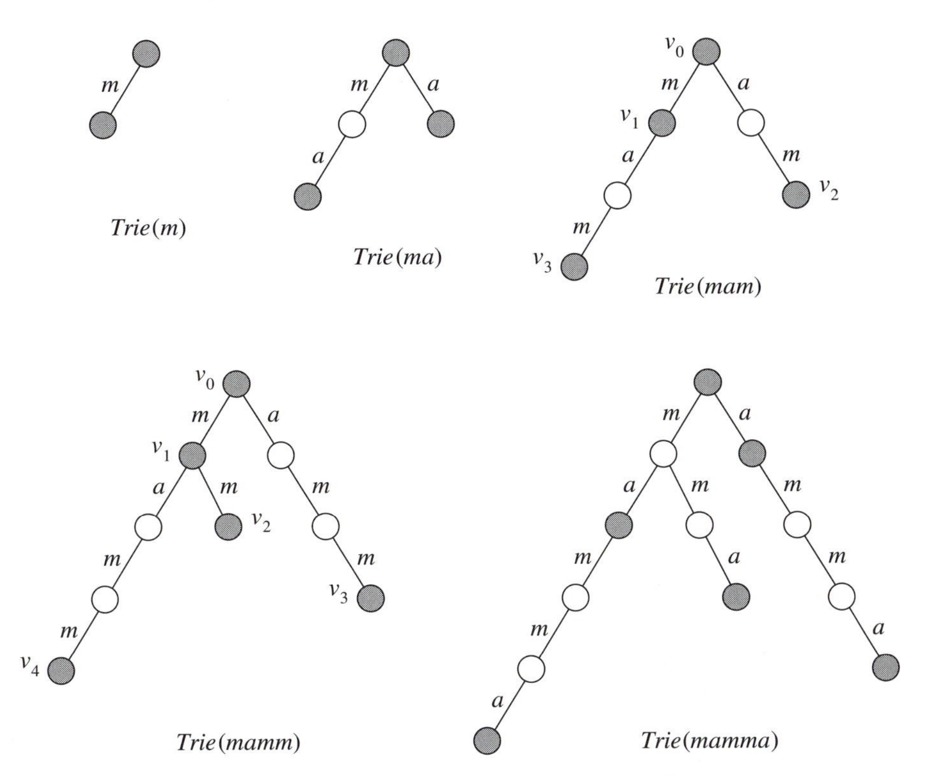

Figure 5.13: The sequence of uncompressed suffix trees for prefixes of text *mamma*. The (compressed) suffix trees result by deleting nodes of out-degree one and composing the labels. Accepting nodes are shaded.

Basic property of the sequence of essential nodes $(**)(v_i, v_{i-1}, ..., v_0) = (v_i, suf[v_i], suf^2[v_i], ..., suf^i[v_i])$.

In order to easily describe the on-line construction of *Trie*(*text*), we introduce the procedure *createson*. Let v be a node of the tree, and let a be a letter. Then, *createson*(v, a) connects a new son of the node v with an edge labeled by letter a. The procedure returns the created son of v. Using properties (*) and (**), one iteration of the algorithm appears as suggested in Figure 5.14. An informal construction is given below. A more concrete description of the on-line algorithm follows.

```
algorithm on_line_trie; { informal version }
begin
    create the two-node tree Trie(a1) with suffix links;
    for i := 2 to m do begin
        vi-1 := deepest leaf of Trie(p^(i-1));
        k := smallest integer such that son(suf^k [vi-1],ai) exists;
```

```
            create a_i-sons for v_{i-1}, suf[v_{i-1}], ..., suf^{k-1}[v_{i-1}],
            creating suffix links (see Figure 5.14);
      end;
   end

   algorithm on_line_trie;
   begin
      create the two-node tree T = Trie(p^1) with suffix links;
      for i := 2 to m do begin
            v := deepest leaf of T;
            q := createson(v, a_i); v := suf[v];
            while son(v,a_i) undefined do begin
                  suf[q] := createson(v, a);
                  q := suf[q]; v := suf[v];
            end;
            if q = son(v,a_i) then { v = root } suf[q] := root
            else suf[q] := son(v,a_i);
      end
   end
```

Theorem 5.6 The algorithm *on_line_trie* builds the tree *Trie*(*text*) of suffixes of *text* in an on-line manner. It works in time proportional to the size of *Trie*(*text*).

Proof The correctness results from properties (*) and (**). The complexity results from the fact that the work performed in one iteration is proportional to the number of created edges (calls to *createson*). This completes the proof. ♦

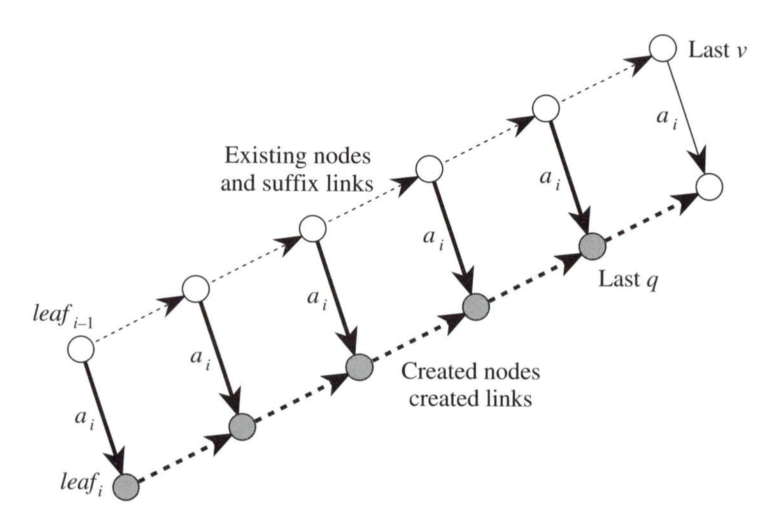

Figure 5.14: One iteration in the construction of *Trie*(*text*). Bold arrows are created at this step.

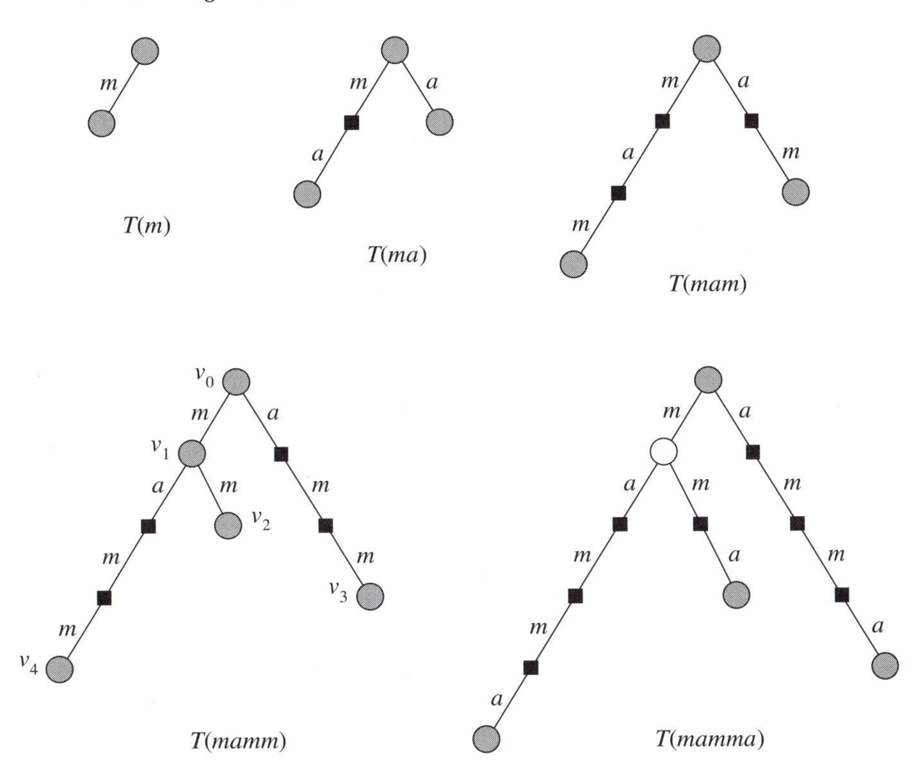

Figure 5.15: The sequence of (compressed) suffix trees (with implicit nodes marked as ■).

The i-th iteration executed in the previous algorithm can be adapted to work on the compressed suffix tree $T(p^{i-1})$. Each node of $Trie(p^{i-1})$ is treated as an implicit node. Hence, the new algorithm simulates the version *on_line_trie* (see Figure 5.15). Here again, the notion of implicit nodes from Section 5.2 is useful. Recall that a pair (v, α) is an implicit node in T iff v is a node of T and α is a prefix of the label of an edge from v to a son of it. The implicit node (v, α) is said to be a "real" node iff α is the empty word.

Let v be a node of the tree T, and let p be a subword of the input word *text* (represented by a pair of integers l, r, where $p = text[l \ldots r]$). The basic function used in McCreight's algorithm is the function *find*(v, p). This function follows a path labeled by p from node v, possibly creating a new node. In the present algorithm we use a similar strategy, except that the creation of a possible node is postponed. The corresponding function is called *normalize* and applies to a pair (v, p). The value of *normalize*(v, p) is the last implicit node (v, α) on the path spelled by p from v. Whenever this function is applied, the word p is fully contained in the followed path, hence, the *fastfind* implementation of Section 5.3 can be adapted to the work.

We have to interpret the suffix links and sons (by single letters) of implicit nodes in the compressed tree. In the algorithm we use the notation *son′* and *create-*

son'. They are interpreted as follows (a is a letter, and (v, α) is an implicit node, identified with v if α is empty):

1. $son'((v, \alpha), a)$ exists iff the path α originated at node v can be extended by the letter a; in this case $son'((v, \alpha), a) = (v, \alpha a)$, which can be a real node.
2. The procedure $createson'((v, \alpha), a)$ first checks if α is non-empty. If so, the procedure first makes (v, α) a real node (applying $break(v, \alpha)$ from Section 5.3, for example). The procedure then creates an a-son of the node.

There is still one important implementation detail. If the node v is a leaf, then there is no need to extend the edge coming from its father by a single symbol. If the label of the edge is a pair of positions (l, r), then, after updating it, it should be $(l, r+1)$. We can omit such updates by setting r to infinity for all leaves. This infinity is automatically understood as the last scanned position i of the pattern. Doing so reduces the amount of work involved in one iteration of the algorithm. If $v_i, v_{i-1}, ..., v_0$ is the sequence of essential nodes, and if we know that $v_i, v_{i-1}, ..., v_k$ are leaves, then we can skip processing them because this is done automatically by the infinity trick. We thus start processing essential nodes from v_{k-1}. In the algorithm we call v the current node that run through essential nodes. Since there is no sense in dealing with the situation when v is a leaf, at a given iteration, the first value of v is the a_i-son of the last value of v (from the preceding iteration), rather than starting from the deepest leaf as in the on-line construction of *Trie*(*text*). All this gives the on-line suffix tree construction following.

```
algorithm of Ukkonen;
begin
   create the two-node tree T(a1) with suffix links;
   (v, a) := (root of the tree, e);
   for i := 2 to m do begin
        if son'((v,a),ai) undefined then createson'((v,a),ai);
        repeat
             (q, β) := son'((v, α), ai);
             (v, α) := normalize(suf[v], α);
             if son'((v,α),ai) undefined then createson'((v,α),ai);
             if β is empty then suf[q] := son'((v,α),ai);
             (v,α) := son'((v,α),ai);
                          { here, the length of α can only be increased by one }
        until no edge created or v = root;
        if v = root then suf[son'(v,ai)] := root;
   end
end
```

Theorem 5.7 Ukkonen algorithm builds the compressed tree $T(text)$ in an on-line manner. It works in linear time (on a fixed alphabet).

Proof The correctness follows from the correctness of the version working on an uncompressed tree. The new algorithm is just a simulation of it.

To prove the $O(|text|)$ time bound, it is sufficient to prove that the total work is proportional to the size of $T(text)$, which is linear. The work is proportional to the work performed by all *normalize* operations. The cost of one *normalize* operation is proportional to the decrease of the length of α. On the other hand, the length of α is increased by at most one per iteration. Hence, the total number of reductions in length of αs is linear. This completes the proof. ♦

5.6 Suffix arrays: An alternative data structure

There is a clever and rather simple way to deal with all suffixes of a text: to arrange their list in increasing lexicographic order in order to perform binary searches on them. The implementation of this idea leads to a data structure called *suffix array*. It is not exactly a "good representation" in the sense defined at the beginning of this chapter. But it is "almost good." This means that it satisfies the following conditions:

1. it has $O(n)$ size,

2'. it can be constructed in $O(n.\log n)$ time,

3'. it allows the computation of $FACTORIN(x, text)$ in $O(|x| + \log n)$ time.

So the time required to construct and use the structure is slightly greater than that needed to compute the suffix tree (it is $O(n.\log|A|)$ for the latter). But suffix arrays have two advantages:

- their construction is rather simple; it is even commonly admitted that, in practice, it behaves better than the construction of suffix trees;
- it consists of two linear size arrays which, in practice again, take little memory space (typically four times less space than suffix trees).

Let $text = a_1a_2\ldots a_n$, and let $p_i = a_ia_{i+1}\ldots a_n$ be the i-th suffix of the text. Let $Pos[k]$ be the position i where the k-th smallest suffix of *text* starts (according to the lexicographic order of the suffixes). In other words, the k-th smallest suffix of *text* is $p_{Pos[k]}$. Denote by $LCPREF[i, j]$ the length of the longest common prefix of suffixes p_i and p_j. For simplicity, we assume that the size of the pattern is a power of two. If it is not, we can add a suitable number of end markers at the end of *text* to meet the condition. Let us call the whole interval $[1\ldots n]$ *regular*, and define regular subintervals by the rule: if an interval is regular, then its halves are also regular. We say that an interval $[i\ldots j]$ is regular iff it can be obtained from the whole interval in this way. Finally, observe that there are exactly $2n-1$ regular intervals.

The suffix array of *text* is the data structure consisting of both the array *Pos*, and the values *LCPREF*[i, j] for each regular interval [i...j]. Hence, the entire data structure has $O(n)$ size, and satisfies condition (1). To show that condition (2') is satisfied we can apply the results of Chapter 9, related to another data structure called the dictionary of basic factors. Its definition has been postponed until this chapter. In fact, this dictionary is an alternative "almost good" representation of the set of factors. Note that condition (2') is rather intuitive because it involves sorting n words having strong mutual dependencies. But, in our opinion, the most interesting property of suffix arrays is that they satisfy condition (3').

Theorem 5.8 Assume that the suffix array of *text* is computed (i.e., are the table *Pos* and the table *LCPREF* for regular intervals). Then, for any word x, we can compute *FACTORIN*(x, *text*) in $O(|x| + \log|text|)$ time.

Proof Let $min_x = \min\{k \,/\, x \leq p_{Pos[k]}\}$ and $max_x = \max\{k \,/\, x \geq p_{Pos[k]}\}$, where ≤ and ≥ denote inequalities in the sense of the lexicographic ordering. Then, all occurrences of the subword x in *text* are at positions given in the table *Pos* between indices min_x and max_x. There is at least one occurrence of x inside *text* iff $min_x \leq max_x$.

We show how to compute min_x. The computation of max_x is symmetrical. We start with an extremely simple algorithm having $O(x.\log n)$ running time. It is a binary search for x inside the sorted sequence of suffixes. Let $suf(k) = p_{Pos[k]}$. Hence, $suf(k)$ is the k-th suffix in the lexicographic numbering. We describe the computation of min_x recursively. The value of min_x is assumed to be found inside the interval [*Left*...*Right*].

```
function find(Left, Right): integer;
begin
    if Left = Right then return Left;
    Middle := right position of the left half of [Left..Right];
    if x ≤ suf(Middle) then return find(Left, Middle)
    else return find(Middle+1, Right)
end
```

Let l = *LCPREF*(*suf*(*Left*), x), and r = *LCPREF*(*suf*(*Right*), x). The basic property of the function *find* is the following fact.

Basic property Let l, r be the respective values of *Left* and *Right* at a given call of the function *find*. And let l', r' be the values of *Left* and *Right* at the next call of *find*. Then, $\max(l, r) \leq \max(l', r')$. If we know the values of l, r, then we can check the inequality $x \leq suf(Middle)$, and compute l', r' in time $O(D)$, where $D = \max(l', r') - \max(l, r)$.

Proof The proof is left to the reader. Several cases are to be considered, depending on the relations between numbers l, r, *LCPREF*(*Left*, *Middle*) and *LCPREF*(*Mid-*

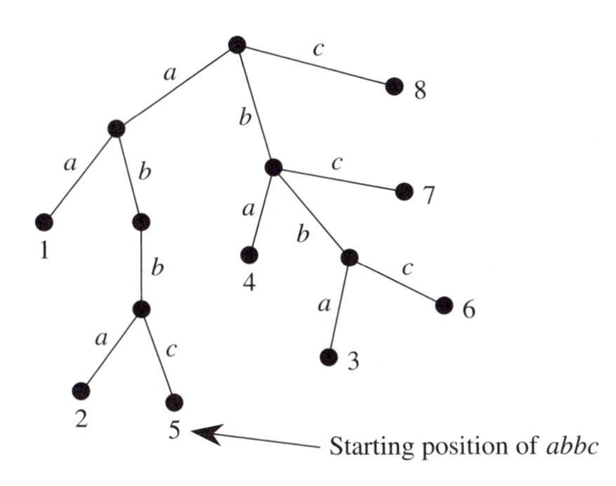

Figure 5.16: The position tree of *text* = *aabbabbc*. Compare with its suffix tree in Figure 5.1.

dle+1, *Right*). The sum of all Δs is $O(|x|)$, since $l, r \leq |x|$. Hence, the total complexity is $O(|x| + \log n)$ because the function *find* makes $\log n$ recursive calls. ♦

The computation of *LCPREF* is discussed at the end of the next section.

5.7 Applications

There is a data structure slightly different from the suffix tree, known as the position tree. It is the tree of position identifiers. The *identifier* of position i on the text is the shortest prefix of $text[i \ldots n]$ that does not occur elsewhere in *text*. Identifiers are well-defined when the last letter of *text* is a marker. Once we have the suffix tree of *text*, computing its position tree is fairly obvious (see Figure 5.16). Moreover, this shows that the construction works in linear time.

Theorem 5.9 The position tree of a given text can be computed in linear time (on a fixed alphabet).

One of the main applications of suffix trees is evident in the situation in which the text *text* is like a dictionary. In this situation, the suffix tree or the position tree act as an index on the text. The index contains virtually all the factors of the text. With the data structure, the problem of locating a word w in the dictionary can be solved efficiently. But we can also perform other operations rapidly, such as computing the number of occurrences of w in *text*.

Theorem 5.10 The suffix tree of *text* can be preprocessed in linear time so that, for a given word w, the following queries can be executed on-line in $O(|w|)$ time:

- find the first occurrence of w in *text*;
- find the last occurrence of w in *text*;
- compute the number of occurrences of w in *text*.

We can list all occurrences of w in *text* which can be listed in time $O(|w|+k)$, where k is the number of occurrences.

Proof We can preprocess the tree, computing bottom-up, for each node, values *first*, *last* and *number*, corresponding respectively to the first position in the subtree, the last position in the subtree, and the number of positions in the subtree. Then, for a given word w, we can (top-down) retrieve this information in $O(|w|)$ time (see Figure 5.17). This gives the answer to the first three queries of the statement.

To list all occurrences, we first access the node corresponding to the word w in $O(|w|)$ time. Then, we traverse all leaves of the subtree to collect the list of positions of w in *text*. Let k be the number of leaves of the subtree. Since all internal nodes of the subtree have at least two sons, the total size of the subtree is less than $2.k$, and the traversal takes $O(k)$ time. This completes the proof. ♦

The longest common factor problem is a natural example of a problem easily solvable in linear time using suffix trees, and very difficult to solve without any essential use of "good" representation of the set of factors. In fact, it has long been believed that no linear time solution to the problem is possible, even if the alphabet is fixed.

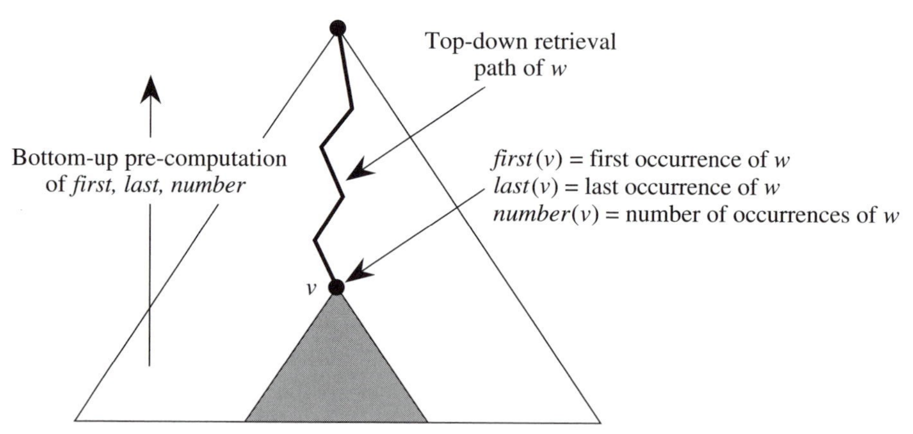

Figure 5.17: We can preprocess the tree to compute (bottom-up), for each node, its corresponding values *first*, *last*, and *number*. Then, for a given word w, we can retrieve this information in $O(|w|)$ time.

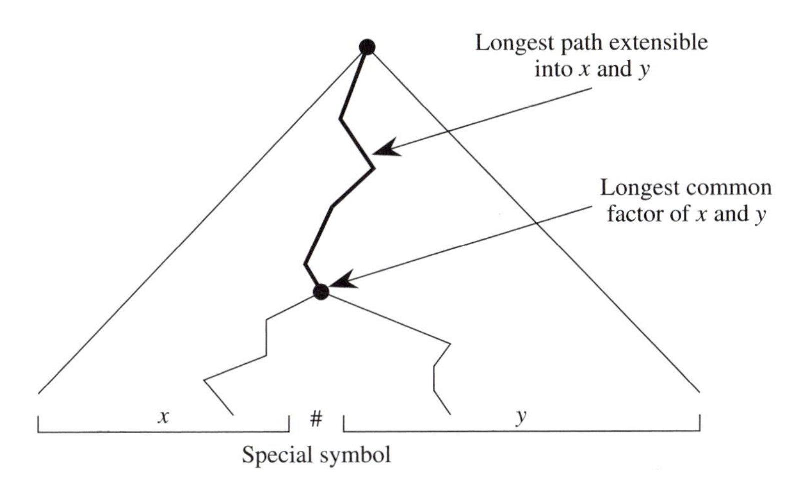

Figure 5.18: Finding longest common factors with suffix trees.

Theorem 5.11 The longest common factor of k words can be found in linear time in the size of the problem, i.e., the total length of words (k is a constant, alphabet is fixed).

Proof The proof for case k=2 is illustrated by Figure 5.18. The general case, with k fixed, is solved similarly. We compute the suffix tree of the text consisting of the given words separated by distinct markers. Then, exploring the tree bottom-up, a vector informing for each node is computed, for each $1 \leq i \leq k$, whether a leaf corresponding to a position inside the i-th subword is in the subtree of this node. The deepest node with positive information of this type for each i ($1 \leq i \leq k$) corresponds to the longest common factor. The total time is linear. This completes the proof. ♦

Another solution to the problem of the longest common factor of two words is given in Chapter 6. It uses the suffix dawg of only one pattern, and processes the other word in an on-line manner.

Figure 5.19 shows an application of suffix trees to the problem of finding the longest repeated factor inside a given text. The solution is analogous to the solution of the longest common factor problem, considering that k=1. Problems of this type (related to regularities in strings) will be treated in more detail in Chapter 8, where we will also give an almost linear-time algorithm ($O(n \log n)$). This latter algorithm is simpler and does not use suffix trees or equivalent data structures. This algorithm will also cover the two-dimensional case, where our "good" representations (considered in the present chapter) are not well suited.

Let $LCPREF(i, j)$ denote, as it did in Section 5.6, the length of the longest common prefix starting at positions i and j in a given text of size n. In Chapter 12 on two-dimensional pattern matching we frequently use the following fact.

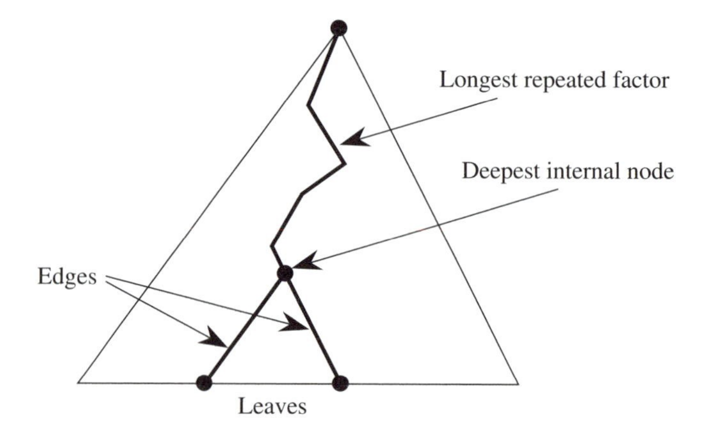

Figure 5.19: Computation of the longest repeated factor.

Theorem 5.12 It is possible to preprocess a given text in linear time so that each query $LCPREF(i, j)$ can be answered in constant time, for any positions i, j. A parallel preprocessing in $O(\log n)$ time with $O(n/\log n)$ processors of an EREW PRAM is also possible (on a fixed alphabet).

Let $LCA(u, v)$ denote the lowest common ancestor of nodes u, v in a given tree T. This proof of Theorem 5.12 easily reduces to preprocessing the suffix tree that enables LCA queries in constant time. The value $LCPREF(i, j)$ can be computed as the size of the string corresponding to the node $LCA(v_i, v_j)$, where v_i, v_j are leaves of the suffix tree corresponding to suffixes starting at positions i and j.

Theorem 5.13 It is possible to preprocess a given tree T in linear time in such a way that each query $LCA(u, v)$ can be answered in constant time. A parallel preprocessing in $O(\log n)$ time with $O(n/\log n)$ processors of an EREW PRAM is also possible (on a fixed alphabet).

The beauty of the proof of Theorem 5.13 is beyond the scope of this book. It is based on a preprocessing corresponding to the following two simple subcases:

- T is single path; then $LCA(u, v)$ is the node (u or v) closer to the root; it is sufficient to associate to each node its distance to the root, to efficiently answer further queries;
- T is the complete binary tree with 2^{k-1} nodes; identify nodes u, v with their numbers in the inorder numbering of the tree; then, the number of $LCA(u, v)$ is given by a simple arithmetic formula of constant size; in this case, the preprocessing consists of computing the inorder number of the nodes of the tree.

In the general case, the proof is based on a transformation F of the tree T into a complete binary tree $F(T)$. The transformation is such that $F^{-1}(q)$ is a path in T, for any node q in $F(T)$.

All applications discussed so far show the power of suffix trees. However, the data structure of the next chapter, dawg, can be applied as well (except maybe for the computation of *LCPREF*). Dawgs are structurally equivalent to suffix trees. We now present explicitly how to use suffix trees or dawgs alternatively on the next problem: compute the number of distinct factors of *text* (cardinality of the set *Fac*(*text*)). This problem also falls into the category of problems for which efficient algorithms without essential use of any of "good" data structures are hardly imaginable.

Lemma 5.14 We can compute the number of factors of a text (cardinality of the set *Fac*(*text*)) in linear time.

Proof 1—suffix tree application Let T be the suffix tree of the text. The weight of an edge in T is defined as the length of its label. Then, the required number is the sum of weights of all labels. ♦

Proof 2—dawg application Let D be the suffix dawg of the text. Compute the number M of all paths in D starting from the root (not necessarily ending at the sink). Then, $M=|Fac(text)|$. The number M is computed bottom-up in linear time. ♦

Illustrated by Lemma 5.14, we conclude this chapter with the following general "metatheorem": suffix trees and dawgs are "good" representations for the set of factors of a text.

Bibliographic notes

The two basic algorithms for suffix tree construction are from Weiner [We 73] and McCreight [McC 76]. Our exposition of Weiner's algorithm uses a version from Chen and Seiferas [CS 85]. This paper also describes the relation between dawgs and suffix trees. An excellent survey on applications of suffix trees has been shown by Apostolico in [Ap 85].

The on-line algorithm of Section 5.5 has been recently discovered by Ukkonen [U 92]. Note that the algorithm of McCreight is not on-line because it has to look ahead of symbols of the text. The notion of suffix array was invented by Manber and Myers [MM 90]. They use a different approach to the construction of suffix arrays than our exposition (which refers to the dictionary of basic factors in Chapter 9). The worst-case complexities of the two approaches are identical, though the average complexity of the original solution proposed by Manber and Myers is better. An equivalent implementation of suffix arrays is considered by Gonnet and Baeza-Yates in [GB 91], and called *Pat trees*.

As reported in [KMP 77], in 1970 Knuth conjectured that a linear-time computation of the longest common factor problem was impossible to achieve. Theorem 5.11 shows that it is indeed possible on a fixed alphabet.

A direct computation of uncompacted position trees, without the use of suffix trees, is given in [AHU 74]. The algorithm is quadratic because uncompacted suffix trees can have quadratic size.

Theorem 5.13, related to the lowest common ancestor problem in trees, is by Schieber and Vishkin [SV 88]. We refer the reader to this paper for more details. In [HT 84] an alternative proof is given. However, their solution is much more complicated, and only a sequential preprocessing is given.

Selected references

[GB 91] Gonnet, G.H., & Baeza-Yates, R., *Handbook of Algorithms and Data Structures*, Addison-Wesley, Reading, Mass., 1991.

[MM 90] Manber, U., & Myers, E., "Suffix arrays: a new method for on-line string searches," in (*Proc. of 1st ACM-SIAM Symposium on Discrete Algorithms*, American Mathematical Society, Providence, 1990): 319–327.

[McC 76] McCreight, E.M., "A space-economical suffix tree construction algorithm," *J. ACM* 232 (1976): 262–272.

[Uk 92] Ukkonen, E., "Constructing suffix trees on-line in linear time," in (*IFIP'*92): 484–492.

[We 73] Weiner, P., "Linear pattern matching algorithms," in (*Proc. 14th IEEE Annual Symposium on Switching and Automata Theory*, Institute of Electrical Electronics Engineers, New York, 1973): 1–11.

6

Subword graphs

The directed acyclic word graph (dawg) is a good data structure representing the set *Fac*(*text*) of all factors of the word *text* of length n. It is an alternative to the suffix trees of Chapter 5. The graph *DAWG*(*text*), called the suffix dawg of *text*, is obtained by identifying isomorphic subtrees of the uncompacted tree *Trie*(*text*) representing *Fac*(*text*) (see Figure 6.1). An advantage of dawgs is that each edge is labeled by a single symbol. It is somehow more convenient to use it when information is associated with edges rather than with nodes. We consider only dawgs representing the set *Fac*(*text*), but it is clear that the approach also applies to other sets of words, such as the set of subsequences of a string.

The linear size of suffix trees is a trivial fact (Chapter 5), but the linear size of suffix dawgs is much less trivial, and it is probably the most surprising fact related to representations of *Fac*(*text*).

Applications of suffix dawgs are essentially the same as applications of suffix trees, such as those presented in Section 5.7. Indexing is the main purpose of these data structures. Theorem 5.11 can also be proven with dawgs. But in the present chapter we present two quite surprising uses of dawgs in Section 6.5. There, the suffix dawg of the pattern serves to search for it inside a text. The second method leads to one of the most efficient algorithms for searching a text.

6.1 Size of suffix DAWGs

A node in the graph *DAWG*(*text*) naturally corresponds to a set of factors of the text: factors having the same right context. It is not difficult to be convinced that all these

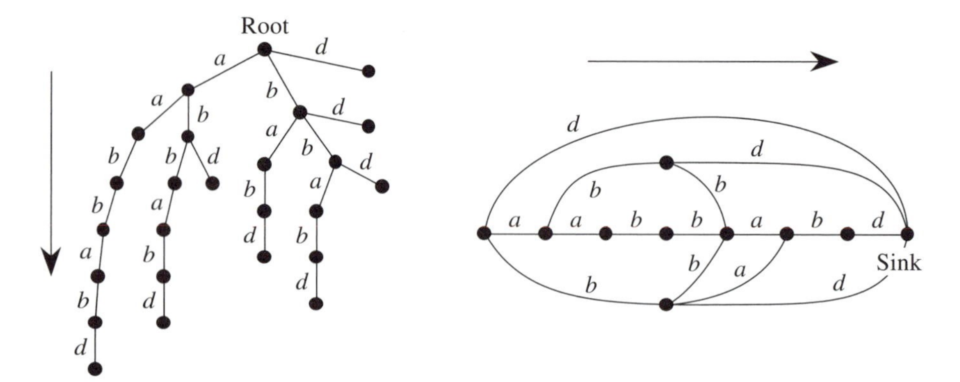

Figure 6.1: The uncompacted suffix tree *Trie*(*aabbabd*), and the direct acyclic word graph *DAWG*(*text*).

factors have the following property: their first occurrences end at the same position in *text*. The converse is not necessarily true, but the remark gives an intuition of the following definition.

Let x be a factor of *text*. We denote by *end-pos*(x) (ending positions) the set of all positions in *text* where an occurrence of x ends. Let y be another factor of *text*. Then, the subtrees of *Trie*(*text*) rooted at x and y (recall that we identify the nodes of *Trie*(*text*) with their labels) are isomorphic iff *end-pos*(x) = *end-pos*(y) (when the text ends with a special end marker). In the graph *DAWG*(*text*), paths x having the same set *end-pos*(x) lead to the same node. Hence, the nodes of G correspond to nonempty sets of the form *end-pos*(x). The root of the dawg corresponds to the whole set of positions $\{0, 1, 2, \ldots, n\}$ on the text. From a theoretical point of view, nodes of G can be identified with such sets (especially when analyzing the construction algorithm). But, from a practical point of view, the sets are never maintained explicitly. The *end-pos* sets, usually large, cannot directly name nodes of G, because the sum of the sizes of all such sets happens to be nonlinear. An explicit representation of each *end-pos* sets by a list of its elements is too large. Figure 6.1 presents both the trie of the text *aabbabd*, and *DAWG*(*text*) for the same word. The graph has 10 nodes and 15 arcs. If we allow labels of edges to have varying lengths then its size can be further reduced to 7 nodes and 12 arcs (by compressing chains).

The small size of dawgs is due to the special structure of the family Φ of sets *end-pos*. We associate each node v of the dawg with its value *val*(v) equal to the longest word leading to it from the root. The nodes of the dawg are, in fact, equivalence classes of nodes of the uncompacted tree *Trie*(*text*), where the equivalence means subtree isomorphism. In this sense, *val*(v) is the longest representative of its equivalence class. It can also be considered that nodes of the dawg are equivalence classes of factors of the text because nodes in *Trie*(*text*) are in one-to-one correspondence with factors of *text*.

The notion of failure function, used intensively in Chapter 3, has an exact counterpart in dawgs. Let v be a node of *DAWG*(*text*) distinct from the *root*. We

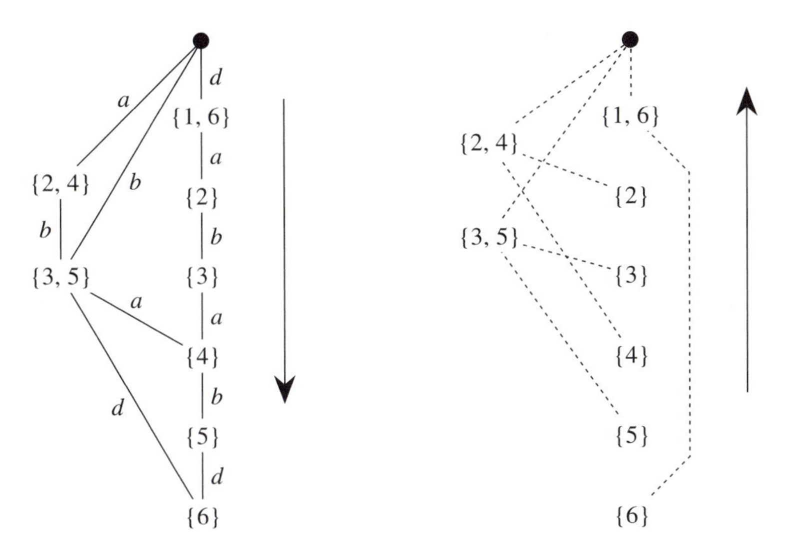

Figure 6.2: *DAWG*(*dababd*) (left) and its suffix links (right). This shows the structure of the family F of *end-pos* sets.

define *suf*[*v*] as the node *w* such that *val*(*w*) is the longest suffix of *val*(*v*) not equivalent to it. In other words, *val*(*w*) is the longest suffix of *val*(*v*) corresponding to a node different from *v*. Note that the definition implies that *val*(*w*) is a proper suffix of *val*(*v*). By convention, we define *suf*[*root*] as *root*. In the implementation this means that *suf* is represented as a table. The table *suf* is analogous to the table *Bord* defined in Chapter 3. We also call the table *suf* the table of *suffix links* (edges (*v*, *suf*[*v*]) are "suffix links").

The suffix links in *DAWG*(*dababd*) are presented in Figure 6.2. Since the value of *suf*[*v*] is a word strictly shorter than *val*(*v*), *suf* induces a tree structure on the set of nodes. The node *suf*[*v*] is interpreted as the father of *v* in the tree. It so happens that the tree of suffix links is the same as the tree structure of sets in Φ induced by the inclusion relation.

Theorem 6.1 The size of *DAWG*(*text*) is linear. More precisely, if *N* is the number of nodes of *DAWG*(*text*) and $n=|text|>0$, then $N < 2n$. Moreover, *DAWG*(*text*) has less than $N+n-1$ edges. This is independent of the size of the alphabet.

Proof The main property used here is derived from the definition of *end-pos* sets: any two subsets of Φ are either disjoint or one is contained in the other. Thus, the family Φ has a tree structure (see Figure 6.2). All leaves are pairwise disjoint subsets of $\{1, 2, \ldots, n\}$ (we do not count position 0 that is associated with the root, because it is not contained in any other *end-pos* set). Hence, there are at most *n* leaves. This does not directly imply the thesis because it can occur that some internal nodes have only one son (as in the example of Figure 6.2).

We partition nodes into two (disjoint) subsets according to the fact that *val*(v) is a prefix of *text* or not. The number of nodes in the first subset is exactly n+1 (number of prefixes of *text*). We now count the number of nodes in the other subset of the partition.

Let v be a node such that *val*(v) is not a prefix of *text*. Then *val*(v) is a non-empty word that occurs in at least two different right contexts in *text*. But we can then deduce that at least two different nodes p and q (corresponding to two different factors of *text*) are such that *suf*[p] = *suf*[q] = v. This show that nodes like v have at least two sons in the tree inferred by *suf*. Since the tree has at most n leaves (corresponding to non-empty prefixes), the number of such nodes is less than n. Additionally note that if *text* contains two different letters, the root has at least two sons but cannot be counted in the second subset because *val*(*root*)=ε is prefix of *text*. If *text* is of the form a^n, the second subset is empty. Therefore, the cardinality of the class is indeed less than n–1. This finally shows that there are less than $(n+1)+(n-1) = 2n$ nodes.

To prove the bound on the number of edges, we consider a spanning tree T over *DAWG*(*text*), and count separately the edges belonging to the tree and the edges outside the tree. The tree T is chosen to contain the branch labeled by the whole text. Since there are N nodes in the tree, there are N–1 edges in the tree. Let us count the other edges of *DAWG*(*text*). Let (v, w) be such an edge. We associate with it the suffix xay of *text* defined by: x is the label of the path in T going from the root to v, a is the label of the edge (v, w), y is any factor of *text* extending xa into a suffix of *text* ($x, y \in A^*$, $a \in A$). It is clear that the correspondence is one-to-one. Moreover, the empty suffix is not considered, nor is *text* itself because it is in the tree. It remains n–1 suffixes, which is the maximum number of edges outside T.

Finally, the number of edges in *DAWG*(*text*) is less than $N+n-1$. ♦

Although the size of *DAWG*(*text*) is linear, it is not always strictly minimal. If minimality is understood in terms of finite automata (number of nodes, for short), then *DAWG*(*text*) is the minimal automaton for the set of suffixes of *text*. The minimal automaton for *Fac*(*text*) can, indeed, be slightly smaller.

6.2 A simple off-line construction

Our first construction of dawgs consists essentially of a transformation of suffix trees. The basic procedure is the computation of equivalent classes of subtrees. It is based on a classical algorithm concerning tree isomorphism (see [AHU 74] for example). We just recall the result without proof.

Lemma 6.2 ([AHU 74]) Let T be a rooted ordered tree in which the edges are labeled by letters (assumed to be of constant size). Then, isomorphic classes of all subtrees of T can be computed in linear time.

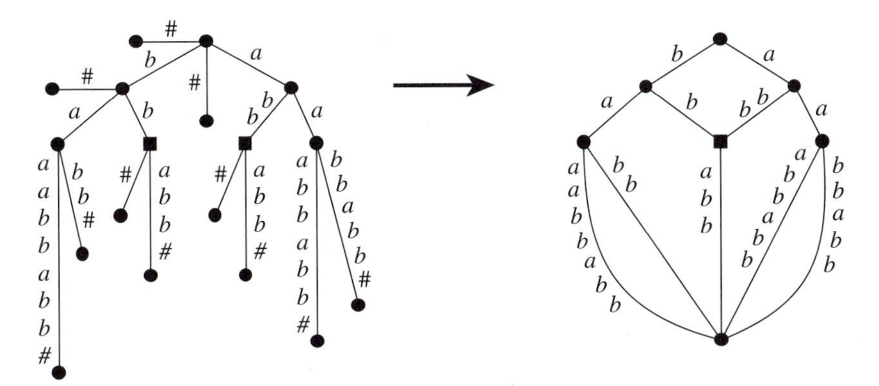

Figure 6.3: Identification of isomorphic classes of nodes of the suffix tree. In the algorithm the labels of the edges are constant-sized names of the corresponding strings.

We call the dawg in which edges are labeled by words, and no node has only one outgoing edge (chains of nodes are compacted) a *compacted dawg*. A simple application of Lemma 6.2 provides a construction of compacted dawgs, that leads to the construction of dawgs.

Theorem 6.3 *DAWG(text)* and its compacted version can be built in linear time (on fixed alphabet).

Proof We illustrate the algorithm with the example text string *w=baaabbabb*. The suffix tree of *w*# is constructed by any method from Chapter 5. For technical reasons inherent to the definition of suffix trees, the end marker # is added to the word *w*. But this end marker is later ignored in the constructed dawg (it is only needed at this stage).

Then, by the algorithm of the preceding lemma, the equivalence classes of isomorphic subtrees are computed. The roots of isomorphic subtrees are identified, and we get the compacted version G' of *DAWG*(w) (see Figure 6.3). The entire process takes linear time.

The difference between the actual structure and the dawg is related to the lengths of strings that are labels of edges. In the dawg each edge is labeled by a single symbol. The "naive" approach could be to break each edge labeled by a string of length k down into k edges. But the resulting graph could have a quadratic number of nodes. We apply such an approach with the following modification. By the weight of an edge we mean the length of its label. For each node v we compute the heaviest (with the largest weight) incoming edge. Denote this edge by *inedge*(v). Then, in parallel for each node v, we perform a local transformation *local_action*(v) (see Figure 6.4). It is crucial that all these local transformations *local_action*(v) are independent and can be performed simultaneously for all v.

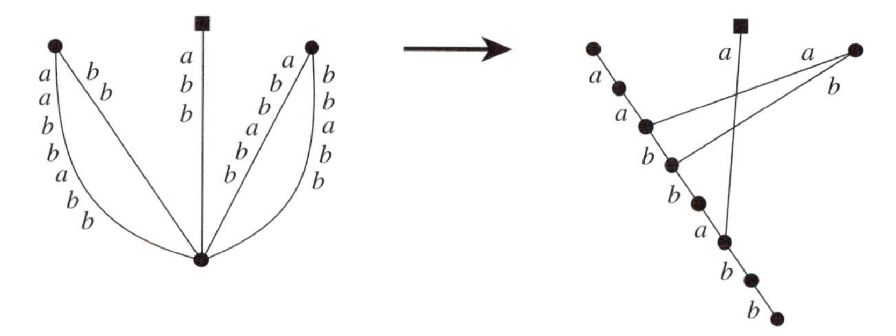

Figure 6.4: Local transformation. New nodes are introduced on the heaviest incoming edge.

The transformation *local_action*(v) consists of breaking the edge *inedge*(v) down into k edges, where k is the length of the string z, label of this edge. The label of the i-th created edge is the i-th letter of z. There are new interleaved k–1 nodes: (v, 1), (v, 2), ..., (v, k–1). The node (v, i) is at the distance i from v. For each other incoming edge of v, e = (v1, v) we perform the following additional action. Suppose that the string, label of e has length p, and that its first letter is a. If p>1 then we remove the edge e and create an edge from v1 to (v, p–1) the label of which is the letter a. This is graphically illustrated in Figure 6.4.

Then we apply the action *local_action*(v) to all nodes v, except the root.

The resulting graph for our example string w is presented in the next figure (see Figure 6.5). This graph is the required *DAWG*(w). This completes the proof. ♦

The proof of the previous statement shows how the dawg *DAWG*(*text*) can be built. The role of the end marker in the preceding proof is to allow the suffix tree construction. Whenever it is removed from the data structure it produces the same graph. In particular, the algorithm of the proof cannot be used directly to build the smallest subword graph accepting *Fac*(*text*), in the sense of automata theory. We do

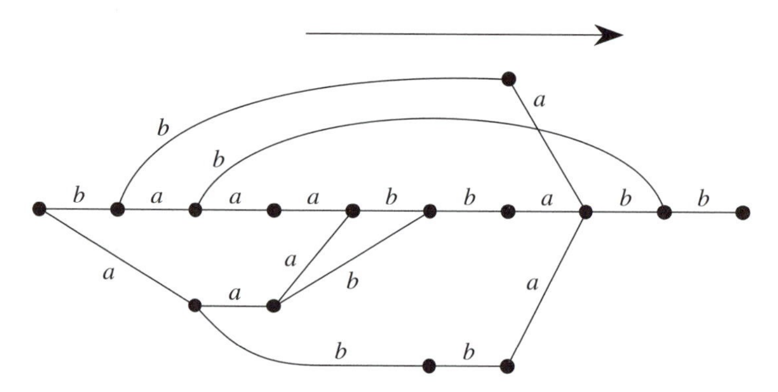

Figure 6.5: The suffix dawg of *baaabbabb*.

not consider these graphs here because their on-line construction is more technical than that of suffix dawgs given in the next section.

6.3 On-line construction

We describe an on-line linear-time algorithm for suffix dawg computation in this section. The algorithm processes the text from left to right. At each step, it reads the next letter of the text and updates the dawg built so far. In the course of the algorithm, two types of edges in the dawg are considered: solid edges (bold in Figure 6.6), and non-solid edges. The solid edges are those contained in the longest paths from the root. In other words, non-solid edges are those creating shortcuts in the dawg. The adjective *solid* means that once these edges are created, they are not modified during the rest of the construction. On the contrary, the target of non-solid edges may change after a while.

In Algorithm *suffix-dawg* we compute successively *DAWG*(*text*[1]), *DAWG* (*text*[1...2]), ..., *DAWG*(*text*[1...n]). Figures 6.6 and 6.7 show an application of the algorithm to *text* = *aabbab*.

The schema of one stage of the algorithm is graphically presented in Figure 6.7. It is the transformation of *DAWG*(*aabba*) into *DAWG*(*aabbab*), which points out a crucial operation of the algorithm. The addition of letter *b* to *aabba* introduces new factors to the set *Fac*(*aabba*). They are all suffixes of *aabbab*. In *DAWG*(*aabba*), nodes corresponding to suffixes of *aabba*, namely, *v*1, *v*2, *v*7, will now have an outgoing *b*-edge. Consider node *v*2 in *DAWG*(*aabba*). It has an outgoing non-solid *b*-edge. The edge is a shortcut between *v*2 and *v*4, compared to the longest path from

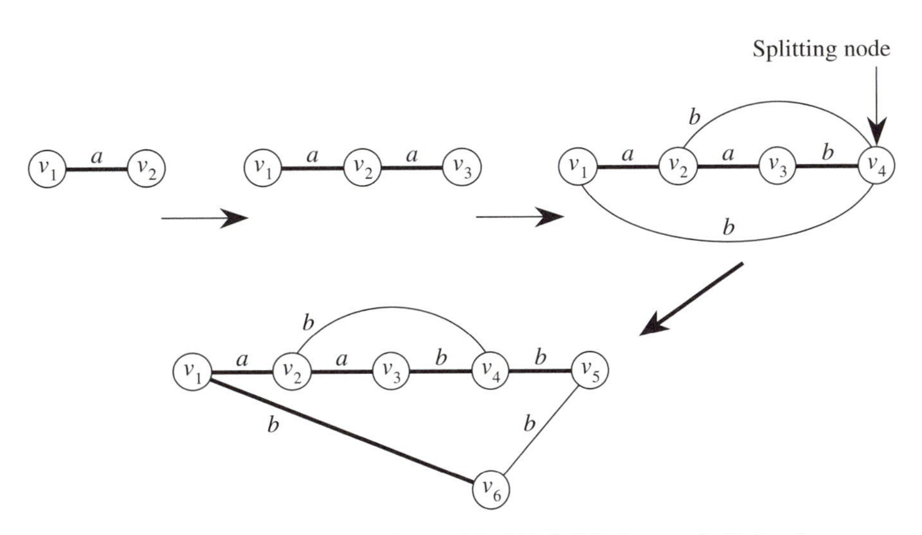

Figure 6.6: Iterative computation of *DAWG*(*aabb*). Solid edges are bold-faced.

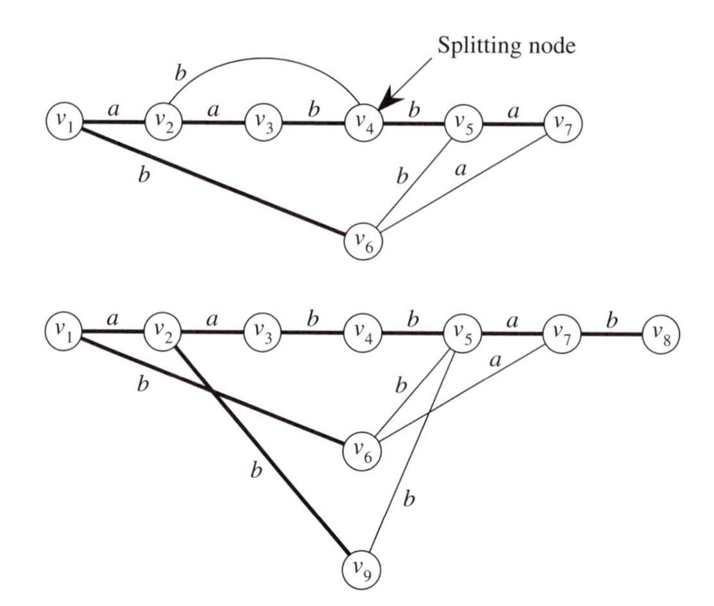

Figure 6.7: Transformation of *DAWG*(*aabba*) into *DAWG*(*aabbab*). The node v_9 is a clone of v_4.

$v2$ to $v4$ which is labeled by *ab*. The two factors *ab* and *aab* are associated with the node $v4$. But in *aabbab*, only *ab* becomes a suffix, not *aab*. This is the reason why the node $v4$ is split into $v4$ and $v9$ in *DAWG*(*aabbab*). Doing so will cause no problem for the rest of the construction, even if $v4$ and $v9$ have different behaviors. But note that the splitting operation is not necessary if the text ends here, because nodes $v4$ and $v9$ of *DAWG*(*aabbab*) have the same outgoing edge.

When splitting a node, it may also occur that some edges need to be redirected to the created node. This situation is illustrated in Figure 6.8 by *DAWG*(*aabbabb*). In *DAWG*(*aabbab*), the node $v5$ corresponds to factors *bb*, *abb*, and *aabb*. In *aabbabb*,

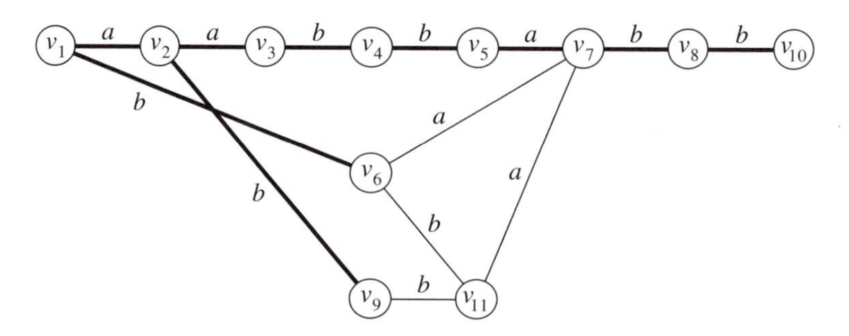

Figure 6.8: The suffix dawg of *aabbabb*.

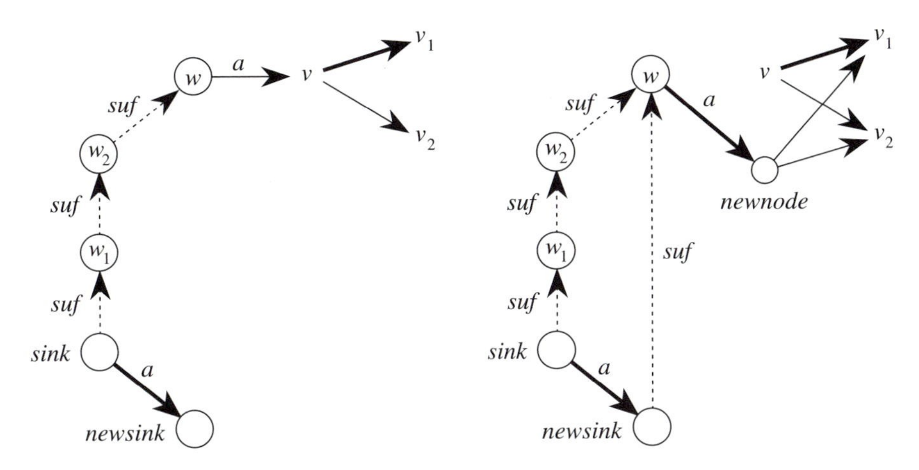

Figure 6.9: One stage of the algorithm with the current letter *a*. The non-solid *a*-edge from *w* is transformed into a solid edge onto a clone of *v*.

only *bb* and *abb* are suffixes. Hence, in *DAWG*(*aabbabb*), paths labeled by *bb* and *abb* should reach the node *v*11, a clone of node *v*5 obtained by the splitting operation.

In the algorithm, we denote the son *v* of node *w* by *son*(*w*, *a*) such that *label*(*w*, *v*)=*a*. We assume that *son*(*w*, *a*)=*nil* if there is no such node *v*.

On nodes of *DAWG*(*text*) a suffix link called *suf* is defined. It creates lists of *working paths* as shown in Figure 6.9. There, the working path is *w*1, *w*2=*suf*[*w*1], *w*3=*suf*[*w*2], *w*=*suf*[*w*3]. Generally, the node *w* is the first node on the path having an outgoing edge (*w*, *v*) labeled by letter *a*. If this edge is non-solid, then *v* is split into two nodes: *v* itself, and *newnode*. The latter is a clone of node *v*, in the sense that outgoing edges and the suffix-pointer for *newnode* are the same as for *v*. The suffix link of *newsink* is set to *newnode*. Otherwise, if edge (*w*, *v*) is solid, the only action performed at this step is to set the suffix link of *newsink* to *v*.

```
Algorithm suffix-dawg;
begin
    create the one-node graph G = DAWG(e);
    root := sink := the node of G; suf [root] := nil;
    for i := 1 to n do begin
        a := text [i];
        create a new node newsink;
        make a solid edge (sink, newsink) labeled by a;
        w := suf [sink];
        while (w≠nil) and (son(w, a)=nil) do begin
            make a non-solid a-edge (w, newsink); w := suf [w];
        end;
        v := son(w, a);
        if w=nil then suf [newsink] := root
```

```
            else if (w, v) is a solid edge then suf [newsink] := v
            else begin { split the node v }
                create a node newnode;
                newnode has the same outgoing edges as v
                except that they are all non-solid;
                change (w, v) into a solid edge (w, newnode);
                suf [newsink] := newnode;
                suf [newnode] := suf [v]; suf [v]:=newnode;
                w := suf [w];
                while (w≠nil) and ((w, v) is a non-solid a-edge) do begin
                {*} redirect this edge to newnode; w :=suf [w];
                end;
            end;
            sink:=newsink;
        end;
    end.
```

Theorem 6.4 Algorithm *suffix-dawg* computes *DAWG*(*text*) in linear time.

Proof We estimate the complexity of the algorithm. We proceed similarly as in the analysis of the suffix tree construction. Let $sink_i$ be the sink of *DAWG*(*text*[1...*i*]). What we call the working path consists of nodes $w1$=suf[*sink*], $w2$=*suf*[$w1$] ... w=*suf*[w_k]. The complexity of one iteration is proportional to the length k_1 of the working path plus the number k_2 of redirections made at step (*) in the algorithm. Let K_i be the value of k_1+k_2 at the i-th iteration. Let *depth*(u) be the depth of node u in the graph of suffix links corresponding to the final dawg (the depth is the number of applications of *suf* needed to reach the root). Then, the following inequality can be proven (we leave it as an exercise):

$$depth(sink_{i+1}) \leq depth(sink_i) - K_i + 2,$$

and this implies

$$K_i \leq depth(sink_i) - depth(sink_{i+1}) + 2.$$

Hence, the sum of all K_is is linear, which in turn implies that the total time complexity is linear. This completes the proof that the algorithm works in linear time. ♦

Remark 1 Algorithm *suffix-dawg* presents another proof that the number of nodes of *DAWG*(*text*) is less than 2|*text*|: at each step i of the algorithm at most 2 nodes are created, except at step 1 and 2 where only one node is created.

Remark 2 We classify the edges of *DAWG*(*text*) as solid and non-solid. Such classification is an important feature of the algorithm, used for knowing when a node needs to be split. However, we can avoid remembering explicitly whether an edge is solid or not. The category of an edge can be tested from the value *length*(v) associated with node v. This information happens to be useful for other purposes as well. We can store in each node v the length *length*(v) of the longest path from the root to v. Then, the edge (v, w) is solid iff $|length(v)|+1=|length(w)|$. Function *length* plays a central role in the algorithm of Section 6.5.

6.4* Further relationships between suffix trees and suffix dawgs

There is a very close relationship between our two "good" representations for the set of factors *Fac*(*text*). This is an interesting feature of these data structures. The relationship is intuitively obvious because suffix trees and dawgs are compacted versions of the same object—the uncompacted tree *Trie*(*text*).

Throughout this section (with one exception) we assume that the pattern *text* starts with a symbol occurring only at the beginning of *text*. In this case the relationship between dawgs and suffix trees is particularly tight and simple.

Remark The uniqueness of the left most symbol is not the crucial point for the relationship between dawgs and suffix trees. Even if the left most symbol is not unique our construction can still essentially be applied. However, the uniqueness case is easier to deal with in the framework of suffix and factor automata.

Two factors of *text*, x and y, are equivalent iff their *end-pos* sets are equal. This means that one of the words is a suffix of the other (say y is a suffix of x), and whenever x appears then y also appears in *text*. However, what happens if, instead of *end-positions*, we consider start-positions? Let us look from the "reverse perspective" and look at the reverse of the pattern. Suffixes then become prefixes, and end-positions become first-positions. Denote by *first-pos*(x) the set of first positions of occurrences of x in *text*.

Recall that a chain in a tree is a maximal path in which all the nodes have out-degree one except the last one. The following obvious lemma is crucial in understanding the relationship between suffix trees and dawgs (see Figure 6.10):

Lemma 6.5 Assume that *text* has a unique left most symbol. Then the following three properties are equivalent:

end-pos(x)=*end-pos*(y) in *text*;
first-pos(x^R)=*first-pos*(y^R) in $text^R$;
x^R, y^R are contained in the same chain of the uncompacted tree *Trie*($text^R$).

Remark The first two equations in the lemma are always true, however, the third property can be damaged if *text* starts with a non-unique symbol. If *text*=*abba* then *text*=$text^R$, and the nodes *a*, *ab* are contained in the same chain. However, *first-*

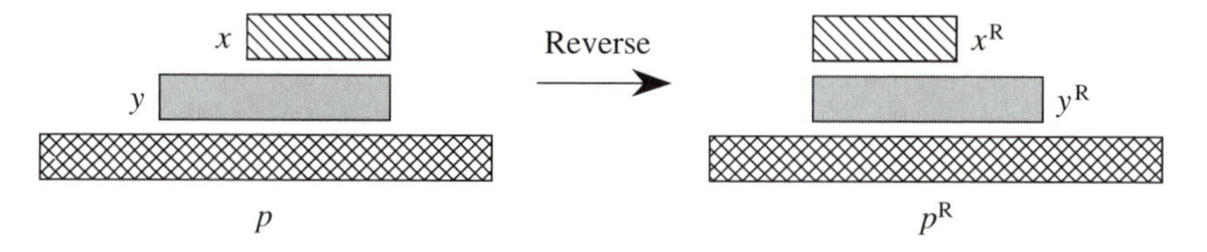

Figure 6.10: When the text has a unique leftmost symbol: *end-pos*(x)=*end-pos*(y) in *text* iff *first-pos*(x^R)=*first-pos*(y^R) in $text^R$ iff x^R, y^R are contained in the same chain of the uncompacted tree *Trie*($text^R$).

pos(a)={1, 4} but *first-pos*(ab)={1}. Fortunately such a situation is impossible with a unique left most symbol.

Corollary 6.6 The reversed values of nodes of the suffix tree $T(text^R)$ are the longest representatives of equivalence classes of factors of *text*. Hence nodes of $T(text^R)$ can be treated as nodes of *DAWG*(*text*).

We consider first the relationship: suffix trees → dawgs. Let T be a suffix tree. We define the shortest extension links (*sext* links, in short) in T, denoted by *sext*[a, v]. The value of *sext*[a, v] is the node w in which the value y=*val*(w) is the shortest word having prefix ax, where x=*val*(v). If there is no such node w, then *sext*[a, v]=*nil*. Observe that *sext*[a, v]≠*nil* iff *test*[a, v]=true. If *link*[a, v]≠*nil* then *sext*[a, v]=*link*[a, v].

Lemma 6.7 If we are given a suffix tree T with tables *link* and *test* computed, then table *sext* can be computed in linear time.

Proof The algorithm is top-down. For the root *sext* links are easy to compute. If table *sext* was computed for entries related to $v1$=*father*(v), then for each symbol a the value of *sext*[a, v] is computed in constant time as follows:

> if *test*[a, v]=false then *sext*[a, v]:=*nil* else
> if *link*[a, $v1$]=*nil* then *sext*[a, v]:=*sext*[a, $v1$] else
> *sext*[a, v]:=the son w of *link*[a, $v1$] such that the label of edge from *link*[a, $v1$] to w has the same first symbol as label of edge ($v1$, v).

The entire computation takes linear time because it takes constant time for each pair (a, v). This completes the proof. ♦

Remark In fact, we could easily modify the algorithm for the construction of suffix trees by removing table *test* and replacing tables *link* and *test* by a more powerful table *sext*. Table *sext* contains essentially the same information as tables *link* and *test* together (see Figure 6.13). This could improve the space complexity of the algorithm

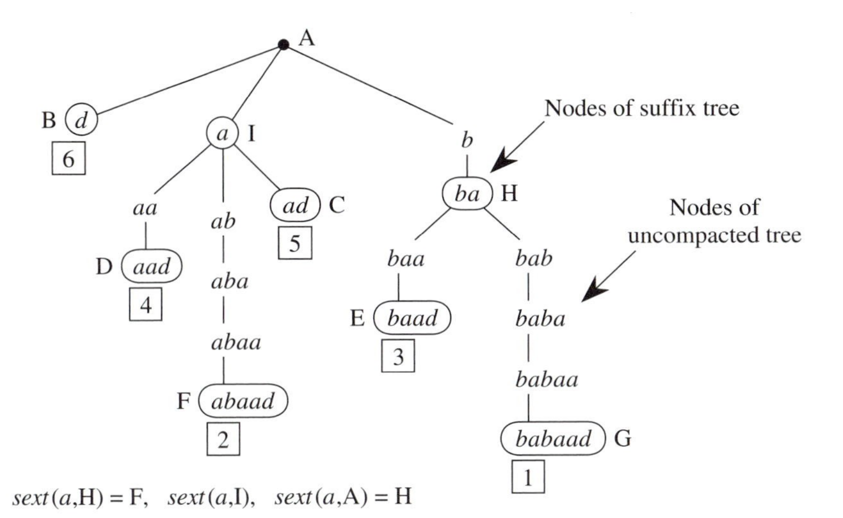

Figure 6.11: The uncompacted node-labeled tree $T1(text)$ for *text*=*babaad*. The nodes of the suffix tree are circled. For example, *sext*[a, H] is the shortest extension of a word $x = (a\ val(\mathrm{H}))$. It is the node F such that x is a prefix of *val*(F).

(by a constant factor). Time complexity does not change much. We leave a suitable modification of the algorithm as an easy exercise.

However, we feel that tables *link* and *test* make the algorithm easier to understand. Moreover, as we shall see later, table *link* gives solid edges of the dawg for reversed pattern and table *sext* (in cases when *link* has value *nil*) gives non-solid edges. Also in most applications it is better to have a choice of several alternative data structures.

Figure 6.11 presents the uncompacted suffix tree for the text *text*=*babaad*.

Figure 6.12 presents the uncompacted suffix tree from the "reverse perspective." The equivalence classes of words of the text of $text^R$=*daabab* are circled.

Theorem 6.8 If *text* has the unique left most symbol then *DAWG*(*text*) equals the graph of sext links of suffix tree T=$T(text^R)$. Solid edges of the dawg of *text* are given by table *link* of T.

Proof The thesis results from the preceding lemma. The corollary from the lemma states that nodes of the dawg correspond to nodes of T. In the dawg for *text*, the edge labeled a goes from the node v with $val(v)=x$ to node w with $val(w)=y$ iff y is the longest representative of the class of factors that contain word xa. If we consider the reversed text, then it means that ax^R is the longest word y^R such that $first\text{-}pos(y^R)=first\text{-}pos(ax^R)$ in $text^R$. This exactly means that $y^R = sext[a, x^R]$. $|y^R|=|x^R|+1$ iff

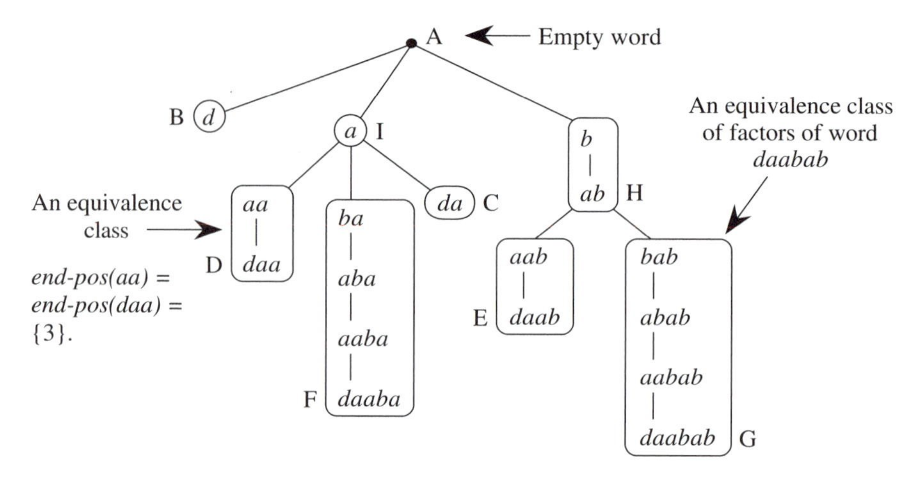

Figure 6.12: The previous uncompacted subword tree with the values of nodes reversed.

$y=xa$ and $link[a, x^R]=y^R$. According to the remark ending the previous section this also means that table *link* gives exactly all solid edges.

This completes the proof. ♦

Remark If one wants to construct a dawg using suffix trees and the string *text* does not start with the left marker, then we can take the string $p' = dp$ which has a left marker d. After constructing the dawg for p' we simply erase the first edge labeled d and delete the nodes inaccessible from the root. The first edge leading to a node of

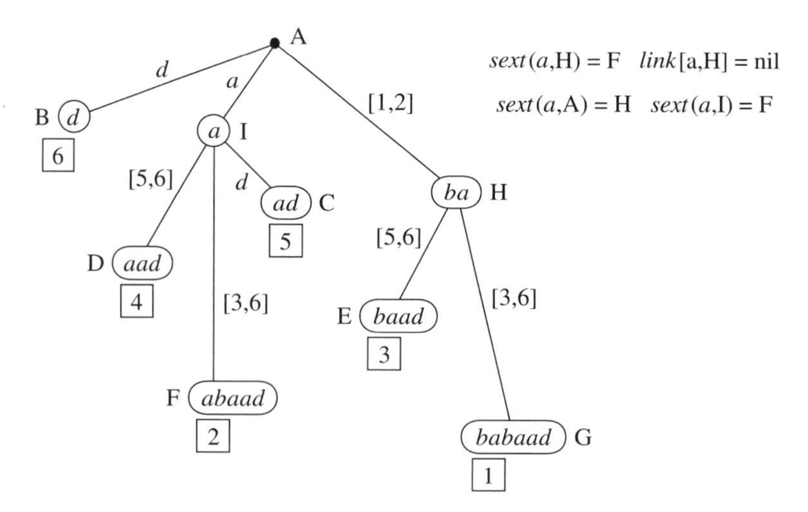

Figure 6.13: The suffix tree $T(babaad)$ with examples of *sext* links.

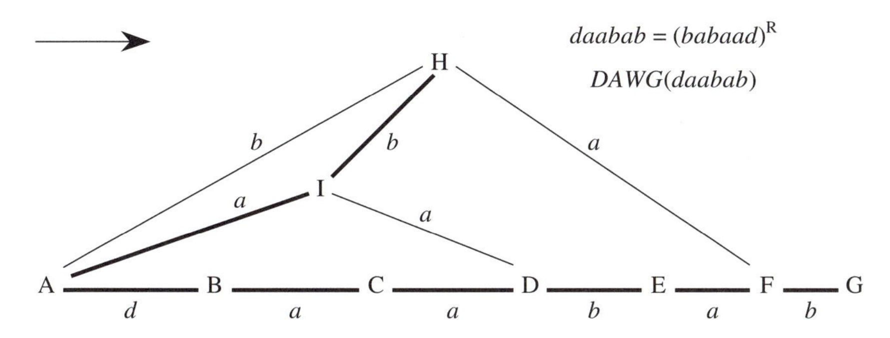

Figure 6.14: *DAWG*(*daabab*) = graph of *sext* links of *T*(*babaad*). Solid edges are given by table link of *T*(*babaad*).

the old longest path becomes solid. The transformation of the dawg from Figure 6.14 is shown as follows.

Let us deal now with the relationship: dawgs → suffix trees. An unexpected property of the dawg construction algorithm is that it also computes the suffix tree of the reversed text if *text* starts with a symbol occurring only at the first position (see Figure 6.15).

Given the dawg G of text *text* we can easily obtain the suffix tree for the reverse of *text*. Let T be a tree of suffix links of G; $suf[v]$ is the father of v. The edge $(v, suf[v])$ is labeled by the word $z=y^R//x^R$, where $x=val(suf[v]$, $y=val(v)$. The operation // consists of cutting the prefix x of y. $z=y^R//x^R$ iff $y^R=x^Rz$. This operation is well-defined in the context $(y=val(v), x=val(suf[v]))$. The tree T is called the tree of suffix links (see Figure 6.16). This tree is automatically constructed during the computation of *DAWG*(*text*).

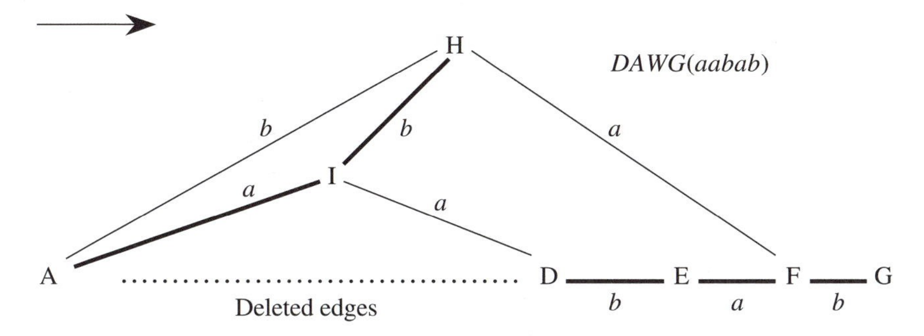

Figure 6.15: *DAWG*(*aabab*) results from *DAWG*(*daabab*), see Figure 6.14. The edge labeled *d* is removed. Then nodes inaccessible from the root are deleted: A and B are deleted. Edge (I, D) becomes solid.

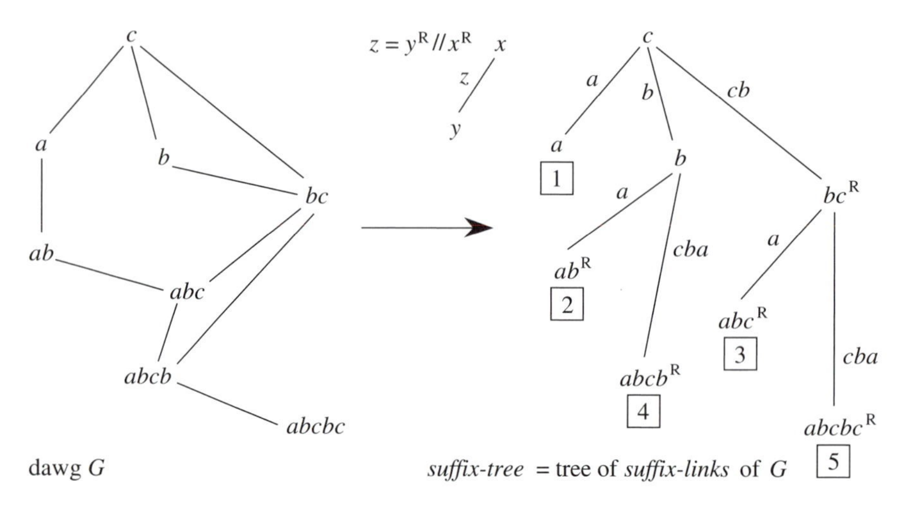

Figure 6.16: Tree of suffix-links of *DAWG*(*abcbc*) = suffix tree *T*(*cbcba*). If text *text* starts with the symbol occurring only at first position then the graph of suffix-links of *DAWG*(*text*) gives the suffix tree $T(text^R)$. We give values of nodes of the dawg and labels of edges of the suffix tree. Hence, the dawg construction also gives an alternative algorithm to construct suffix trees.

Theorem 6.9 If *text* starts with a unique left most symbol then the tree of suffix links of *DAWG*(*text*) equals the suffix tree $T(text^R)$.

Proof The same arguments used in the proof of the previous theorem can be applied. ♦

Remark There is a functional symmetry between suffix trees and dawgs: essentially they have the same range of applications. They are closely related: the structure of one of them is "contained" in the structure of the other. However, one of these data structures is a tree while the other is not. As is written in [CS 85]: "It is remarkable that such functional symmetry is efficiently achieved by such an integrated asymmetric construction."

From the point of view of applications the dawg can be compressed. This will result in a considerable savings in space. Such compression allows us to see a relationship between *G*=*DAWG*(*text*) and $G'=DAWG(text^R)$. It is also possible to make a symmetric version of a dawg: a good data structure *SymRep*(*text*) that can efficiently represent simultaneously *Fac*(*text*) and $Fac(text^R)$.

Let us look at a relationship between *G* and *G*′. Take the example text *text*=*aabbabd*. *G* is as presented in Figure 6.1 and it has 10 nodes. The graph *G*′ is presented in Figure 6.17 and it has 11 nodes.

At first glance it is difficult to see any relationship between *G* and *G*′. Let *root* and *sink* be the names of the root and the sink of *G*, and let *root*′, *sink*′ be the root and

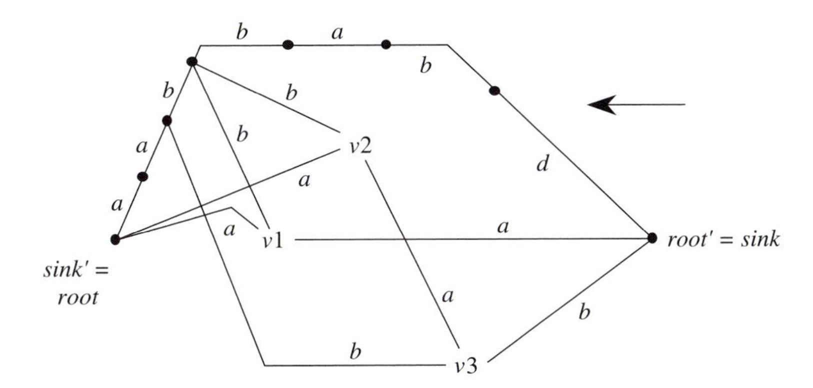

Figure 6.17: The graph G'=*DAWG*($text^R$). There are three nodes of out-degree greater than 1 (edges go right-to-left).

the sink of G'. We have *root*=*sink'* and *sink*=*root'* (we can call these nodes external, while other nodes can be called internal). Let us look at internal nodes with out-degree larger than one. Call such nodes essential. There are three essential nodes $v1$, $v2$, $v3$ in G, and also three essential nodes in G'. It so happens that there is a one-to-one correspondence between these nodes. Let values of nodes in G' be words spelled right-to-left (words from the "reverse perspective"). Then we have the same set of values of essential nodes in G as in G'. We leave the general proof of this fact to the reader. We can identify internal nodes of G and G'. Now we define the compressed versions of G and G' to be graphs consisting only of essential and external nodes (all other nodes disappear by compressing chains). Hence, the compressed versions are two graphs with the same set of nodes. Therefore, these graphs can be combined (by taking union of edges of both graphs) and we get a labeled graph that is a symmetric representation *SymRep*(*text*). There will be two kinds of edges in such graph: left-to-right and right-to-left edges. *SymRep* is a good representation of *Fac*(*text*) and satisfies: *SymRep*(*text*)=*SymRep*($text^R$).

More precisely: a compressed version *compr*(G) of the dawg G is the graph in which the nodes are essential (internal) and external nodes of G. The graph *compr*(G) results from compressing all chains of G—each internal node of out-degree one is removed and its incoming edges are redirected. Figures 6.18 and 6.19 show the compressed versions of dawgs for the word *aabbabd* and its reverse. The nodes of *compr*(G) correspond to so-called prime factors (see [BBHME 87], page 580, for a definition). Intuitively these are factors that appear in two distinct left contexts (if extendable to the left) and two distinct right contexts (if extendable to the right).

The constructed data structure can be used to "travel" in the dawg in two directions. We can find some useful information for the factor x, then for its left extension ax, and then for the right extension axb.

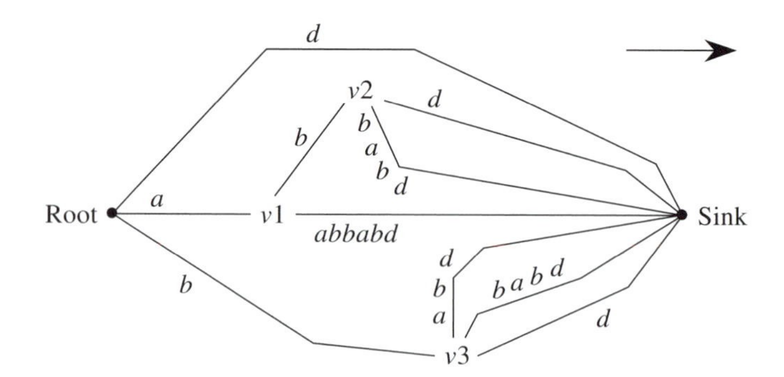

Figure 6.18: The graph *compr*(*G*) for *text*=*aabbabd*.

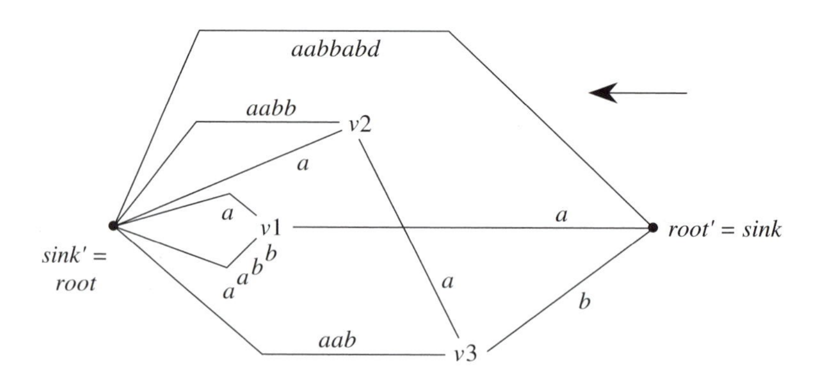

Figure 6.19: The graph *compr*(*G*′) for $text^R$=*dbabbaa*. The direction of edges is right-to-left. The labels of edges are also to be read right-to-left. *SymRep*(*text*) = *compr*(*G*′) ∪ *compr*(*G*).

6.5 String matching using suffix dawgs

In this section, we describe two other string-matching algorithms. They apply the strategy developed in Chapters 3 and 4, respectively, but are based on suffix dawgs built on the pattern. For fixed alphabets, the first algorithm yields a real-time search, while the second searching algorithm is optimal on the average.

A simple application of subword dawgs produces an efficient algorithm to compute the longest common factor of two strings. We have already discussed an algorithm for that purpose in Chapter 5. There, the solution is based on suffix trees. The present solution is quite different in several aspects. First, it computes the longest common factor while processing one of the two input words (a priori, the longer word) in an on-line way. Second, it uses the dawg of only one word (a priori, the shorter word), which makes the whole process cheaper according to the space used. The overall algorithm can be adapted from the string-matching algorithm presented below.

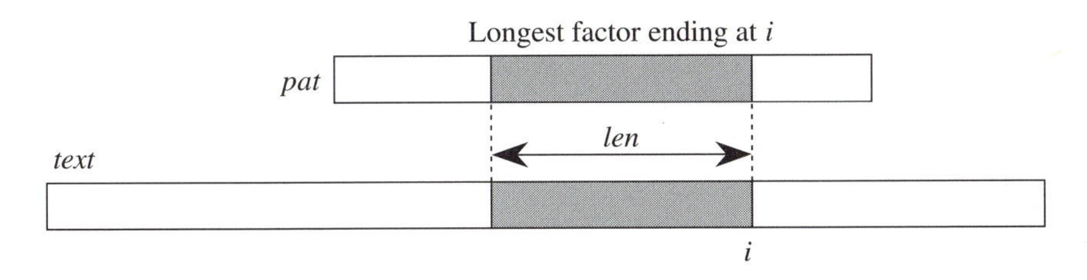

Figure 6.20: Current configuration in FDM algorithm.

The first string-matching algorithm of the present section searches the text for factors of the pattern. It computes the longest factor of the pattern ending at each position in the text. It is called the *forward-dawg-matching* algorithm (FDM for short) because factors are scanned from left to right (see Figure 6.20). The algorithm can be viewed as another variation of algorithm MP (Chapter 3). In this latter algorithm, the text is searched for prefixes of the pattern, and shifts are executed immediately when a mismatch is found. In contrast, scans performed by the FDM algorithm continue, while the match is a factor of the pattern. We explain the principles of FDM algorithm.

Let *pat* and *text* be the two words. Let *DAWG*(*pat*) be the dawg of the pattern. The text is scanned from left to right. At a given step, let *p* be the prefix of *text* that has just been processed. Then, we know the longest suffix *s* of *p* that is a factor of *pat*, and we have to compute the same value associated with the next prefix *pa* of *text*. It is clear that *DAWG*(*pat*) can help to compute this value efficiently if *sa* is a factor of *pat*. But, this is still true if *sa* is not a factor of *pat* because in this situation we can use suffix links of the dawg in a manner similar to the failure function used in MP algorithm (Chapter 3).

In a current situation of FDM algorithm, we memorize a state of *DAWG*(*pat*). But, the detection of an occurrence of *pat* in *text* cannot be achieved based only on the state of *DAWG*(*pat*), because states are usually reachable through different paths. To cope with this problem, in addition to the current state of *DAWG*(*pat*), we compute the length *len* of the longest factor of *pat* ending at the present position in the *text*. When suffix links are used to compute the next factor, the algorithm resets the value of *len* with the help of the function *length* defined on states of *DAWG*(*pat*). Recall that *length*(w) is the length of the longest path from the root of *DAWG*(*pat*) to the node *w*. The following property shows why the computation of *len* is valid when going through a suffix link.

Property of *DAWG*(*pat*) Let x be the label of any path from the root of *DAWG*(*pat*) to *w*. Let y be the label of the longest path from the root of *DAWG*(*pat*) to *v*. If $suf[w] = v$, the word y is a suffix of x.

The property is an invariant of the dawg construction of Section 6.3. Note that each time a suffix link is created in algorithm *suffix-dawg* the target of the suffix link is also the target of a solid edge. This is assumed by the splitting operation. Therefore the proof follows by induction.

```
function FDM: boolean; { forward-dawg-matching algorithm }
{ it computes longest factors of pat occurring in text }
begin
   D := DAWG(pat);
   len := 0; w := root of D; i := 0;
   while not i ≤ |text| do begin
       if there is an edge (w, v) labeled by text[i] then begin
          len := len+1; w := v;
       end else begin
           repeat w := suf[w];
           until (w undefined) or (there is (w,v) labeled by text[i]);
           if w undefined then begin
              len := 0; w := root of D;
           end else begin
              let (w, v) be the edge labeled by text[i] from w;
              len := length(w)+1; w := v;
           end;
       { len = larger length of factor of pat ending at i in text }
         if len=|pat| then return(true);
         i := i+1;
   end;
   return(false);
end;
```

Algorithm FDM can still be improved. The modification is similar to the transformation of algorithm MP into algorithm KMP. It is effected through a modified version of the failure function. The criterion to be considered is the delay of the search, that is, the time spent on a given letter of the text. Algorithm FDM has delay $O(|pat|)$ (consider the pattern a^n). The transformation of FDM is realized on the suffix link *suf* as explained below.

Assume that during the search, the current letter of the text is a and the current node of the dawg is w. If no a-edge starts from w, then we would expect that some a-edge starts from $v=suf[w]$. But it sometimes happens that outgoing edges on v have the same label as outgoing edges on w. Then, node v does not play its presumed role in the computation, and *suf* has to be iterated at the next step. This iteration can in fact be preprocessed because it is independent of the text.

We define the context of node w of *DAWG(pat)* as the set of letters that are labels of edges starting from w:

$$Context(w) = \{a\text{: there is an } a\text{-edge } (w, w') \text{ in } DAWG(pat)\}.$$

Then, the improved suffix link is *SUF* defined as ($v=suf[w]$)

$SUF(w) = v$, if $Context(w) \neq Context(v)$,
$SUF(w) = SUF[v]$, otherwise.

Note that $SUF(w)$ can be left undefined in this definition because *suf*[*root*] is undefined.

In practice, for the computation of *SUF*, no explicit context of node is necessary. This is due to a special feature of dawgs. When v=*suf*[w], *Context*(w) is included in *Context*(v), because the value of *suf*[w] is a suffix of the value of w. Thus, tests to compute *SUF* can be conducted only on the sizes of context sets. The following is an equivalent definition of *SUF*:

$SUF(w) = v$, if $|Context(w)| \neq |Context(v)|$,
$SUF(w) = SUF[v]$, otherwise.

Then, computing *SUF* is executed with a standard breadth-first exploration of *DAWG*(*pat*). Quantities |*Context*(w)| can be stored in each node during the construction of *DAWG*(*pat*), to help with computing *SUF* afterwards.

It is clear that the number of times *SUF* can be applied consecutively is bounded by the size of the alphabet. This leads to the next theorem that summarizes the overall concept.

Theorem 6.10 Algorithm FDM searches for *pat* in *text*, and computes longest factors of *pat* ending at each position in *text*, in linear time (on fixed alphabet). The delay of the search can be bounded by $O(|A|)$.

Proof The correctness comes from the discussion above. If we modify the algorithm FDM, replacing *suf* by *SUF*, then the delay of the new algorithm is obviously $O(|A|)$. ♦

We now present the second string-searching of the section. It is called *backward-dawg-matching* algorithm (BDM for short), and is somehow a variant of BM algorithm (Chapter 4). BM Algorithm searches for a suffix of the pattern *pat* at the current position of the text *text*. In BDM algorithm we search for a factor x of *pat* ending at the current position in the text (see Figure 6.21). Shifts use a pre-computed function on x.

The advantage of the approach is to produce an algorithm on the average very quick, not only on large alphabets, but also for small alphabets. The strategy of BDM algorithm is more "optimal" than that of BM algorithm: the smaller the cost of a given step, the bigger the shift. In practice, the match (and the cost) at a given iteration is usually small, hence, the algorithm, in which the shifts are reversely proportional to local matches, is close to optimal. If the alphabet is of size at least 2, then the average time complexity of BDM algorithm is $O(n \log(m)/m)$. This reaches the lower bound for the problem. But, it is quadratic in the worst case. However, rather

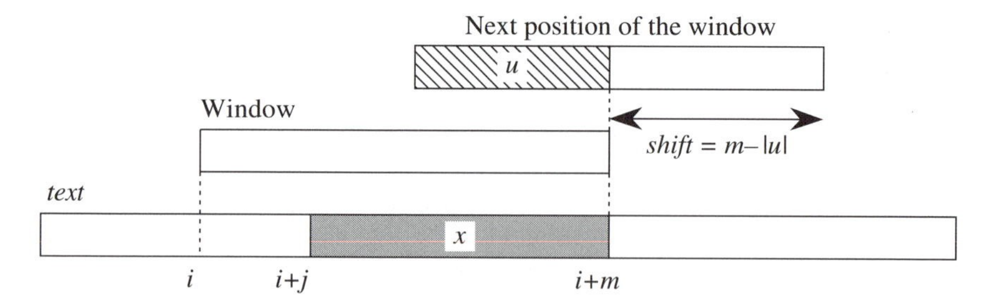

Figure 6.21: One iteration of Algorithm BDM. Word x is the longest factor of *pat* ending at $i+m$. Word u is the longest prefix of the pattern that is a suffix of x.

standard techniques, such as those applied on BM algorithm (see Chapter 4), lead to an algorithm that is both linear in the worst case and optimal on the average.

BDM Algorithm makes a central use of the dawg structure. The current factor x of *pat* found in *text*, is identified with the node corresponding to x^R in $DAWG(pat^R)$. We use the reverse pattern because we scan the window on the text from right to left. A constant-sized memory is sufficient to identify x. On the current letter a of the text, we further try to match ax with a factor of *pat*. What remains is to find an a-edge in $DAWG(pat^R)$ from the current node. If no appropriate a-edge is found, a shift occurs (see Figure 6.21).

```
Algorithm BDM; { backward-dawg-matching algorithm }
{ use the suffix dawg DAWG(pat^R) }
begin
   i := 0;
   while i ≤ n-m do begin
      j:=m;
      while j>1 and text[i+j..i+m]∈ Fac(pat) do j:=j-1;
      if j = 0 then report a match at position i;
      shift := BDM_shift [node of DAWG(pat^R)];
      i := i+shift;
   end;
end.
```

To each node of $DAWG(pat^R)$ we add information specifying if words corresponding to that node are suffixes of the reversed pattern pat^R (i.e., prefixes of *pat*). We traverse this graph when scanning the text right-to-left in BDM algorithm. Again, let x be the longest word which is a factor of *pat*, found at a given iteration. The time spent to scan x is proportional to $|x|$. The multiplicative factor is constant if a matrix representation is used for transitions in the data structure. Otherwise, it is $O(\log|A|)$ (where A can be restricted to the pattern alphabet), which applies for arbitrary alphabets.

We now define the shift *BDM_shift*. Let u be the longest suffix of x that is a proper prefix of the pattern *pat*. We can assume that we always know the current value of u. It is associated with the last node on the scanned path in *DAWG*(*pat*) corresponding to a suffix of pat^R. Then, the shift *BDM_shift*[x] is $m-|u|$ (see Figure 6.21).

The expected time complexity of BDM algorithm is given in the next statement. We analyze the average running time of BDM algorithm, considering the situation in which the text is random. The probability that a specified letter occurs at any position in the text is $1/|A|$, independent of the position of the letter.

Theorem 6.11 Let $c = |A| > 1$. Under independent equiprobability conditions, the expected number of inspections done by BDM algorithm is $O((n\log_c m)/m)$.

Proof We first count the expected number of symbol inspections necessary to shift the pattern $m/2$ places to the right. We show that $O(\log_c m)$ inspections are sufficient to achieve that goal. Since there are $2n/m$ segments of text of length $m/2$, we get the expected time $O((n\log_c m)/m)$. Let $r = 3.\lfloor\log_c m\rfloor$. There are more than m^3 possible values of the suffix *text*[*j-r...j*]. However, the number of subwords of length $r+1$ ending in the right half of the pattern is at most $m/2$ (provided m is large enough). The probability that *text*[*j-r...j*] matches a subword of the pattern ending in its right half is then $1/(2m^2)$. This is also the probability that the corresponding shift has length less than $m/2$. In this case, we bound the number of inspections by $m(m/2)$ (worst behavior of the algorithm making shifts of length 1). In the other case, the number of inspections is bounded by $3.\log_c m$.

The expected number of inspections that lead to a shift of length at least $m/2$ is thus less than

$$3.\log_c m\,(1-\frac{1}{2m^2}) + m(m/2)\,\frac{i}{2m^2},$$

that is $O(\log_c m)$. This achieves the proof. ♦

Theorem 6.11 shows that BDM algorithm realizes an optimal search of the pattern in the text, when branching in the dawg takes constant time. However, the worst-case time of the search can be quadratic.

To speed up BDM algorithm we use the prefix-memorization technique. The prefix u of size *m-shift* of the pattern (see Figure 6.22) that matches the text at the current position is kept in memory. The scan, between the part v of the pattern and the part of the text aligned with it (Figure 6.22), is done from right to left. When we arrive at the boundary between u and v in a successful scan (all comparisons positive), then we are at a *decision point*. Now, instead of scanning u until a mismatch is found, we need only scan (again) a part of u, due to combinatorial properties of primitive words. The part of u which is scanned again in the worst case has size *period*(u) The crucial point is that if we scan successfully, v and the suffix of u of size *per*(u), then we know what shift to execute without any further calculation:

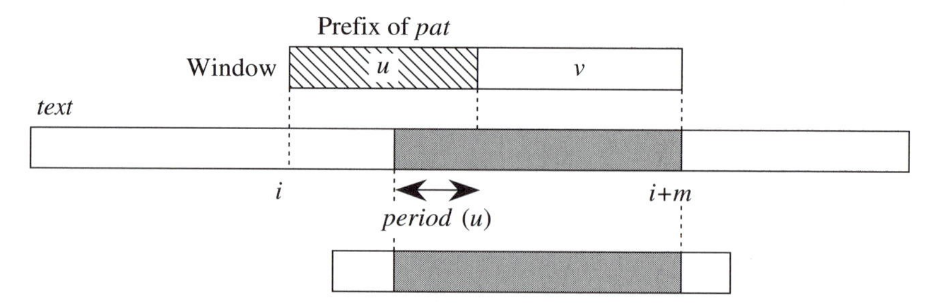

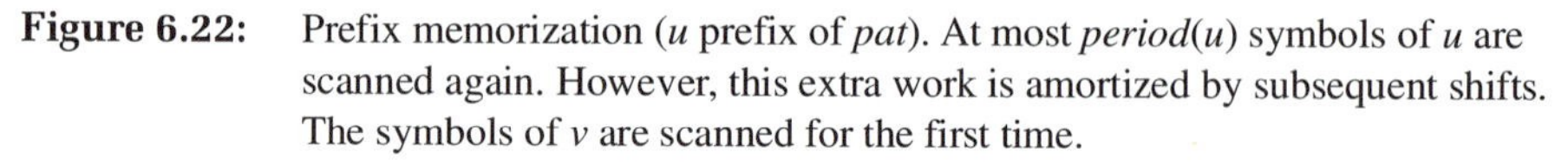

Figure 6.22: Prefix memorization (u prefix of *pat*). At most *period*(u) symbols of u are scanned again. However, this extra work is amortized by subsequent shifts. The symbols of v are scanned for the first time.

many comparisons are saved at precisely this moment, and this speeds up BDM algorithm.

```
Algorithm Turbo_BDM;
{ linear-time backward-dawg-matching algorithm }
begin
   i := 0; u := empty;
   while i ≤ n-m do begin
      j := m;
      while j>|u|-period(u) and text[i+j..i+m]∈ Fac(pat) do
         j := j-1;
      if j=|u|-period(u) and text[i+j..i+m] suffix of pat
      then report a match at position i;
      shift := BDM_shift[text [i+j+1..i+m]]
      i := i+shift; u := prefix of pattern of length m-shift;
   end;
end.
```

The proof of *Turbo_BDM* algorithm essentially relies on the next lemma. It is stated with the notion of *displacement* of factors of *pat*, which is incorporated in BDM shifts. If $y \in Fac(pat)$, we denote by *displ*(y) the smallest integer d such that $y = pat[m-d-|y|+1..m-d]$ (see Figure 6.23).

Lemma 6.12 (key lemma) Let u, v be as in Figure 6.22. Let z be the suffix of u of length *period*(u), and let x be the longest suffix of uv that belongs to *Fac*(*pat*). Then

$$zv \in FACT(p) \text{ implies } BDM\_shift(x)=displ(zv).$$

Proof It follows from the definition of *period*(u), as the smallest period of u, that z is a primitive word. The primitivity of z implies that occurrences of z can appear from the end of u only at distances which are multiples of *period*(u). Hence, *displ*(zv) should be a multiple of *period*(u), and this naturally implies that the smallest proper

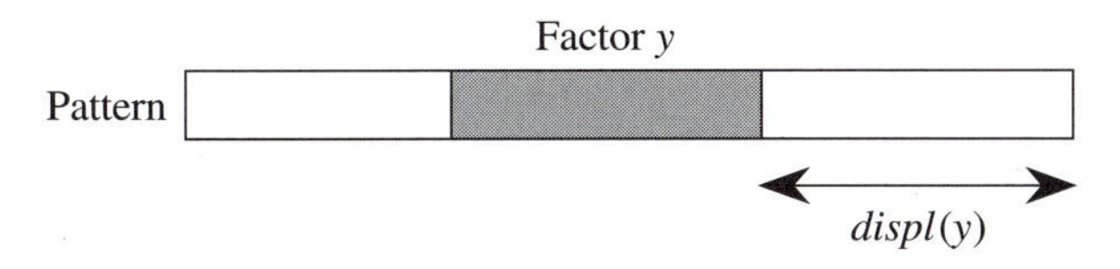

Figure 6.23: Displacement of factor y in the pattern.

suffix of uv that is a prefix of *pat* has size $|uv|$-$displ(zv)$. Hence, the next shift executed by the algorithm, *BDM_shift*(x), should have length $displ(zv)$. This proves the claim. ♦

The transformation of BDM algorithm into *Turbo_BDM* algorithm can be executed as follows. The variable u is specified by only one pointer to the text. The values of displacements are incorporated into the data structure *DAWG*(*pat*), and similarly the values of periods *period*(u) for all prefixes of the pattern can be precomputed and stored in *DAWG*(*pat*). In fact, the table of periods can be removed and values of *period*(u) can be computed dynamically inside *Turbo_BDM* algorithm using constant additional memory. This is quite technical and, therefore, has been omitted here.

Theorem 6.13 On a text of length n, *Turbo_BDM* algorithm runs in linear time (for a fixed alphabet). It makes at most $2n$ inspections of text symbols.

Proof In *Turbo_BDM* algorithm at most *period*(u) symbols of factor u are scanned. Let *extra_cost* be the number of symbols of u scanned at the actual stage. Since the next shift has length of at least *period*(u), *extra_cost* ≤ *next_shift*. Hence, all extra inspections of symbols are amortized by the total sum of shifts, which gives together at most n inspections. The symbols in parts v (see Figure 6.22) are scanned for the first time at a given stage. These parts are disjoint in distinct stages. Hence, they also give together at most n inspections. The work spent inside segments u, and inside segments v altogether is thus bounded by $2n$. This completes the proof. ♦

Bibliographic notes

Probably the most impressive feature of this chapter is the linear size of dawgs first discovered by Blumer, Blumer, Ehrenfeucht, Haussler, and McConnell [BBEHC 83]. They have simultaneously designed a linear-time algorithm for the construction of dawgs. After marking nodes of the dawg associated with suffixes of the text as terminal states, the dawg becomes an automaton recognizing the suffixes of the text. In fact, this is indeed the minimal automaton for the set of suffixes of the text. This point is discussed in [Cr 85] where an algorithm to build the minimal automaton recognizing the set of all factors of the text is also presented (the factor automaton can be slightly smaller than *DAWG(text)*). Constructions of *DAWG(text)* by Blumer et al. [BBEHCS 85] and Crochemore [Cr 86] are essentially the same.

The standard application of dawgs is for building efficient dictionaries of factors of many strings, see [BBHME 87]. The relationship between dawgs and suffix trees is adeptly described in [CS 85].

The application of dawgs to find the longest common factor of two words (Section 6.5) is presented in [Cr 86] (see also [Cr 87]).

The average-optimal algorithm BDM is from [C-R 92]. The paper also contains several variations on the implementation of the algorithm.

Selected references

[BBEHCS 85] Blumer, A., Blumer, J., Ehrenfeucht, A., Haussler, D., Chen, M.T.,& Seiferas, J., "The smallest automaton recognizing the subwords of a text," *Theoret. Comput. Sci.* 40 (1985): 31–55.

[CS 84] Chen, M.T., & Seiferas, J., "Efficient and elegant subword tree construction," in (A. Apostolico, Z. Galil, editors, *Combinatorial Algorithms on Words*, NATO Advanced Science Institutes, Series F, vol. 12, Springer-Verlag, Berlin, 1985): 97–107.

[Cr 86] Crochemore, M., "Transducers and repetitions," *Theoret. Comput. Sci.* 45 (1986): 63–86.

[C-R 92] Crochemore, M., Czumaj, A., Gasieniec, L., Jarominek, S., Lecroq, T., Plandowski, W., & Rytter, W., "Speeding up two string matching algorithms," in (A. Finkel and M. Jantzen editors, *9th Annual Symposium on Theoretical Aspects of Computer Science*, Springer-Verlag, Berlin, 1992): 589–600.

7

Automata-theoretic approach

Finite automata can be considered both as simplified models of machine and as mechanisms used to specify languages. As machines, their only memory is composed of a finite set of states. In the present chapter, both aspects are considered and lead to different approaches to pattern matching. Formally, a (deterministic) automaton G is a sequence $(A, Q, init, \delta, T)$, where A is an input alphabet, Q is a finite set of states, $init$ is the initial state (an element of Q), and δ is the transition function. For reasons of economy we allow δ to be a partial function. The value $\delta(q, a)$ is the state reached from state q by the transition labeled by input symbol a. The transition function extends in a natural way to all words, and, for the word x, $\delta(q, x)$ denotes, if it exists, the state reached after reading the word x in the automaton from the state q. The set T is the set of accepting states, or terminal states of the automaton.

The automaton G accepts the language:

$$L(G) = \{x: \delta(init, x) \text{ is defined and belongs to } T\}.$$

The size of G, denoted by $size(G)$ is the number of all transitions of G: number of all pairs (q, a) (q is a state, a is a single symbol) for which $\delta(q, a)$ is defined. Another example of a useful size of G is the number of states denoted by $statesize(G)$.

Probably the most fundamental problem in this chapter is to construct in linear time a (deterministic) finite automaton G accepting the words ending by one pattern among a finite set of patterns, and gives a representation of G of linear size, independently on the size of the alphabet. Another very important and useful algorithm solves the question of matching a regular expression.

7.1 Aho-Corasick automaton

We denote by *SMA(pat)*, for *String-Matching Automaton*, a (deterministic) finite automaton G accepting the set of all words containing *pat* as a suffix, and denote by $SMA(\Pi)$ an automaton accepting the set of all words containing a word of the finite set of words Π as a suffix. In other words, noting A^* the set of all words on the alphabet A,

$$L(SMA(pat)) = A^*pat \text{ and } L(SMA(\Pi)) = A^*\Pi.$$

In this section we present a construction of the minimal finite automaton $G = SMA(pat)$, where minimality is understood based on the number of states of G. Unfortunately, the total size of G depends heavily upon the size of the alphabet. We show how to construct these automata in linear time (with respect to the output).

With such an automaton, the following real-time algorithm *SMA* can be applied to the string-matching problem (for one or many patterns). The algorithm outputs a string of 0s and 1s that locates all occurrences of the pattern in the input text (1s mark the end position of occurrences of the pattern). The algorithm does not use the same model of computation as the algorithms of Chapters 3 and 4 do. There, the elementary operation used by algorithms (MP, BM, and their variations) is letter comparison. Here, the basic operation is branching (computation of a transition).

```
Algorithm SMA { real-time transducer for string-matching }
begin
    state := init; read(symbol);
    while symbol ≠ end marker do begin
        state := δ(state, symbol);
        if state in T then
            write(1) { reports an occurrence of the pattern }
        else write(0);
        read(symbol);
    end;
end.
```

We start with the case of only one pattern *pat*. We show how to build the minimal automaton *SMA(pat)*. The function *build_SMA* builds *SMA(pat)* sequentially. The core of the construction consists, for each letter a of the pattern, of unfolding the a-transition from the last created state t. This is illustrated by pattern *abaaba* in Figure 7.1.

```
function build_SMA (pat) : automaton;
begin
    create a state init; terminal := init;
    for all b in A do δ(init, b) := init;
```

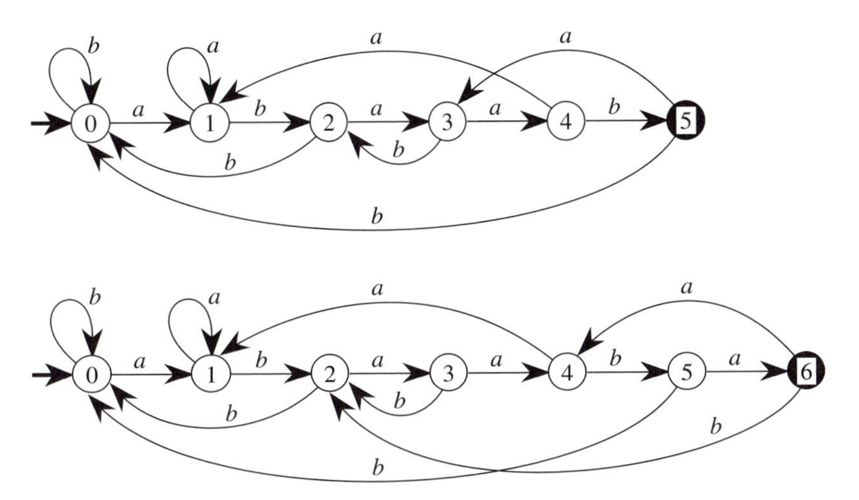

Figure 7.1: One step in the construction of an sma—from *SMA*(*abaab*) to *SMA*(*abaaba*)—unfolding the *a*-transition from state 5. Terminal states are black.

```
    for a := first to last letter of pat do begin
        temp := δ(terminal, a); δ(terminal, a) := new state x;
        for all b in A do δ(x, b) := δ(temp, b);
        terminal := x;
    end;
    return (A, set of created states, δ, init, {terminal});
end;
```

Lemma 7.1 Algorithm *build_SMA* constructs the (minimal) automaton *SMA*(*pat*) in time $O(m.|A|)$. The automaton *SMA*(*pat*) has $(m+1)|A|$ transitions, and $m+1$ states.

Proof The proof of Lemma 7.1 is left as an exercise. ♦

There is an alternative construction of the automaton *SMA*(*pat*) that shows the relation between SMAs and the MP-like algorithm of Chapter 3. Once we have computed the failure table *Bord* for pattern *pat*, the automaton *SMA*(*pat*) can be constructed as follows. We first define $Q = \{0, 1, \ldots, m\}$, $T = \{m\}$, $init = 0$. The transition function (table) δ is computed by the algorithm below. Figure 7.2 simultaneously displays the failure links (arrows going to the left) and *SMA* (*abaaba*).

```
Algorithm { computes the transition function of SMA(pat) }
          { assuming that table Bord is already computed }
begin
    for all a in A do δ(0, a) := 0;
    if m>0 then δ(0, pat[1]) := 1;
    for i := 1 to m do
```

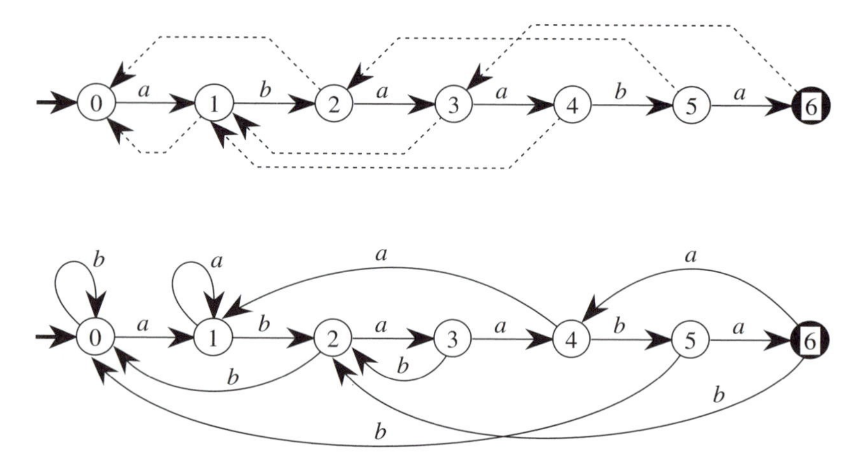

Figure 7.2: The function *Bord*, and the automaton *SMA*(*abaaba*).

```
        for all a in A do
            if i<m and a=pat[i+1] then δ(i, a) := i+1
            else δ(i, a) := δ(Bord[i], a);
    end.
```

In some sense, we can consider that table *Bord* represents the transition function δ of the automaton *SMA*(*pat*). Then, MP algorithm becomes a mere simulation of the SMA algorithm above. In the simulation, branching operations are substituted by letter comparisons. This remark is indeed the basis of Simon algorithm (MPS algorithm in Chapter 3). The representation of *SMA*(*pat*) by a failure function makes the size of the representation independent of the alphabet without increasing the total time complexity of the search phase.

The algorithm above shows that the transition function of *SMA*(*pat*) can be computed from the failure table *Bord*.

We next apply the same strategy for the recognition of a finite set of patterns. Assume that we have a set Π of r patterns. The i-th pattern is denoted by p_i. Let m be the total length (sum) of all patterns. We no longer try to build the minimal string-matching automaton corresponding to the problem. Therefore, *SMA*(Π) is not necessarily the minimal (deterministic) automaton of the language $A^*\Pi$, as it is when Π contains only one pattern.

To construct *SMA*(Π), we first consider a tree (a word trie) *Tree*(Π) in which the branches are labeled by elements of Π. The nodes of *Tree*(Π) are identified with prefixes of words in Π. The root is the empty word ε. The father of a non-empty prefix xa is the prefix x. We write *father*(xa) = x, and *son*(x, a) = xa. Nodes of *Tree*(Π) are considered as states of an automaton, and they are marked as terminal or non-terminal. A node is marked as terminal if the word it represents is in the set Π. All

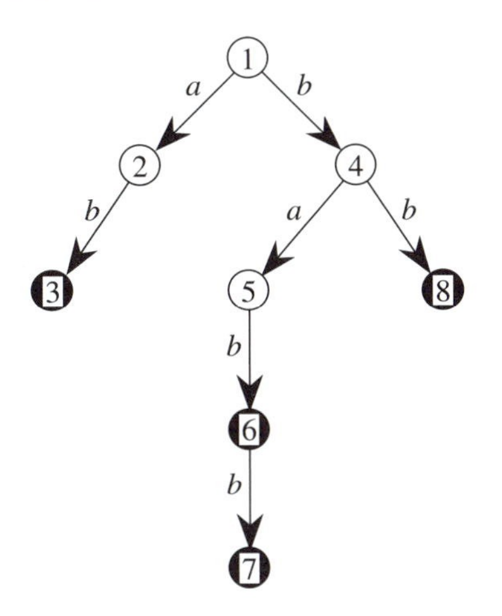

Figure 7.3: The word tree of set $\{ab, babb, bb\}$. Terminal nodes are black.

leaves are terminal states, but it may also occur that some internal node is also a terminal state. This happens when a pattern is a proper prefix of another pattern. Figure 7.3 displays *Tree*($\{ab, babb, bb\}$).

When applied to a set of strings Π, the algorithm *build_SMA* below builds an automaton *SMA*(Π) from the tree *Tree*(Π). The states of the automaton are the nodes of the tree. The algorithm essentially transforms and completes the relation *son* into the transition δ. The algorithm is very similar to the case of a single word. Here, no state is created, because the states are taken from the existing tree.

There is, however, a delicate point involved in the computation of terminal states. It may occur that some word of Π is an internal factor of another word of the set. This is the case for the set $\{ab, babb, bb\}$ considered in Figure 7.3. Node 6 of *Tree*($\{ab, babb, bb\}$) becomes a terminal state in *SMA*($\{ab, babb, bb\}$), because *bab* ends with the word *ab* of the set (see Figure 7.4). More generally, this happens during the construction when the clone of node x, namely node *temp* in the algorithm, is itself a terminal node.

```
function build_SMA (Π) : automaton;
begin
{ father and son links refer to the tree Tree(Π) }
{ states of SMA(Π) are the nodes of Tree(Π) }
   if root terminal node then T := {root} else T := ∅;
   for all b in A do δ(root, b) := root;
   for all non-root nodes x of Tree(Π) in bfs order do begin
      t := father(x); a := the letter such that x=son(t, a);
      temp := δ(t, a); δ(t, a) := x;
      if x or temp terminal nodes then add x to T;
```

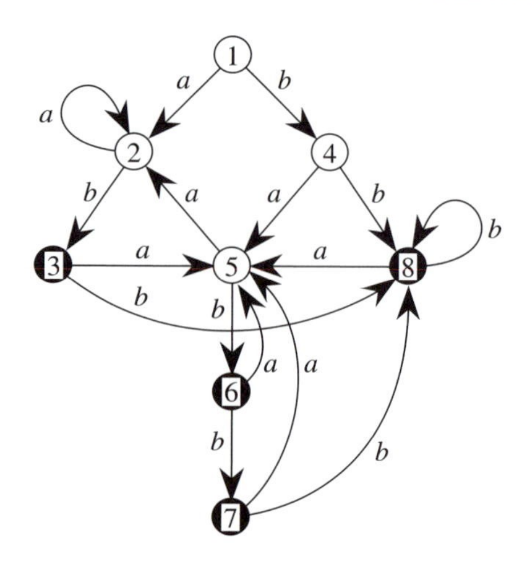

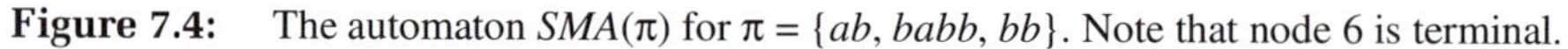

Figure 7.4: The automaton $SMA(\pi)$ for $\pi = \{ab, babb, bb\}$. Note that node 6 is terminal.

```
            for all b in A do
                if son(temp, b) defined then δ(x, b) := son(temp, b)
                else δ(x, b) := δ(temp, b);
        end;
        return (A, nodes of Tree(Π), δ, root, T);
    end;
```

Lemma 7.2 The algorithm *build_SMA* builds a deterministic automaton $SMA(\Pi)$ having the same set of nodes as the word trie $Tree(\Pi)$. It runs in time $O(statesize(Tree(\Pi)).|A|)$.

The automaton $SMA(\Pi)$ can also be built from scratch without a previous computation of $Tree(\Pi)$. Branches are unfolded as in the case of one pattern. Patterns are processed simultaneously, all prefixes of the same length at a time, which correspond to the breath-first-seach order applied on $Tree(\Pi)$ by the algorithm *build_SMA*. Figure 7.5 illustrates this alternative strategy for building $SMA(\Pi)$.

With the automaton $SMA(\Pi)$ built by the previous algorithm, searching a text for occurrences of the patterns in Π can be realized by the algorithm SMA. The search is then performed in real time, and the space required to store the automaton is $O(statesize(Tree(\Pi)).|A|)$. Again, it is possible to represent the automaton $SMA(\Pi)$ with a failure function. The advantage is to represent the automaton within space $O(statesize(Tree(\Pi)))$, which is independent of the alphabet. The search then becomes analogous to MP algorithm of Chapter 3.

The function *Bord* related to Π is defined, for a non-empty word u, by

$$Bord(u) = \text{longest proper suffix of } u \text{ that is prefix of a pattern in } \Pi.$$

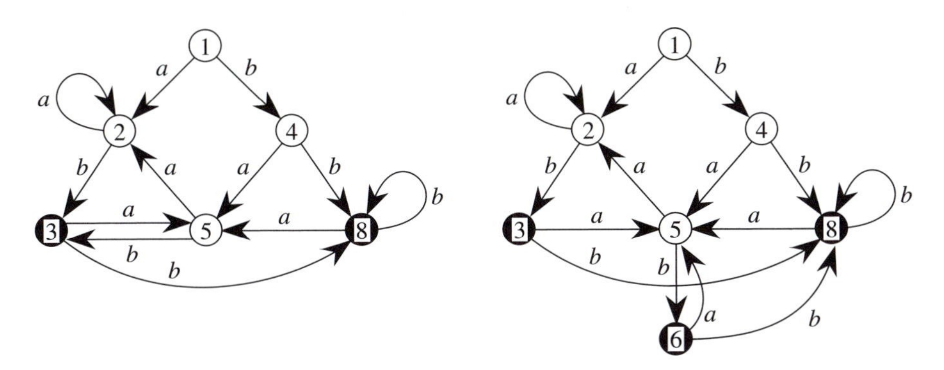

Figure 7.5: One step of another possible algorithm for the construction of *SMA*({*ab*, *babb*, *bb*}); Node 6 is a clone of d(5, *b*)=3 (on the left). Transitions on node 6 are the same as for node 3.

We also denote by *Bord* the failure table defined on nodes of *Tree*(Π) (except on the root) (see Figure 7.6). The relation used by the following algorithm that computes table *Bord* is (*t* is a node of *Tree*(Π) different from the *root*):

$Bord[ta] = Bord^k[t]a$, for the smallest k such that $Bord^k[t]a \in Trie(\Pi)$,
$Bord[ta] = \varepsilon$, otherwise.

Note that the algorithm following also marks nodes as terminal in the same situation explained for the direct construction of *SMA*(Π).

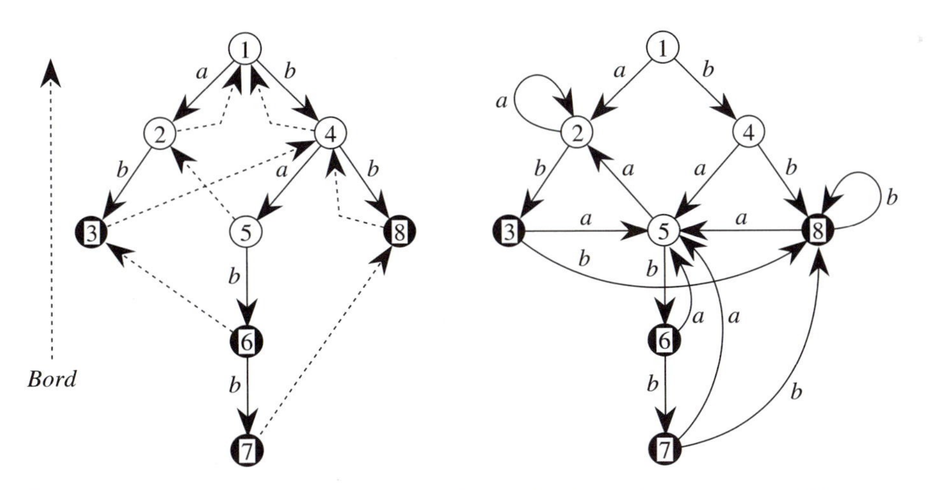

Figure 7.6: The tree *Tree*(π) with suffix links *Bord* (left), and the automaton *SMA*(π) (right), for π = {*ab*, *babb*, *bb*}.

```
procedure compute_Bord; { failure table on Tree(Π) }
begin
{ father and son refer to the tree Tree(Π) }
{ Bord is defined on nodes of Tree(Π) }
    set Bord [ε] as undefined;
    for all a in Tree(Π) do Bord [a] := ε;
    for all nodes x of Tree(Π), |x|>2, in bfs order do begin
        t := father(x); a := the letter such that x=son(t, a);
        z := Bord [t];
        while z defined and za not in Tree(Π) do z := Bord [z];
        if za is in Tree(Π) then Bord [x] := za else Bord [x] := ε;
        if Bord [u] is terminal then mark x as terminal;
    end;
end.
```

The complexity of the algorithm above is proportional to m. The analysis is similar to that of computing *Bord* for a single pattern. It is sufficient to estimate the total number of all executed statements "$z := Bord[z]$". This statement can again be treated as deleting some items from a store and z as the number of items. Let us fix a path π of length k from the root to a leaf. Using the store principle it is easy to prove that the total number of insertions (increases of z) into the store for nodes of π is bounded by k, hence, the total number of deletions (executing "$z := Bord[z]$") is also bounded by k. If we sum this over all paths we then get the total length m of all patterns in Π.

We can again base the construction of an automaton $SMA(\Pi)$ on the failure table of $Tree(\Pi)$. The transition function is defined on nodes of the tree as shown by the following algorithm.

```
Algorithm { computes transition function for SMA(Π) }
begin
    for all a not in Tree(Π) do δ(ε , a) := ε;
    for all nodes x of Tree(Π),|u|>0, in bfs order do
        for all a in A do
            if xa is in Tree(Π) then δ(x, a) := xa
            else δ(x, a) := δ(Bord [x], a);
end.
```

```
function AC(Tree(Π); text) : boolean;
{ Aho-Corasick multi-pattern matching }
{ uses the table Bord on Tree(Π) }
begin
    state := root; read(symbol);
    while symbol ≠ end marker do begin
        while state defined and son(state, symbol) undefined do
            state := Bord [state];
        if state undefined then state := root
        else state := son(state, symbol);
```

```
            if state is terminal then return true;
            read(symbol);
        end;
        return false;
    end;
```

Algorithm AC is the version of algorithm MP for several patterns. It is a straightforward application of the notion of the failure table *Bord*. The preprocessing phase of AC algorithm is the procedure *compute_Bord*.

Terminal nodes of *SMA*(Π) can have numbers corresponding to the numbering of patterns in Π. Hence, the automaton can produce in real-time numbers that correspond to those patterns ending at the last scanned position of the text. If no pattern occurs, 0 is written. This proves the following statement.

Theorem 7.3 The string-matching problem for a finite number of patterns can be solved in time $O(n+m)$. The additional memory is of size $O(m)$.

The table *Bord* used in AC algorithm can be improved in the following manner. Assume, for example, that during the search, node 6 of Figure 7.6 has been reached, and that the next letter is *a*. Since node 6 has no *a*-son in the tree, AC algorithm iterates function *Bord* from node 6. It successively finds nodes 3 and node 4 where the iteration stops. It is clear that the test on node 3 is useless in any situation because it has no son at all. On the contrary, node 4 plays its role because it has an *a*-son. Some iterations of the function *Bord* can be precomputed on the tree. The transformation of *Bord* is similar to the transformation of *suf* into *SUF* based on the contexts of nodes described in Section 6.5. Figure 7.7 shows the result of the transformation. The optimization does not change the worst-case behavior of AC algorithm.

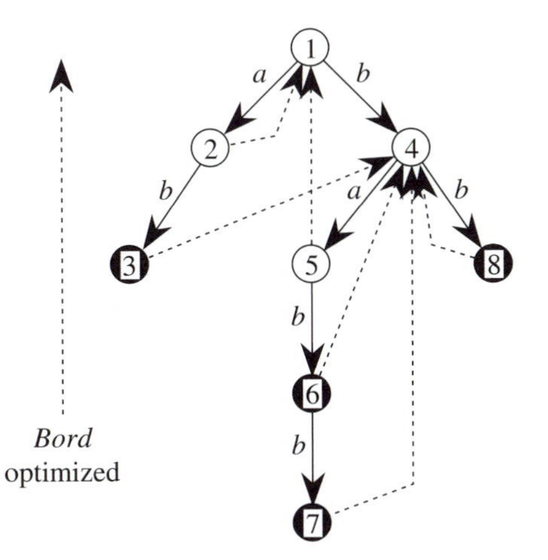

Figure 7.7 Optimized table *Bord* on *Tree*({*ab*, *babb*, *bb*}). Note that the table is undefined on node 4.

7.2 Faster multi-pattern matching

The method presented in the preceding section is widely used, but its efficiency is not entirely satisfactory. It is similar to the method described in Chapter 3. The minimum time complexity of the search phase in all these algorithms is still linear, which is rather large compared to the performance of the BM-type algorithms of Chapter 4. The idea then is to have the counterpart of BM algorithm to search for several patterns. It is possible to extend the BM strategy to the multi-pattern-matching problem, but the natural extension becomes very technical. We present another solution here based on the use of dawgs.

The search for several patterns combines the techniques used in AC algorithm and in the backward-dawg-matching algorithm (BDM in Section 6.5). The dawg related to the set of pattern Π is used to avoid searching certain segments of the text. It must be considered as an extra mechanism added to the AC automaton $SMA(\Pi)$ to speed up the search. The search is executed through a window which slides along the text. The size of the window here is the minimal length of pattern in Π. In the extreme case where the minimal length is very small, the dawg is not useful, and the AC algorithm is efficient enough on its own.

The idea for speeding up AC algorithm is the same as the central idea used in BM algorithm. On the average, only a short suffix of the window needs to be scanned to discover that no occurrence of a pattern in Π occurs at the current position (see Section 4.6). In order to achieve an average-optimal procedure we use the strategy of BDM algorithm. This approach has the further advantage of simplifying the whole algorithm compared to direct extensions of BM algorithm to the problem.

The algorithm of the present section, used for matching a finite set of patterns Π, is called *multi_BDM* (see Figure 7.8). The search uses both the automaton $SMA(\Pi)$ and the suffix dawg of reverse patterns $DAWG(\Pi^R)$. The basic work is supplied by the former structure. The role of the latter structure is to provide large shifts over the text.

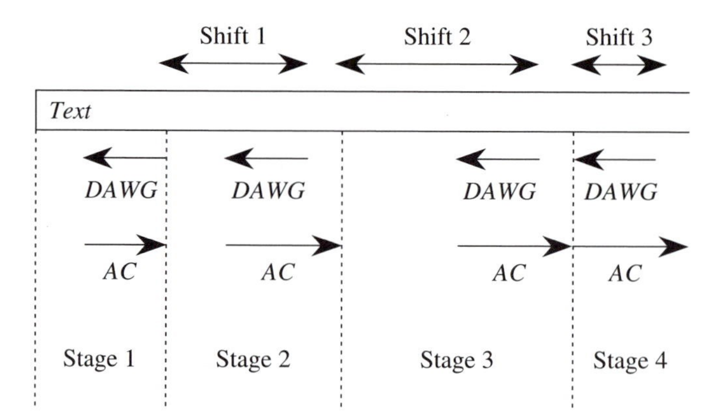

Figure 7.8: Series of stages in *multi_BDM* search phase.

In the current configuration of *multi_BDM* algorithm, the window is at position i on the text. A prefix u of the window has already been scanned through the AC automaton $SMA(\Pi)$. The word u is known as a prefix of some pattern. Both the length lp of that prefix and the corresponding state *state* of $SMA(\Pi)$ are stored. They function when extending the prefix during the present stage, if necessary. They also serve to compute the length of the next shift in some situation. Each stage is broken down into two sub-stages (see Figure 7.9). We assume that the window contains the word uv (u as above).

- **Substage I**, scan with $DAWG(\Pi^R)$. The part v of the window, not already scanned with $SMA(\Pi)$, is scanned from right to left with $DAWG(\Pi^R)$. Note that the process may stop before v has been entirely scanned. This occurs when v is not a factor of the patterns. This is likely to happen with high probability (at least if the window is large enough). The longest prefix u' of Π found during the scan is memorized to be used precisely in such a situation.
- **Substage II**, scan with $SMA(\Pi)$. If the part v of the window is entirely scanned at substage I, then it is scanned again, but with $SMA(\Pi)$ and in the reverse direction (from left of right). Doing so, the possible matches are reported, and we get the new prefix of patterns on which the window is adjusted. If v is not scanned entirely at substage I, the scan with $SMA(\Pi)$ starts from the beginning of u', and the rest is similar to the previous situation. This is the situation in which a part of the text is ignored by the search, which leads to a faster search than with AC algorithm alone.

The basic assumption of Substage I is that u is a common prefix of the window and the patterns. Since the size of the window is fixed and equals the minimal length of patterns, in particular, u is shorter than the shortest pattern. To meet the condition at the end of Substage II, the scan continues until the current prefix u is shorter than the required length.

The algorithm is given below. The current position of the window is i. Position *critpos* ($\geq i$) is the position of the input head of automaton $SMA(\Pi)$. The corresponding state of the automaton is denoted by *state*. The first internal while loop executes Substage I. The test "$text[i+j..i+min] \in Fac(\Pi^R)$" is assumed to be executed with the help of the suffix dawg $DAWG(\Pi^R)$. The second internal while loop executes Substage II.

```
Algorithm multi_BDM;
{ searches text for the finite set of patterns Π }
{ G is SMA(Π) (AC automaton of Section 7.1), }
{ D is DAWG(Π^R) for the reverse patterns (see Chapter 6) }
begin
    min := minimal length of patterns;
    i := 0;            { i is the position of the window on text}
    critpos := 0; state := initial state of G;
```

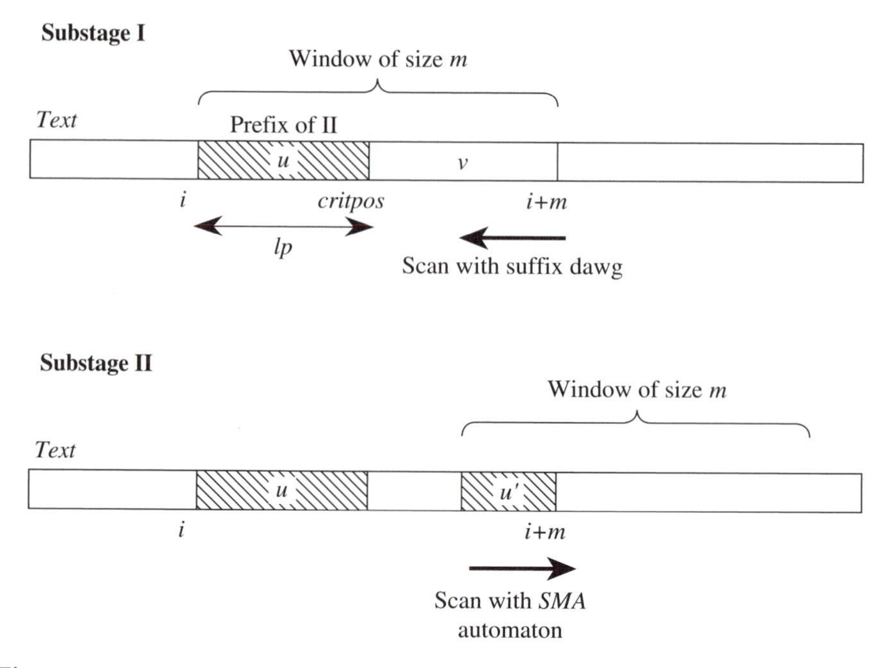

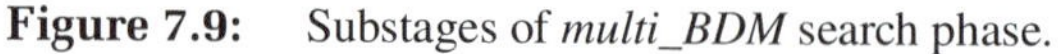

Figure 7.9: Substages of *multi_BDM* search phase.

```
while i ≤ n-min do begin
    j := min;
    while i+j > critpos and text[i+j..i+min] ∈ Fac(Π^R) do
        j := j-1;
    if i+j > critpos then begin
        lp := length of the longest prefix of Π found by D
                  during the preceding while loop;
        state := initial state of G;
        critpos := i+min-lp;
    end;
    while not end of text and (critpos < i+min or length of
            the longest prefix of Π found by G ≥ min) do begin
        apply G to text[critpos+1..n], reporting matches,
        incrementing critpos;
    end;
    state := state reached by G;
    lp := length of the longest prefix of Π found by G;
    i := critpos-lp;
  end;
end.
```

Example Consider the set of patterns $\Pi = \{abaabaab, aabb, baaba\}$, and the text $text = abacbaaabaabaabbccc$..... Let $D = DAWG(\Pi)$, and $G = SMA(\Pi)$.

- Stage 1. The window contains *abac*. The dawg D discovers that the longest suffix of the window that is prefix of Π is the empty word. Then, *critpos* is set 4, which leads to a shift of length 4.
- Stage 2. The window contains *baaa*. The dawg D finds the suffix *aa* as factor of Π. The factor *aa* is scanned with G. The length of the shift is 2. Letter b is not scanned at all.
- Stage 3. The window contains *aaba*. The dawg D scans the suffix *ba* of the window backward, reaching the point where the head of G is located. Then, G scans the same suffix of the window forward, and finds that the suffix *aba* of the window is the longest prefix of Π found at this position. The resulting shift has length 1.
- Stage 4. The window contains *abaa*. The dawg scans only its suffix of length 1. Then, G scans it again. But, since the entire window is a prefix of some pattern, the scan continues along the factor *abaabb* of the text. Two occurrences (of *baaba* and *abaabaab*) are reported. The longest prefix of Π at this point is b. The next contents of the window is *bccc*. ♦

The following statement shows that *multi_BDM* algorithm makes a linear search of patterns, provided branching in the automata is realized in constant time. Otherwise, the search takes $O(n.\log|A|)$, where A can be restricted to the alphabet of the patterns in Π. The second statement gives the average behavior of *multi_BDM* algorithm. The theorem proves that the longer the smallest pattern, the faster the search.

Lemma 7.4 The algorithm *multi_BDM* executes at most $2n$ inspections of text characters.

Proof Text characters are not scanned more than once by $D = DAWG(\Pi^R)$, nor more than once by $G = SMA(\Pi)$. Therefore, the search executes at most $2n$ inspections of text characters. ♦

Theorem 7.5 Let $c=|A|>1$, and $m>1$ be the length of the shortest pattern of Π. Under independent equiprobability conditions, the expected number of inspections executed by *multi_BDM* algorithm on a text of length n is $O((n.\log_c m)/m)$.

Proof The proof is similar to the corresponding proof for BDM algorithm in Chapter 6. ♦

7.3 Regular expression matching

In this section we consider a problem more general than string searching. The pattern is a set of strings specified by a regular expression. The problem, called *regular expression matching*, or sometimes only *pattern matching*, contains both the string-matching and the multiple string-matching problems. However, since the patterns are more general, their recognition becomes more difficult, and it so happens that solutions to pattern matching are not as efficient as those to string matching.

The pattern is specified by a regular expression on the alphabet A with the *regular operations*: concatenation, union, and star. For example, $(a+b)^*$ represents the set of all strings on the alphabet $\{a, b\}$, and the expression ab^*+a^*b represents the set $\{a, ab, abb, abbb, \ldots, b, aab, aaab, aaaab, \ldots\}$.

Let us recall the meaning of regular operations. Assume that e and f are regular expressions representing the sets of strings (regular languages) $L(e)$ and $L(f)$ respectively. Then,

- $e.f$ (also denoted by ef) represents $L(e)L(f) = \{uv : u \in L(e) \text{ and } v \in L(f)\}$,
- $e+f$ represents the union $L(e) \cup L(f)$,
- e^* represents the set $L(e)^* = \{u^k : k \geq 0 \text{ and } u \in L(e)\}$.

Elementary regular expressions are 0, 1, and a (for $a \in A$), which represents the empty set and the singletons $\{\varepsilon\}$ and $\{a\}$, respectively.

The formalism of regular expressions is equivalent to that of finite automata. It is the latter that is used for the recognition of the specified patterns.

The pattern-matching algorithm of this section consists of two phases. First, the regular expression is transformed into an equivalent automaton. Second, the search on the text is realized with the help of the automaton.

We consider only specific automata having good properties according to the space used for their representations. We call them *normalized automata*. They are not necessarily deterministic, as is in previous sections of the chapter. Such an automaton G is given by the tuple (Q, i, F, t), assuming that the alphabet is A. The set Q is the finite set of states, i is the only one initial state, t is the only one terminal state, and F is the set of edges (or arcs) labeled by letters of the alphabet A or by the empty word ε. The language of the automaton is denoted by $L(G)$, and is the set of words, labels of paths in G starting at i and ending at t. Finally, the automaton G and the regular expression e are said to be equivalent if $L(G) = L(e)$. Figure 7.10 displays an automaton equivalent to ab^*+a^*b.

A normalized automaton $G = (Q, i, F, t)$ has specific properties listed here:

- it has a unique initial state i, a unique terminal state t, and $i \neq t$
- no edge starts from t, and no edge points on i;

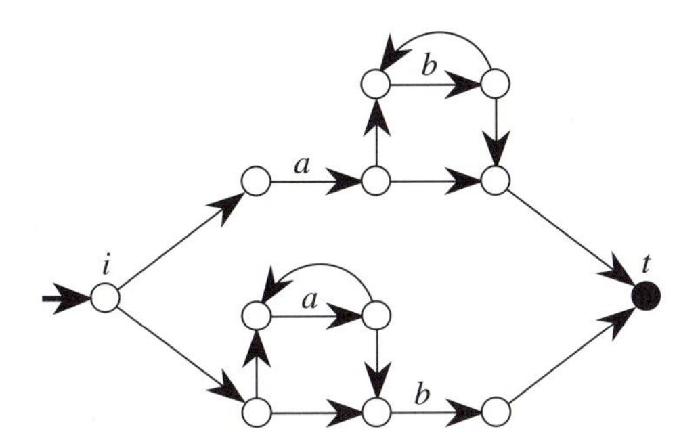

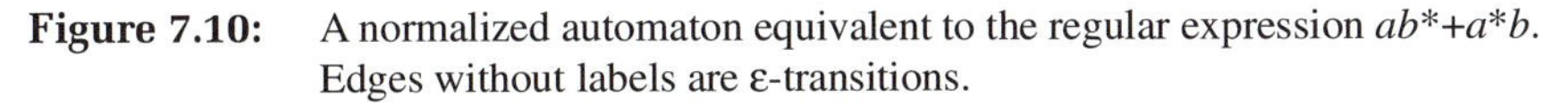

Figure 7.10: A normalized automaton equivalent to the regular expression ab^*+a^*b. Edges without labels are ε-transitions.

- on every state, there is either only one outgoing edge labeled by a letter of A, or there are at most two outgoing edges labeled by ε;
- on every state, there is either only one in-going edge labeled by a letter of A, or there are at most two in-going edges labeled by ε.

Note that properties satisfied by G have their dual counterparts satisfied by the reverse automaton G^R obtained by exchanging i and t and reversing directions of edges.

The pattern-matching preprocessing is based on the following theorem. Its proof is given in the form of an algorithm that builds a normalized automaton equivalent to the regular expression. We can even bound the size of the automaton with respect of the size of the regular expression. We define $size(e)$ as the total number of letters (including 0 and 1), number of plus signs, and number of stars occurring in the expression e. Note that parentheses and dots that may occur in e are not reckoned with in the definition of $size(e)$.

Theorem 7.6 Let e be a regular expression. There is a normalized automaton (Q, i, F, t) equivalent to e, that satisfies $|Q| \leq 2size(e)$ and $|F| \leq 4size(e)$.

Proof The proof is by induction on the size of e. The automaton associated with e by the construction is denoted by $G(e)$. If e is an elementary expression, its equivalent automaton is given in Figure 7.11.

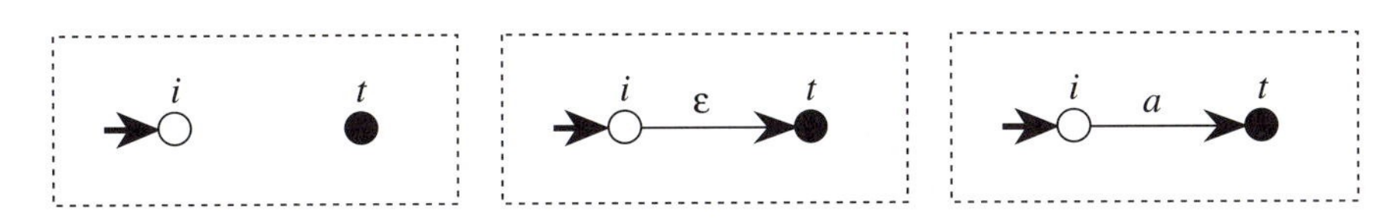

Figure 7.11: Normalized automata equivalent to elementary expressions 0 (empty set), 1 (set {ε}), and a (a letter) from left to right.

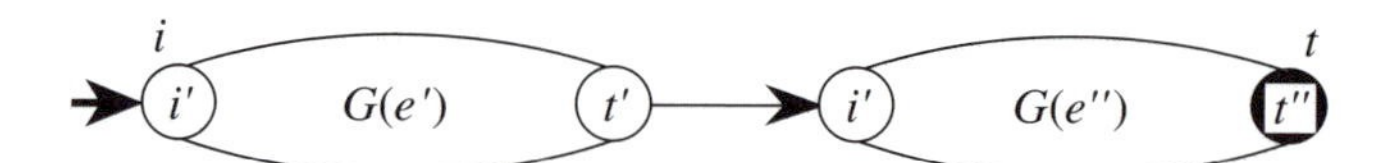

Figure 7.12: Concatenation of two automata.

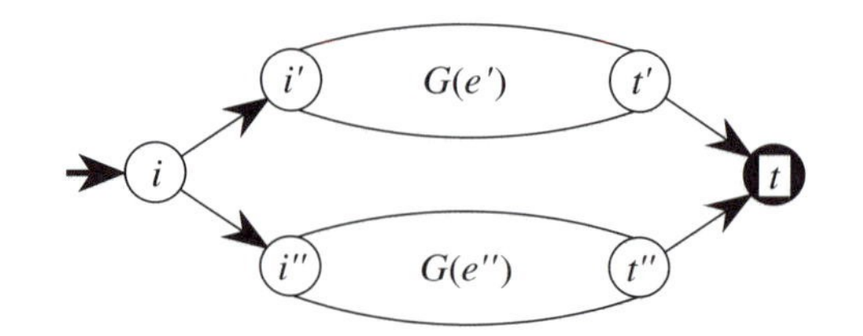

Figure 7.13: Union of two automata.

Let $G(e') = (Q', i', F', t')$, and $G(e'') = (Q'', e'', F'', t'')$ for two expressions e' and e''. We assume that the set of states are disjoint ($Q' \cap Q'' = \emptyset$). If e is a composed expression of the form $e'e''$, $e'+e''$, or $e'*$, then $G(e)$ is defined respectively by *concat*($G(e')$, $G(e'')$), *union*($G(e')$, $G(e'')$), and *star*($G(e')$). The operations *concat*, *union*, and *star* on normalized automata are defined in Figure 7.12, Figure 7.13, and Figure 7.14. In these figures, the design of automata displays their unique initial and terminal states.

We leave as an exercise the fact that the constructed automaton is normalized. This is, by the way, clear from the figures.

In the construction, exactly two nodes are created for each element of the regular expression accounting for its size. This proves, by induction, that the number of states of $G(e)$ is $2size(e)$.

Then the bound on the number of edges in $G(e)$ comes easily from the properties of $G(e)$. This completes the proof of Theorem 7.6. ♦

Figure 7.15 shows the automaton built by the algorithm of Theorem 7.6 from expression $(a+b^*)^*c$.

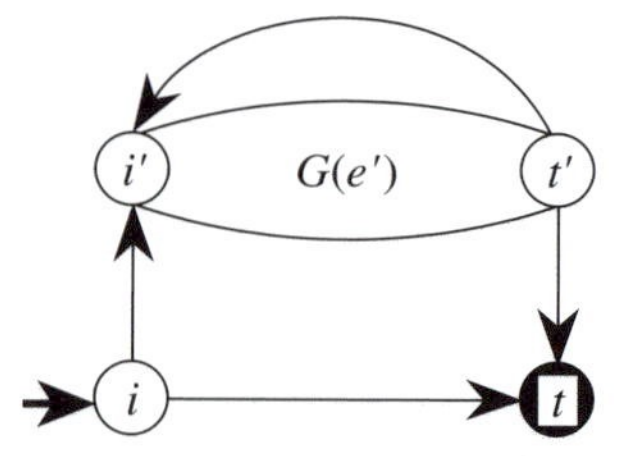

Figure 7.14: Star of an automaton.

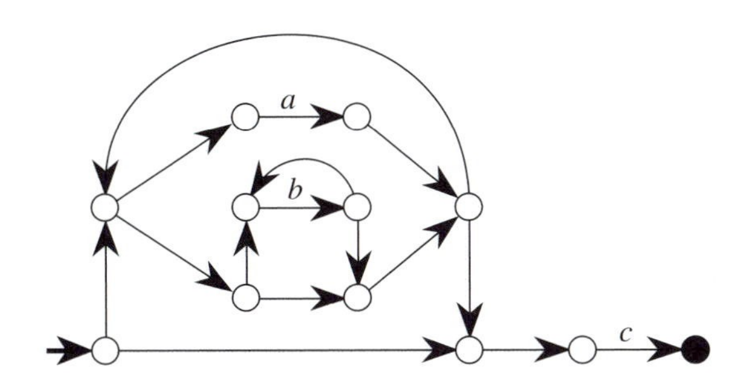

Figure 7.15: The normalized automaton for $(a+b^*)^*c$ built by the algorithm.

To evaluate the time complexity of algorithms, it is necessary to give some hints about the data structures involved in the machine representation of normalized automata. States are simply indices on an array that stores the transitions. The array has three columns. The first corresponds to a possible edge labeled by a letter. The two others correspond to possible edges labeled by the empty word (the three columns can even be compressed into two columns because of properties of outgoing edges in normalized automata). Initial and terminal states are stored apart. This shows that the space required to represent a normalized automaton $G = (Q, i, F, t)$ is $O(|Q|)$. It is then $O(size(e))$ if $G = G(e)$.

The algorithm that proves Theorem 7.6 is given in the form of four functions below. Overall, it integrates the analysis of expression e. It is implemented here as a standard predictive analyzer with one look-ahead symbol (*char*). The effect of procedure *error* is to stop the analysis, in which case no automaton is produced. This occurs when the input expression is not written correctly. Function *FACTOR* contains a small improvement on the construction described in the proof of Theorem 7.6: several consecutive symbols "*" are considered as just one symbol "*". This is because $(f^*)^*$ represents the same set of strings as f^*, for any regular expression f.

Theorem 7.7 A normalized automaton $G(e)$ equivalent to a regular expression e can be built in time and space $O(size(e))$.

Proof Operations *union*, *concat*, and *star* can be implemented to work in constant time with the array representation of normalized automata. Then, the algorithm *automaton* spends a constant time on each symbol of the input expression. This proves that the entire running time is $O(size(e))$. The statement on space complexity is essentially a consequence of Theorem 7.6, after noting that the number of recursive calls is less than $size(e)$. ♦

```
function automaton(e regular expression) : normalized automaton;
{ returns the normalized automaton equivalent to e, }
{ which is defined in the proof of Theorem 7.6 }
```

```
begin
    char := first symbol of e;
    G := EXPR;
    if not end of e then error else return G;
end;

function EXPR : normalized automaton;
begin
    G := TERM;
    while char = '+' do begin
        char := next symbol of e;
        H := TERM; G := union(G, H);
    end;
    return G;
end;

function TERM : normalized automaton;
begin
    G := FACTOR;
    while char in A ∪ {'0', '1', '('} do begin
        H := FACTOR; G := concat(G, H);
    end;
    return G;
end;

function FACTOR : normalized automaton;
begin
    if char in A ∪ {'0', '1'} then begin
        G := the corresponding elementary automaton;
        char := next symbol of e;
    end else if char = '(' then begin
        char := next symbol of e;
        G := EXPR;
        if char = ')' then char := next symbol of e else error;
    end else error;
    if char = '*' then begin
        G := star(G);
        repeat char := next symbol of e; until char ≠ '*';
    end;
    return G;
end;
```

The automaton $G(e)$ can be used to test if a word x belongs to the set $L(e)$ of patterns specified the regular expression e. The usual procedure for this is to spell a path labeled by x in the automaton. But there can be many such paths, and even an infinite number of them. This is the case, for example, with the automaton of Figure 7.15 and $x = bc$. An algorithm using a backtracking mechanism is possible, but the usual solution is based on a dynamic kind of programming technique. The idea is to cope with the path labeled by the empty word. Then the notion of *closure* of a set of states as those states accessible from them by ε-path occurs. For S, a subset of states of the automaton $G(e) = (Q, i, F, t)$, we define

$$closure(S) = \{q \in Q\text{: there exists an } \varepsilon\text{-path in } G(e) \text{ from a state of } S \text{ to } q\}.$$

We also use the following notation for transitions in the automaton ($a \in A$):

$$trans(S, a) = \{q \in Q\text{: there exists an } a\text{-edge in } G(e) \text{ from a state of } S \text{ to } q\}.$$

Testing whether the word x belong to $L(e)$ is implemented by the algorithm below.

```
function test(x word) : boolean;
{ test(x) = true iff x belongs to the language L(Q, i, F, t) }
begin
   S := closure({i});
   while not end of x do begin
        a := next letter of x;
        S := closure(trans(S, a));
   end;
   if t is in S then return true else return false;
end;
```

Functions *closure* and *trans* on sets of states are realized by the algorithms below. With careful implementation, based on standard manipulation of lists, the running times to compute *closure*(S) and *trans*(S, a) are proportional to the sets involved. This is independent of the alphabet.

```
function closure(S set of states) : set of states;
{ closure of a set of states in the automaton (Q, i, F, t) }
begin
   T := S; File := S;
   while File not empty do begin
        delete state p from File;
        for any ε-edge (p, q) in F do
             if q is not in T then begin
                  add q to T; add q to File;
             end;
   end;
   return T;
end;
```

```
function trans(S set of states; a letter): set of states;
{ transition of a set of states in the automaton (Q, i, F, t) }
begin
   T := ∅;
   for all p in S do
        if there is an a-edge (p, q) in F then
             add q to T;
   return T;
end;
```

Lemma 7.8 Testing whether the word x belongs to the language described by the normalized automaton (Q, i, F, t) can be realized in time $O(|Q|.|x|)$ and space $O(|Q|)$.

Proof Each computation of $closure(S)$ and of $trans(S, a)$ takes a time that is bounded by $O(|Q|)$. The total time is then $O(|Q|.|x|)$.

The array representation of the automaton requires $O(|Q|)$ memory space. ♦

The recognition of expression e in the text *text* also uses the automaton $G(e)$. But the problem is slightly different, because the test is done on factors of *text* and not only on *text* itself. Indeed we could consider the expression A^*e, and its associated automaton, to locate all occurrences of patterns like prefixes of *text*. But no transformation of $G(e)$ is even necessary, because it is integrated into the searching algorithm. At each iteration of the algorithm below, the initial state is incorporated into the current set of state, the same as if we restart the automaton at the new position in the text. Moreover, after each computation of set, a test is performed to report a possible match ending at the current position in the text.

Theorem 7.9 The recognition of patterns specified by a regular expression e in a text *text* can be realized in time $O(size(e).|text|)$ and space $O(size(e))$.

Proof We admit that the construction of $G(e)$ can be realized in time $O(size(e))$.

This recognition is given by PM algorithm below. The detailed implementation is similar to that of function *test*. The statement is essentially a corollary of Lemma 7.8, and a consequence of Theorem 7.6, asserting that the number of states of $G(e)$ is $|Q| = 2size(e)$. ♦

```
Algorithm PM; { Regular expression matching algorithm }
{ searches text for the regular expression e represented }
{ by the normalized automaton (Q, i, F, t) }
begin
    S := closure({i});
    while not end of text do begin
        a := next letter of text;
        S := closure(trans(S, a) ∪ {i});
        if t is in S then report an occurrence;
    end;
end;
```

There is an alternative to the pattern-matching procedure presented in this section. It is known that any (normalized finite) automaton is equivalent to a deterministic automaton (defined by a transition function). More precisely, any automaton G can be transformed into a deterministic automaton D such that $L(G) = L(D)$. Indeed, algorithm PM simulates moves in a deterministic automaton equivalent to $G(e)$. But states are computed again and again without any memorization.

It is clear that if $G(e)$ is deterministic, algorithms PM and *test* become much more simple, similar to algorithm SMA of Section 7.1. Moreover, with a deterministic automaton, the time of the search becomes independent of the size of e with an appropriate representation of $G(e)$. Therefore, there seem to be only advantages to making the automaton $G(e)$ deterministic, if it is not already so. The main restriction is that the size of the automaton may grow exponentially, going from s to 2^s. This point is discussed in Section 7.4. We summarize these remarks in the following theorem.

Theorem 7.10 The recognition of patterns specified by a regular expression e in a text *text* can be realized in time $O(|A|.2^{size(e)} + |text|)$ and space $O(|A|.2^{size(e)})$, where A is the set of letters effectively occurring in e.

Proof Build the deterministic version $D(e)$ of $G(e)$, and use it to scan the text. Use a matrix to implement the transition function of $D(e)$. The branching in $D(e)$ thus takes constant time, and the entire search takes $O(|text|)$. ♦

7.4* Determinization of finite automata: Efficient instances

We consider a certain set S of patterns represented in a succinct way by a deterministic automaton G with n states. The set S is the language $L(G)$ accepted by G. Typical examples of sets of patterns are $S = \{pat\}$, a singleton, and $S = \{pat_1, pat_2, \ldots, pat_r\}$, a finite set of words. In the first example, the structure of G is the line of consecutive positions in *pat*. The right most state (position) is marked as terminal. In the second example, the structure of G is the tree of prefixes of patterns. All leaves of the tree are terminal states, and some internal nodes can also be terminal if a pattern is a prefix of another pattern of the set.

We can transform G into a non-deterministic pattern-matching automaton, noted $loop(G)$, that accepts the language of all words having a suffix in $L(G)$. The automaton $loop(G)$ is obtained by adding a loop on the initial state, for each letter of the alphabet. The automata $loop(G)$ for the two examples of the cases mentioned above are presented in Figure 7.16 and Figure 7.17. The actual non-determinism of the automata appears only on the initial state.

We can apply the classical powerset construction (see [HU 79], for example) to a non-deterministic automaton $loop(G)$. It appears that, in these two cases of one pattern and of a finite set of patterns, by doing so we obtain efficient deterministic string-matching automata. In the powerset construction, only those subsets that are accessible from the initial subset are considered.

However, the idea of using $loop(G)$ and the powerset construction on it altogether is not always efficient. In Figure 7.18 a non-efficient case is presented. It is not hard to become convinced that the deterministic version of $loop(G)$, which has a tree-like structure, cannot have less than 2^7 states. Extending the example shows that

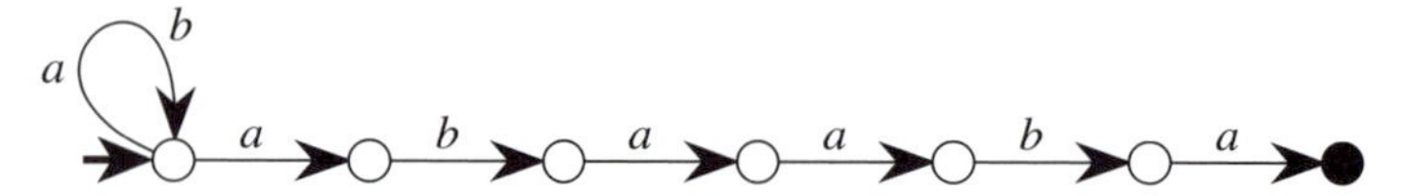

Figure 7.16: The non-deterministic pattern-matching automaton for *abaaba*. The powerset construction gives the automaton *SMA*(*abaaba*) of Figure 7.2. An efficient case of the powerset construction.

a non-deterministic automaton, even of the form *loop*(G) with n+1 states, can be transformed into an equivalent deterministic automaton having 2^n states.

Similarly we can transform the deterministic automaton G accepting one word *pat* into a non-deterministic automaton *FAC*(G) accepting the set *Fac*(*pat*) (see Figure 7.19). The transformation could be done simply by saying that all states are simultaneously initial and terminal states, but we prefer creating additional edges from the single initial state to all other states and labeled by all letters. It again happens that we find an efficient case of the powerset construction if we start with the deterministic automaton accepting just a single word. The powerset construction gives the smallest deterministic automaton accepting the suffixes of *pat*.

The powerset construction applied on automata of the form *FAC*(G) is not always efficient (see Figure 7.20). If we take as G the deterministic automaton with $2n$+3 states accepting the set $S = (a+b)^n a(a+b)^n c$, then the automaton *FAC*(G) also has $2n$+3 states. But the smallest deterministic automaton accepting the set of all factors of words in S has an exponential number of states.

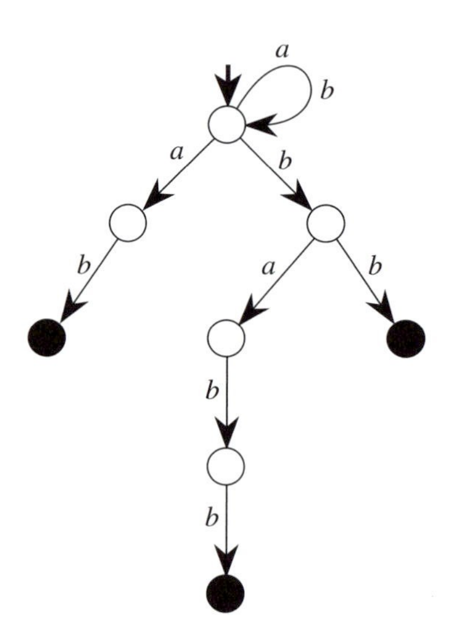

Figure 7.17: The non-deterministic pattern-matching automaton for the set {*ab*, *babb*, *bb*}. The powerset construction gives the automaton *SMA*({*ab*, *babb*, *bb*}) of Figure 7.4. An efficient case of the powerset construction.

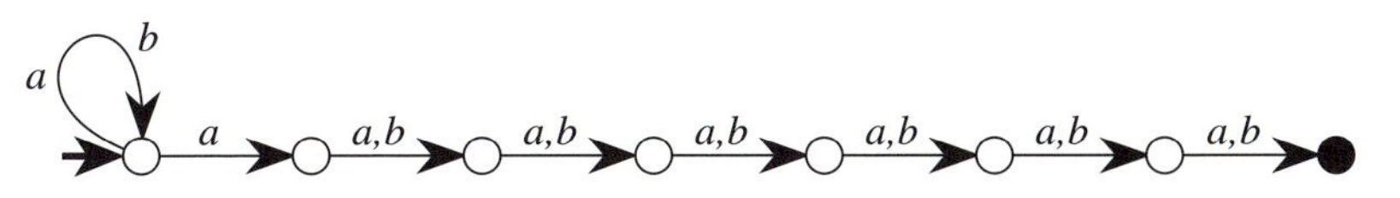

Figure 7.18: A non-efficient case of the powerset construction. The automaton presented results by adding the loop to the initial state of the deterministic automaton accepting words of length 7 that start with letter a. The smallest equivalent deterministic automaton has 128 states.

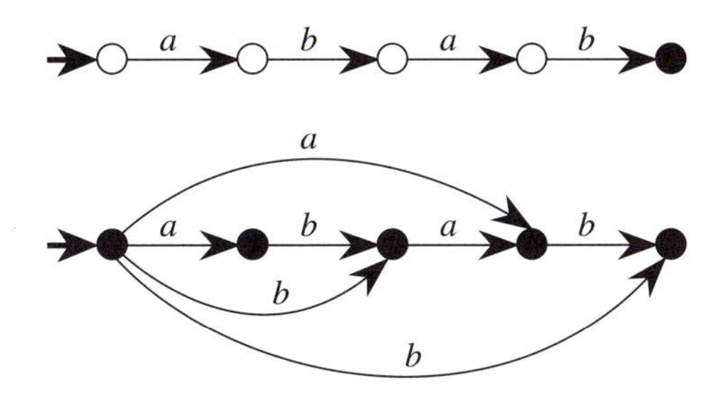

Figure 7.19: An efficient case of the powerset construction. Applied to *FAC*(G) accepting *Fac*(*text*), it gives the smallest automaton for the suffixes of *pat* with a linear number of states.

Combining *loop* and *FAC* sometimes yields an efficient powerset construction. This is done implicitly in Section 6.5. There, the failure function defined on the automaton *DAWG*(*pat*) serves to represent the automaton *loop*(*DAWG*(*pat*)). The overall result is efficient because both steps—from *pat* to *DAWG*(*pat*), and from *DAWG*(*pat*) to *loop*(*DAWG*(*pat*))—are. However, in general, the whole determinization process is inefficient.

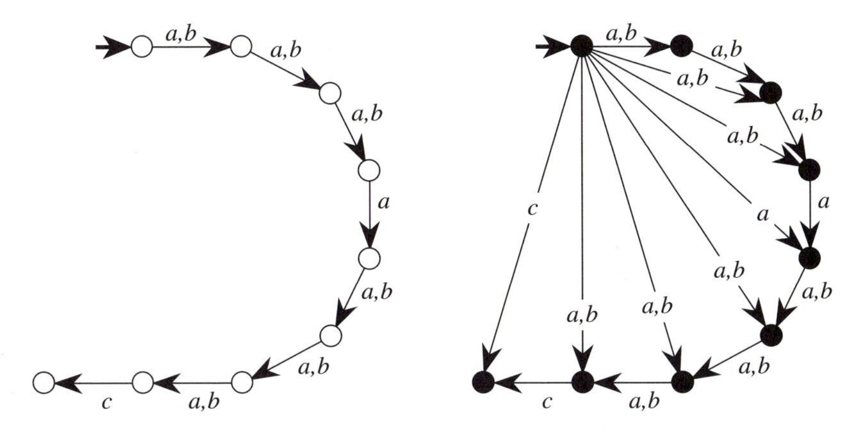

Figure 7.20: A non-efficient case of the powerset construction. G (on the left) accepts $(a{+}b)^n a(a{+}b)^n c$, for $n = 3$. A deterministic version of *FAC*(G) (on the right) has, in this case, an exponential number of states.

7.5 Two-way pushdown automata

The two-way deterministic pushdown automaton (2dpda) is an abstract model of linear-time decision computations on texts. A 2dpda G is essentially a usually deterministic pushdown finite-state machine (see [HU 79]) that differs from the standard model in its ability to move the input head in two directions.

The possible actions of the automaton are: changing the current state of the finite control, moving the head by one position, and locally changing the contents of the stack "near" its top. For simplicity, we assume that each change of the contents of the stack is of one of two types: push(a)—pushing a symbol a onto the stack; pop—popping one symbol off the stack. The automaton has access to the top symbol of the stack and to the symbol in front of the two-way head. We also assume that there are special left and right end markers on both ends of the input word. The output of such an abstract algorithm is "true" iff the automaton stops in an accepting state. Otherwise, the output is "false." Initially the stack contains a special "bottom" symbol, and we assume that in the final moment of acceptance the stack also contains one element. When the stack is empty the automaton stops (the next move is undefined).

The problem solved by a 2dpda can be treated abstractly as a formal language L consisting of all words for which the answer of the automaton is "true" (G stops in an accepting state). We say also that G accepts the language L. The string-matching problem, for a fixed input alphabet, can be also interpreted as the formal language:

$$L_{sm} = \{\ pat\&text\text{: } pat \text{ is a factor of } text\}.$$

This language is accepted by some 2dpda. This gives an automata-theoretic approach to linear-time string-matching, because, as we shall see later, 2dpda can be simulated in linear time. Historically, in fact, it was one of the first methods used for the (theoretical) design of a linear-time string-matching algorithm.

Lemma 7.11 There is a 2dpda accepting the language L_{sm}.

Proof We define a 2dpda G for L_{sm}. It simulates the naive string-matching algorithm *brute-force1* (see Chapter 3). At a given stage of the algorithm, we start at a position i in the text *text*, and at the position $j = 1$ in the pattern *pat*. The pair (i, j) is replaced by the pair (stack, j), where j is the position of the input head in the pattern (and has exactly the same role as j in algorithm *brute-force1*). The contents of the stack is $text[i...n]$ with $text[i+1]$ at the top. The automaton tries to match the pattern starting from position $i+1$ in the text. It checks if the top of the stack equals the current symbol at position j in the pattern; if so, then ($j = j+1$; *pop*). This action is repeated until a mismatch is found, or until the input head points on the marker "&." In the latter case, G accepts. Otherwise, it goes on to the next stage. The stack should now correspond to $text[i+1...n]$ and $j = 1$. It is not difficult to reach such a configuration. The stack is reconstructed by shifting the input head back while simultaneously pushing scanned symbols of *pat* (that have been matched successfully against the pattern). This shows that the algorithm *brute-force1* can be simulated by a 2dpda, and completes the proof. ♦

Similarly the problem of finding the prefix palindromes of a string, and several other problems related to palindromes, can be interpreted as formal languages. Here is a sample:

$L_{prefpal} = \{ww^R u\colon u, w \in A^*, w \text{ is non-empty}\}$,
$L_{prefpal3} = \{ww^R uu^R vv^R z\colon w, u, v \in A^*, w, u, v \text{ non-empty}\}$,
$L_{pal2} = \{ww^R uu^R\colon w, u, v \in A^*, w, u \text{ are non-empty}\}$.

All these languages related to symmetries are accepted by 2dpdas. We leave it as an exercise for the reader to construct the appropriate 2dpdas.

The main feature of 2dpdas is that they correspond to some simple linear-time algorithms. Assume that we are given a 2dpda G and an input word w of length n. The size of the problem is n (the static description of G has constant size). Then, it is proven below that testing whether w is in $L(G)$ takes linear time. We present the concepts that lead to the proof of the result.

The key concept for 2dpda is the consideration of *top configurations*, also called *surface configurations*. The top configuration of a 2dpda retains its top element from the stack only. The first basic property of top configurations is that they contain sufficient information for the 2dpda to choose the next move. The entire configuration consists of the current state, the present contents of the stack, and the position of the two-way input head. Unfortunately, there are too many such configurations (potentially infinite, as the automaton can loop while making push operations). It is easy to see that in every accepting computation the height of the stack is linearly bounded. This does not help very much because there is an exponential number of possible contents of the stack of linear height. The second basic property of top configurations follows: their number is linear in the size of the problem n.

Formally, a top configuration is a triple C = (*state*, *top symbol*, *position*). The linearity of the set of top configurations obviously follows from the fact that the number of states and the number of top symbols (elements of the stack alphabet) is bounded by a constant; it does not depend on n.

We can classify top configurations according to the type of next move of the 2dpda *pop configurations* and *push configurations* (see Figure 7.21). For a top configuration C, we define *Term*[C] as a pop configuration C' accessible from C after some number (possibly zero) of moves that are not allowed to pop the top symbol C. It is as if we started with a stack of height one, and at the end the height of the stack equals one. If there is no such C' then *Term*[C] is undefined. Assume (w. l. o. g.) that the accepting configuration is a pop configuration and the stack is a simple element.

The theorem below is the main result from 2dpdas. It is surprising in view of the fact that the number of moves is usually bigger than that of top configurations. In

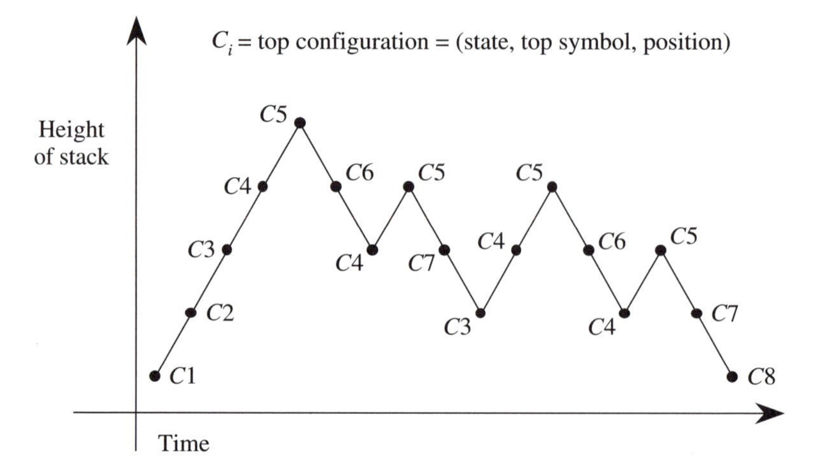

Figure 7.21: The diagram representing the history of the computation of a 2dpda. $Term[C3] = C7$, $Term[C2] = Term[P2(C2, Term[P1(C2)])]$, where $P1(C2) = C3$, $Term[C3] = C7$ and $P2(C2, C7) = C3$, $Term[C3] = C7$.

fact, a 2dpda can make an exponential number of moves and halt. But the result simply shows that shortcuts are possible.

Theorem 7.12 If the language L is accepted by a 2dpda, then there is a linear-time algorithm to test whether x belongs L (for fixed size alphabets).

Proof Let G be a 2dpda, and let w be a given input word of length n. Let us introduce two functions acting on top configurations:

- $P1(C) = C'$, where C' results from C by a push move; this is defined only for push configurations;
- $P2(C1, C2) = C'$, where C' results from $C2$ by a pop move, and the top symbol of C' is the same as in $C1$ ($C2$ determines only the state and the position).

Let POP be the Boolean function defined by: $POP(C)$ = true iff C is a pop configuration. All these functions can be computed in constant time by a random access machine using the (constant-sized) description of the automaton.

It is sufficient to compute in linear time the value of $Term[C0]$ (or find that it is undefined), where $C0$ is an initial top configuration. According to our assumptions simplifying 2dpdas, if G accepts, then $Term[C0]$ is defined. Assume that initially all entries of the table $Term$ contain a special value "not computed."

We start with an assumption that G never loops and ends with a one-element stack. In fact, if the move of G is at some moment undefined we can assume that it goes to a state in which all the symbols in the stack are successively popped.

```
Algorithm; { linear-time simulation of the halting 2dpda }
begin
    for all configuration C do onstack [C] := false;
    return Comp(C0);
end.

function Comp(C) ; { returns Term [C] if defined, else 'false' }
begin
    if Term [C] = "not computed" then
        Term [C] := if POP(C) then C else Comp(P2(C, Comp(P1(C))));
    return Term [C]
end;
```

The correctness and time linearity of the algorithm above are obvious in the case of a halting automaton. The statement "*Term*[*C*] := ... " is executed at most once for each *C*. Hence, only a linear number of push moves (using function *P*1) are applied.

```
Algorithm Simulate;
{ a version of the previous algorithm that detects loops }
begin
    for all configuration C do onstack [C] := false;
    return Comp(C0);
end.

function Comp(C) ;
{ returns Term [C] if defined, 'false' otherwise }
begin { C1 is a local variable }
    if Term [C] = "not computed" then begin
        C1 := P1(C);
        if onstack [C1] then return false { loop }
        else onstack [C1] := true;
        C1 := Comp(C1); onstack [C1] := false; { pop move };
        C1 := P2(C, C1);
        if onstack [C1] then return false { loop } else begin
            onstack [C] := false; onstack [C1] := true
        end;
        Term [C] := Comp(C1)
    end;
    return Term [C]
end;
```

The algorithm *Simulate* above is for a general case: it also has to detect a possible looping of *G* for a given input. We use the table *onstack* initially consisting of "false" values. Whenever we make a push move we then set *onstack*[*C*1] to true for the current top configuration *C*1, and whenever a pop move is made we set *onstack*[*C*1] to false. The looping is detected if we try to put a configuration that is already on the (imaginary) stack of top configurations. If there is a loop then such a situation occurs.

The algorithm *Simulate* has also a linear-time complexity by the same argument as in the case of halting 2dpdas. This completes the proof. ♦

Sometimes it is quite difficult to design a 2dpda for a given language *L*, even if we know that such 2dpda exists. An example of such a language *L* is:

$$L = \{\ 1^n : n \text{ is the square of an integer}\}.$$

A 2dpda for this language can be constructed by a method presented in [Mo 85].

It is much easier to construct 2dpdas for the following languages:

$$\{a^n b^m : m = 2^n\},\ \{a^n b^m : m = n^4\ \},\ \{a^n b^m : m = \log^*(n)\}.$$

But it is not known whether there is a 2dpda accepting the set of even palstars, or the set of all palstars. In general, there is no good technique for proving that a specific language is not accepted by any 2dpda. In fact the "P = NP?" problem can be reduced to the question: does a 2dpda exist for a specific language *L*? There are several examples of such languages. Generally, 2dpdas are used to give alternative formulations of many important problems in complexity theory.

Bibliographic notes

Two "simple" pattern-matching machines are discussed by Aho, Hopcroft, and Ullman in [AHU 74] and, in the case of many patterns, by Aho and Corasick in [AC 75]. The algorithms first compute failure functions, and then the automata. The constructions given in Section 7.1 are direct constructions of the pattern-matching machines. The algorithm of Aho and Corasick is implemented by the command "fgrep" of the UNIX system.

A version of BM algorithm adapted to the search for a finite set of pattern was first sketched by Commentz-Walter [Co 79]. The algorithm was completed by Aho [Ah 90]. Another version of Commentz-Walter's algorithm is presented in [BR 90]. The version of multiple search of Section 7.2 is from Crochemore et al. [C-R 93], where experiments on the real behavior of the algorithm are presented.

The recognition of regular expressions is a classical subject treated in many books. The first algorithm is from Thompson [Th 68]. The equivalence between regular expressions and finite automata is attributed to Kleene [Kl 56]. The equivalence between deterministic automata and non-deterministic automata is from Rabin and Scott [RS 59]. The very useful command "grep" of UNIX implements a variant of the pattern-matching algorithm based on non-deterministic automata (Theorem 7.9). The command "egrep" makes the search with a deterministic automaton (Theorem

7.10). But the states of the automaton are computed only when they are reached effectively during the search (see [Ah 90]).

The determinization of automata can be found in the standard textbook of Hopcroft and Ullman [HU 79]. The question of efficient determinization of automata is from Perrin [Pe 90]. This paper is a survey on the main properties and most recent discoveries about automata theory.

The reader can refer to [KMP 77] for a discussion about the 2dpda approach to string matching (see also [KP 71]).

The linear-time simulation of 2dpdas is from Cook [Co 71] (see also the presentation by Jones [Jo 77]). Our presentation is from Rytter [Ry 85]. This paper also gives a different algorithm similar to the efficient recognition of unambiguous context-free languages. We refer to Galil [Ga 77] for the theoretical significance of 2dpdas, and other interesting theoretical questions. An example of interesting 2dpda computation can be found in [Mo 84].

Selected references

[Ah 90] Aho, A.V., "Algorithms for finding patterns in strings," in (J. van Leeuwen, editor, *Handbook of Theoretical Computer Science*, vol A, *Algorithms and complexity*, Elsevier, Amsterdam, 1990): 255–300.

[AC 75] Aho, A.V., & Corasick, M., "Efficient string matching: an aid to bibliographic search," *Comm. ACM* 18 (1975): 333–340.

[AHU 74] Aho, A.V., Hopcroft, J. E., & Ullman, J. D., *The Design and Analysis of Computer Algorithms*, Addison-Wesley, Reading, Mass. 1974.

[C-R 93a] Crochemore, M., Czumaj, L., Gasieniec, S., Jarominek, S., Lecroq, T., Plandowski, W., & Rytter, W., "Fast multi-pattern matching," Rapport I.G.M. 93-3, Université de Marne la Vallée, 1993.

[Ga 77] Galil, Z., "Some open problems in the theory of computations as questions about two-way deterministic pushdown automaton languages," *Math. Syst. Theory* 10 (1977): 211–228.

[HU 79] Hopcroft, J. E., & Ullman, J. D., *Introduction to Automata, Languages and Computations*, Addison-Wesley, Reading, Mass. (1979)

8

Regularities in texts: Symmetries and repetitions

This chapter presents algorithms dealing with regularities in texts. By regularity we mean a similarity between one factor and some other factors of the text. Such a similarity can be of two kinds: one factor is an exact copy of the other, or it may be a symmetric copy of the other. Algorithmically, an interesting exercise is to detect situations in which similar factors are consecutive in the text. For exact copies we get repetitions of the same factor of the form xx (squares). In the case of symmetric copies we have words of the form xx^R, called even palindromes. Odd palindromes are also interesting regularities (words of the form xax^R, where x is a non-empty word). The compositions of very regular words are in some sense also regular: a palstar is a composition of palindromes and analogously a squarestar is a composition of squares. Algorithms of this chapter are aimed at discovering regularities in text.

8.1 KMR algorithm: Longest repeated factors in words and arrays

We start with a search for copies of the same factor inside a text. Let us denote by *lrepfac*(*text*) the length of the *longest repeated factor* of *text*. It is the longest word occurring at least twice in *text* (occurrences are not necessarily consecutive). When there are several such longest repeated factors, we compute any of them. Let also $lrepfac_k(text)$ be the longest factor that occurs at least k times in *text*.

The central notion used in algorithms of the section is called *naming* or *numbering*. It serves to compute longest repeated factors, but its range of application is much wider. The algorithm that attributes numbers is called *KMR*, and described as follows.

To explain the technique of numbering, we begin with the search for repetitions of factors of a given length r. The algorithm will be an example of the reduction of a string problem to a sorting problem. Suppose that there are k distinct factors of length r (in fact k is computed by the algorithm) in the *text*. Let $fac(1), fac(2),\ldots, fac(k)$ be the sequence of these k different factors in lexicographic order. The algorithm computes the vector NUM_r of length $n\text{–}r$, in which the i-th component is (for $i = 1\ldots n\text{-}r\text{+}1$):

$$NUM_r[i] = j \text{ iff } text[i\ldots i{+}r{-}1] = fac(j).$$

In other words, for each position i ($i = 1\ldots n\text{-}r$), NUM_r identifies the rank or index j, inside the ordered list, of the factor of length r that starts at this position. For example, if $text = ababab$ and $r = 4$, we have $fac(1) = abab$, $fac(2) = baba$ and $NUM_4 = [1, 2, 1]$. In this sense, we compute equivalence classes of positions: two positions are r-equivalent iff the factors of length r starting at these positions are equal. We denote the equivalence on positions by $\equiv_r$. Then, for two positions i and $j (1 \le i, j \le n\text{-}r\text{+}1)$,

$$i \equiv_r j \text{ iff } NUM_r[i] = NUM_r[j].$$

The equivalences on *text* are strongly related to the suffix tree of *text*.

Remark String matching can be easily reduced to the computation of the table NUM_m defined on *text*, for a pattern *pat* of length m. Consider the string w = *pat&text*, where & is a special symbol not in the alphabet. If $NUM_m[1] = q$ then the pattern *pat* starts at all positions i in *text* (relative to w) such that $NUM_m[i] = q$.

The algorithm KMR described following is aimed at computing NUM_r. For simplicity, we start with the assumption that r is a power of two. Later this assumption is dropped. We describe a special function RENUMBER(x) used at each step of KMR algorithm. The value of RENUMBER applied to the vector x is the vector NUM_1 associated with x that is treated as a text. Doing so, the alphabet of x consists of all distinct entries of x; its size can be large, but it does not exceed the length of x. The function RENUMBER realizes one step of KMR algorithm.

Let x be a vector (an array) of size n containing elements of some linearly ordered set. The function RENUMBER assigns to x a vector x'. Let $val(x)$ be the set of values of all entries of x. The vector x' satisfies:

1. if $x[i] = x[j]$ then $x'[i] = x'[j]$, and
2. $val(x')$ is a subset of $[1\ldots n]$.

There are two variations of this procedure depending on whether the following condition is also satisfied:

3. $x'[i]$ is the rank of $x[i]$ in the ordered list of elements of $val(x)$.

Note that condition (3) implies the other two, and can be used to define x'. But the real values $x'[i]$s are of no importance because their role is to identify equivalence classes. We also require that the procedure computes the vector *POS* as a side effect: if q is in val(x'), then $POS[q]$ is any position i such that $x'[i] = q$.

The main part of the algorithm RENUMBER (satisfying conditions (a-c)) is the lexicographic sort. We explain the action of RENUMBER on the following example:

$$x = [(1, 2), (3, 1), (2, 2), (1, 1), (2, 3), (1, 2)].$$

We present a method to compute a vector x' satisfying conditions (1) and (2). We first create the vector y of composite entries $y[i] = (x[i], i)$. Then, the entries of y are lexicographically sorted. Therefore, we get the ordered sequence

$$((1, 1), 4), ((1, 2), 1), ((1, 2), 6), ((2, 2), 3), ((2, 3), 5), ((3, 1), 2).$$

Next, we partition this sequence into groups of elements having the same first component. These groups are consecutively numbered starting from 1. The last component i of each element is used to define the output vector n' : $x'[i]$ is the number associated with the group of the element. Therefore, in the example, we get

$$x'[4] = 1, x'[1] = 2, x'[6] = 2, x'[3] = 3, x'[5] = 4, x'[2] = 5.$$

Doing so is equivalent to computing the vector NUM_1 for x. A vector *POS* associated with x' is

$$POS[1] = 4, POS[2] = 1, POS[3] = 3, POS[4] = 5, POS[5] = 2.$$

Lemma 8.1 If the vector x has size n, and its components are pairs of integers in the range $(1, 2, \ldots, n)$, RENUMBER(x) can be computed in time $O(n)$.

Proof The linear-time lexicographic sorting based on bucket sort can be applied (see [AHU 83], for example). In this way, procedure RENUMBER has the same complexity as sorting n elements of a special form. ♦

The crucial fact that yields the performance of algorithm KMR is the following simple observation:

(*) $NUM_{2p}[i] = NUM_{2p}[j]$ iff
($NUM_p[i] = NUM_p[j]$) and ($NUM_p[i+p] = NUM_p[j+p]$).

This fact is used in KMR algorithm to compute vectors NUM_r. Let us look at how the algorithm works for the example string *text* = *abbababba* and $r = 4$. The first vector NUM_1 identifies the letters in *text*:

$$NUM_1 = [1, 2, 2, 1, 2, 1, 2, 2, 1].$$

Then, the algorithm successively computes vectors NUM_2, and finally NUM_4.

$NUM_2 = [1, 3, 2, 1, 2, 1, 3, 2]$;
$NUM_4 = [2, 5, 3, 1, 4, 2]$.

```
Algorithm KMR;
{ relatively to text, computes the vector NUMr (r is a power }
{ of two) and, as a side effect, vectors NUMP (p = 2^q < r) }
begin
   NUM1 := x := RENUMBER(text); { text considered as a vector }
   p := 1;
   while p < r do begin
        for i := 1 to n-2p+1 do y[i] := (x[i], x[i+p]);
        NUM2p := RENUMBER(y);
        p := 2.p;
   end;
end.
```

Once all vectors NUM_P, for all powers of two smaller than r, are computed, we can easily compute the vector NUM_q in linear time for each integer $q < r$. Let q' be the greatest power of two not greater than q. We can compute NUM_q using a fact similar to (*):

$NUM_q[i] = NUM_q[j]$ iff
$(NUM_q'[i] = NUM_q'[j])$ and $(NUM_q'[i+q-q'] = NUM_q'[j+q-q'])$.

Equivalences on positions of a text (represented by vectors NUM_P) computed by KMR algorithm capture the repetitions in the text. The algorithm is a general strategy that can serve several purposes. A first application provides a longest factor repeated at least k times in the text (recall that its length is denoted by $lrepfac_k(text)$). And it can produce the position of such a factor at the same cost. Let us call it $REP_k(r, text)$. If r is its length, and if the vector $NUM_r[i]$ on $text$ is known, then $lrepfac_r(text)$ can be trivially computed in linear time.

Theorem 8.2 The function $lrepfac_k(text)$ can be computed in $O(n.\log n)$ time for alphabets of size $O(n)$.

Proof We can assume that length n of the text is a power of two; otherwise a suitable number of "dummy" symbols can be appended to text. The algorithm KMR can be used to compute all vectors NUM_p for powers of two not exceeding n. We then apply a kind of binary search using function $REP_k(r, text)$: the binary search looks for the maximum r such that $REP_k(r, text) \neq$ nil. If the search is successful we then return the longest (k times) repeated factor. Otherwise, we report that there is no such repetition. The sequence of values of $REP_k(r, text)$ (for $r = 1, 2, \ldots, n-1$) is "mono-

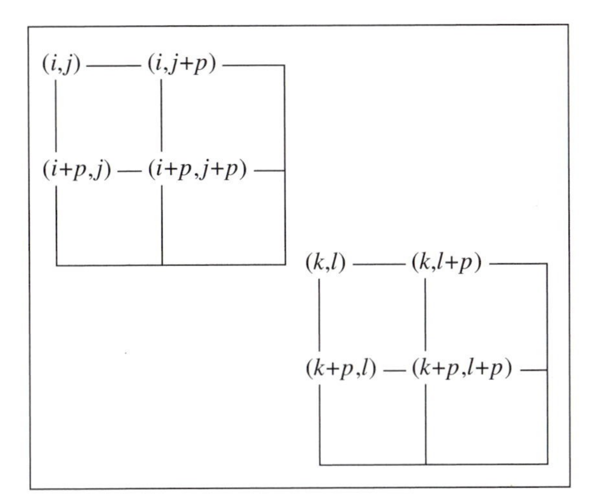

Figure 8.1: A repeated sub-array of size $2p \times 2p$. The occurrences can overlap.

tonic" in the following sense: if $r1 < r2$ and $REP_k(r2, text) \neq$ nil, then $REP_k(r1, text) \neq$ nil. The binary search behaves similarly to searching an element in a monotonic sequence. It has $\log n$ stages; in each stage the value $REP_k(r, text)$ is computed in linear time. Altogether the computation takes $O(n.\log n)$ time. This completes the proof. ♦

Remark Later in Section 8.2 we give a linear-time algorithm for this problem. However, it will be more sophisticated, and it uses one of the structures from Chapters 5 and 6.

The longest repeated factor problem generalizes an equivalent two-dimensional problem in a straightforward way. It is called the *longest repeated sub-array* problem (see Figure 8.1). We are given an $n \times n$ array T. The size of the problem is $N = n^2$. For this problem KMR algorithm gives $O(N.\log N)$ time complexity, which is the best upper bound known up to now. The algorithm works for finding repetitions in trees as well.

Still, numbering is used on sub-arrays. Let $NUM_r[i, j]$ be the number of the $r \times r$ sub-array of the array T having its upper-left corner at position (i, j). There is a fact, analogous to (*), illustrated by Figure 8.1:

(**) $NUM_{2p}[i, j] = NUM_{2p}[k, l]$ iff
$NUM_p[i, j] = NUM_p[k, l]$ and $NUM_p[i+p, j] = NUM_p[k+p, l]$ and
$NUM_p[i, j+p] = NUM_p[k, l+p]$ and $NUM_p[i+p, j+p] = NUM_p[k+p, l+p]$.

Using fact (**), all the computation of repeating $2p \times 2p$ sub-array reduces to the computation of repeating $p \times p$ sub-arrays. The matrix NUM_{2p} is computed from

NUM_p using a procedure analogous to RENUMBER. Now, the internal lexicographic sorting is executed on elements having four components (instead of two for texts). But, Lemma 8.1 still holds in this case, such as KMR algorithm that is deduced from RENUMBER. This proves the following.

Theorem 8.3 The size of a longest repeated sub-array of an $n \times n$ array of symbols can be computed in $O(N.\log N)$ time, where $N = n^2$.

The longest repeated factor problem for texts can be solved in a straightforward way if we have already constructed the suffix tree *T*(*tree*) (see Section 5.7). The value *lrepfac*(*text*) is the longest path (in the sense of length of word corresponding to the path) leading from the root of *T*(*tree*) to an internal node. Generally, $lrepfac_k(text)$ is the length of a longest path leading from the root to an internal node in which the subtree contains at least k leaves. The computation of such a path can be accomplished easily in linear time. The preprocessing needed to construct the suffix tree makes the whole algorithm much more complicated than the one applying the strategy of KMR algorithm. Nevertheless, this proves that the computation takes linear time.

Theorem 8.4 The function $lrepfac_k$ can be computed in time linear in the length of the input text, for fixed-size alphabets.

8.2 Finding squares

It is a nontrivial problem to find a square factor in linear time, that is, a non-empty factor of the form *xx*. A naive algorithm gives cubic bound on the number of steps. A simple application of failure function gives a quadratic algorithm. For that purpose, we can compute a failure function $Bord_i$ for each suffix $text[i...n]$ of the text. Then, there is a square starting at position i in the text iff $Bord_i[j] \geq j/2$, for some j ($1 \leq j \leq n-i+1$). Since each failure function is computed in linear time, the whole algorithm takes quadratic time.

The notions introduced in Section 8.1 are useful for finding a square of a given length r. Using NUM_r the test takes linear time, giving an overall $O(n.\log n)$ time algorithm. In fact, values of NUM_r, for all sensible values of r, can help with finding squares at a similar cost, but KMR algorithm does not provide all these vectors.

We now develop an $O(n.\log n)$ algorithm that tests the squarefreeness of texts, and afterward design a linear-time algorithm (for fixed alphabets). The first method is based on a divide-and-conquer approach. The main feature of both algorithms is a fast implementation of the Boolean function *test*(u, v) that tests whether the word uv contains a square for two squarefree words u and v. Then, if uv contains a square, it must begin in u and ends in v. Thus, the operation *test* is a composition of two smaller Boolean functions: *righttest* and *lefttest*. The first one searches for a square

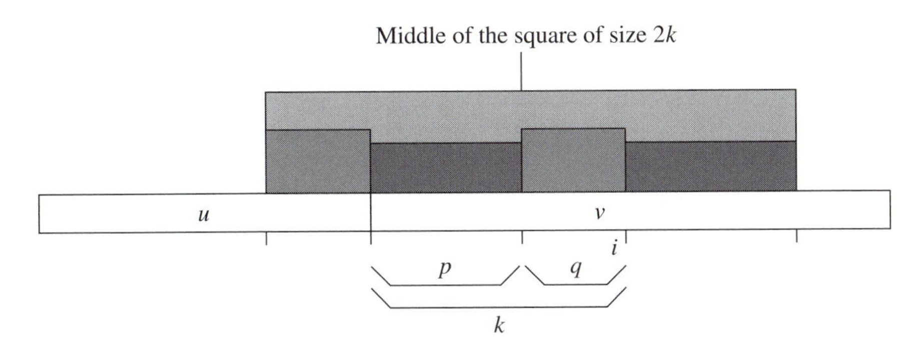

Figure 8.2: A square xx of size $2k$ occurs in uv. The suffix of $v[1\ldots k]$ of size q is also a suffix of u; the prefix of $v[k+1]\ldots$ *of size p* is a prefix of v. $PREF[k]+ SUF_u[k] \geq k$.

in which the center is in v, while the second searches for a square in which the center is in u.

We will now describe how *righttest* works on words u and v. We use two auxiliary tables related to string matching. The first table *PREF* is defined on the word v. For a position k on v, $PREF[k]$ is the size of longest prefix of v occurring at position k (it is a prefix of $v[k+1\ldots]$. The second table is called *SUF*. The value $SUF_u[k]$ (k is still a position on v) is the size of longest suffix of $v[1\ldots k]$ that is also a suffix of u. Table SUF_u is a generalization of table S discussed in Section 4.6. These tables can be computed in linear time with respect to $|v|$ (see Section 4.6). With the two tables, the existence of a square in uv centered in v reduces to a simple test on each position of v, as shown by Figure 8.2.

Lemma 8.5 The Boolean value *righttest*(u, v) can be computed in $O(|v|)$ time (independently of the size of u).

Proof The computation of tables *PREF* and SUF_u takes $O(|v|)$ time. It is clear (see Figure 8.2) that there exists a square centered in v iff for some position k $PREF[k]+ SUF_u[k] \geq k$. All these tests again take $O(|v|)$ time. ♦

Corollary 8.6 The Boolean value *test*(u, v) can be computed in $O(|u|+|v|)$ time.

Proof Compute *righttest*(u, v) and *lefttest*(u, v). The value *test*(u, v) is the Boolean value of *righttest*(u, v) or *lefttest*(u, v). The computation takes $O(|v|)$ (Lemma 8.5), and $O(|u|)$ time (by symmetry from Lemma 8.5), respectively. The result follows. ♦

We present another very interesting algorithm computing *righttest*(u, v) in linear time according to v, but requiring only a constant additional memory space. It is based on the following combinatorial fact:

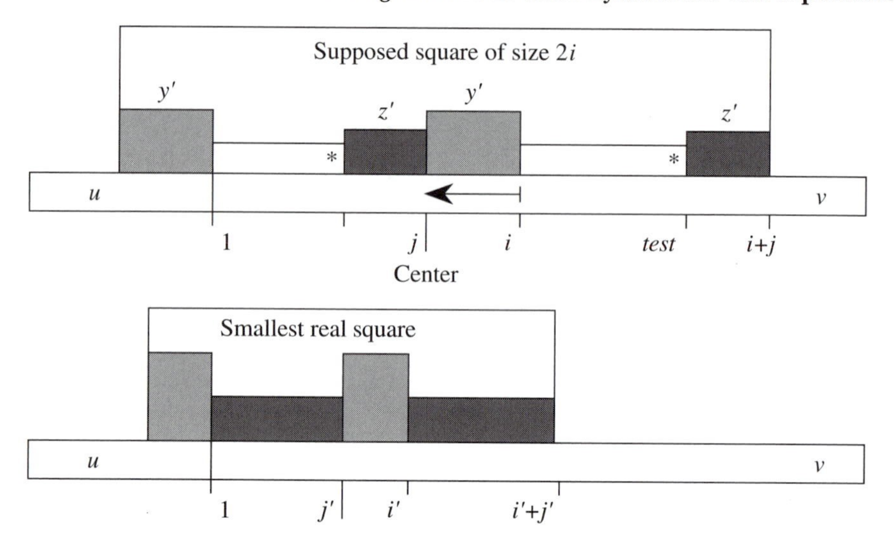

Figure 8.3: One stage of an algorithm to compute *righttest*(u, v) start at i and check right-to-left until a mismatch at j; then, assume the middle of a square at j, and check from $i+j$ right-to-left until a mismatch occurs at position *test*. Parts of text shaded with the same pattern match. Each part may be empty, but not all together.

- if u, v are squarefree, and we have a situation as in Figure 8.3 (where segments shaded with the same pattern indicate which factors are equal) then $i' < \max(j, i/2)$ and $i'+j' < test$. The proof is rather technical and has been left to the reader as an excellent exercise on word combinatorics.

One stage of the algorithm consists of finding the largest matching segments shaded with a light screen, and then, (if $i+j < test$) of finding the largest matching segments shaded with a darker screen. If we reach position 1, then a square is found. Otherwise, we set $i = i'$, where i' can be taken small according to the combinatorial fact above. Then, we start the next (similar) stage.

```
function righttest(u, v) : boolean;
{ return true iff uv contains a square centered in v }
begin
   i := |v|; test := |v|+1;
            { initially, no reasonable bound on i+j is known }
   while i ≥ 1 do begin
      j := i;
      { traversing right-to-left segments painted with a light screen}
      while j ≥ 1 and |u|-i+j ≥ 1 and u[|u|-i+j] = v [j] do
         j := j-1;
```

```
            if j = 0 then return true;
            k := i+j;
            if k < test then begin
                test := k;
            { traversing right-to-left segments painted with a dark screen}
                while test > i and v[test] = v[j-k+test] do
                        test := test-1;
                if test = i then return true;
            end; { if }
            i := max(j, ⌈i/2⌉)-1;
        end; { main while loop }
        return false; { no square centered in v }
    end;
```

We leave the proof that the algorithm makes only a linear number of steps as an exercise. Symmetrically *lefttest* can be accomplished in linear time also using a constant-sized memory. Both implementations of functions *lefttest* and *righttest* yield an implementation of function *test*.

Lemma 8.7 The function *test* can be computed in linear time using only O(1) auxiliary memory.

The recursive algorithm *SQUARE* for testing occurrences of squares in a text is designed with the divide-and-conquer method, for which function *test* has been written.

Theorem 8.8 Algorithm *SQUARE* tests whether the text *text* of length n contains a square in time $O(n.\log n)$ with $O(1)$ additional memory space.

Proof The algorithm has $O(n.\log n)$ time complexity, because *test* can be computed in linear time. The recursion has a highly regular character, and can be organized without using a stack. It is sufficient to remember the depth of recursion and the current interval. The tree of recursion can be computed in a bottom-up manner, level-by-level. This completes the proof. ♦

```
function SQUARE(text) : boolean;
{ checks if text contains a square, n = |text| }
begin
    if n > 1 then begin
        if SQUARE(text[1..⌊n/2⌋]) then return true;
        if SQUARE(text[⌊n/2⌋+1..n]) then return true;
        if test(text[1..⌊n/2⌋], text[⌊n/2⌋+1..n]) then return true;
        return false;       { if value true not already returned }
end;
```

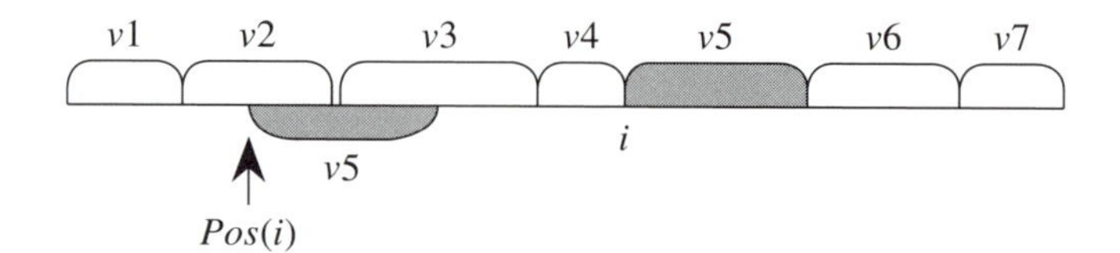

Figure 8.4: Efficient factorization of the source. v_5 is the longest factor occurring before.

The algorithm *SQUARE* inherently runs in $O(n.\log n)$ time. This is due to the divide-and-conquer strategy for the problem. But, it can be shown that the algorithm extends to the detection of all squares in a text. And this shows that the algorithm becomes optimal, because some texts may contain exactly $O(n.\log n)$ squares, namely Fibonacci words.

We show that the question of testing the squarefreeness of a word can be answered in linear time on a fixed alphabet. This contrasts with the above problem. The strategy will again use a kind of divide-and-conquer, but with an unbalanced nature. It is based on a special factorization of texts. Our interest is mainly theoretical because of a rather large overhead in memory usage: the extra memory space is of linear size (instead of $O(1)$ for *SQUARE*), and an efficient data structure for factors of the text has to be used. But the algorithm, or more exactly the factorization defined for the purpose, is related to data compression methods based on the elimination of repetitions of factors. The algorithm shows another profound application of data structures developed in Chapters 5 and 6 for storing all factors of a text.

We first define the *f*-factorization of *text* (the *f* stands for factors) (see Figure 8.4). It is a sequence of non-empty words $(v_1, v_2, \ldots, v_m)$, where

- $v_1 = text[1]$, and
- for $k > 1$, v_k is defined as follows. If $|v_1v_2\ldots v_{k-1}| = i$ then v_k is the longest prefix u of $text[i+1\ldots n]$ that occurs at least twice in $text[1\ldots i]u$. If there is no such u then $v_k = text[i+1]$. Denote by $pos(v_k)$ the smallest position $l < i$ (where $i = |v_1v_2\ldots v_{k-1}|$) such that an occurrence of v_k starts at l. If there is no such position then take $pos(v_k) = 0$.

The *f*-factorization of a *text*, and the computation of values $pos(v_k)$ can be executed in linear time with the directed acyclic word graphs $G = DAWG(text)$ (also with the suffix tree $T(text)$). Indeed, the factorization is computed while the dawg is built. The overall procedure has the same asymptotic linear bound as just building the dawg. This leads to the final result of the section, the consequence of the following observation in which the proof is left as an exercise.

Lemma 8.9 Let $(v_1, v_2, \ldots, v_m)$ be the *f*-factorization of *text*. Then, *text* contains a square iff for some k at least one of the following conditions holds:

1. $pos(v_k)+|v_k|\geq|v_1v_2\ldots v_{k-1}|$ (selfoverlapping of v_k),
2. $lefttest(v_{k-1}, v_k)$ or $righttest(v_{k-1}, v_k)$,
3. $righttest(v_1v_2\ldots v_{k-2}, v_{k-1}v_k)$.

```
function linear-SQUARE(text) : boolean;
{ checks if text contains a square, n = |text| }
begin
   compute the f-factorization (v1, v2, .., vm) of text;
   for k :=1 to m do
      if (1) or (2) or (3) hold then { Lemma 8.9 }
         return true;
   return false;
end;
```

Theorem 8.10 Function *linear-SQUARE* tests whether a text of length n contains a square in time $O(n)$ (on a fixed alphabet), with $O(n)$ additional memory space.

Proof The key point is that the computation of $righttest(v_1v_2\ldots v_{k-2}, v_{k-1}v_k)$ can be executed in $O(|v_{k-1}v_k|)$ time. Thus the total time is proportional to the sum of length of all v_ks; hence, it is linear. This completes the proof. ♦

8.3 Symmetries in texts

We start with a decision problem that consists of verifying if a given text has a prefix that is a palindrome (such palindromes are called prefix palindromes). There is a very simple linear-time algorithm used to search for a prefix palindrome:

- compute the failure function *Bord* for $text\&text^R$ (of length $2n+1$),
- then, text has a prefix palindrome iff $Bord(2n+1) \neq 0$.

However, this algorithm has two drawbacks: it is not an on-line algorithm, and moreover, we could expect to have an algorithm that computes the smallest prefix palindrome in time proportional to the length of this palindrome (if a prefix palindrome exists). Later in the section it will be seen (when testing palstars) why we impose such requirements.

For the time being, we restrict ourselves to even palindromes. We proceed in a similar way as in the derivation of KMR algorithm. An efficient algorithm will be derived from an initial *brute_force* algorithm by examining its invariants. However more combinatorics on words are used here (as compared with KMR algorithm).

Observe a strong similarity between this algorithm and the first algorithm for string matching, *brute_force1*. The time complexity of the algorithm is quadratic (in worst case). A simple instance of a worst text for the algorithm is $text = ab^n$. On the other hand, the same analysis as for algorithm *brute_force1* (string matching) shows that the expected number of comparisons (if symbols of text are randomly chosen) is linear.

```
Algorithm brute_force;
{ looking for prefix even palindromes }
begin
   i := 1;
   while i ≤ ⌊n/2⌋ do begin
        { check if text[1..2i] is a palindrome }
        j := 0;
        while j < i and text[i-j] = text[i+1+j] do j := j+1;
        if j = i then return true;
        { inv(i, j) : text[i-j] ≠ text[i+1+j] }
        i := i+1;
   end;
   return false;
end.
```

The key to the improvement is to make appropriate use of the information gathered by the algorithm. This information is expressed by invariant

$w(i, j)$: $text\ [i\text{-}j\ldots i] = text\ [i+1\ldots i+j]$.

The maximum value of j satisfying $w(i, j)$ for a given position i is called the radius of the palindrome centered at i, and denoted by $Rad[i]$. Hence, algorithm *brute_force* computes values of $Rad[i]$ but does not profit from their computation. The information is wasted. At the next iteration, the value of j is reset to 0. As an alternative, we try to make use of all possible information, and, for that purpose, the computed values of $Rad[i]$ are stored in a table for further use. The computation of prefix palindromes easily reduces to the computation of table *Rad*. Hence, we convert the decision version of algorithm *brute_force* into its optimized version computing *Rad*. For simplicity, assume that the text starts with a special symbol. A palindrome, then, centered in i is actually a prefix palindrome iff $Rad[i] = i-1$.

The key to the improvement is not only a mere recording of table *Rad*, but also a surprising combinatorial fact about symmetries in words. Suppose that we have already computed $Rad[1]$, $Rad[2]$,...., $Rad[i]$. It so happens that we can sometimes compute many new entries of table *Rad* without comparing any symbols. The following fact enables us to do so.

Lemma 8.11 If $1 \leq k \leq Rad[i]$ and $Rad[i\text{-}k] \neq Rad[i]\text{-}k$, then $Rad[i+k] = \min(Rad[i-k], Rad[i]-k)$.

Proof Two cases are considered.

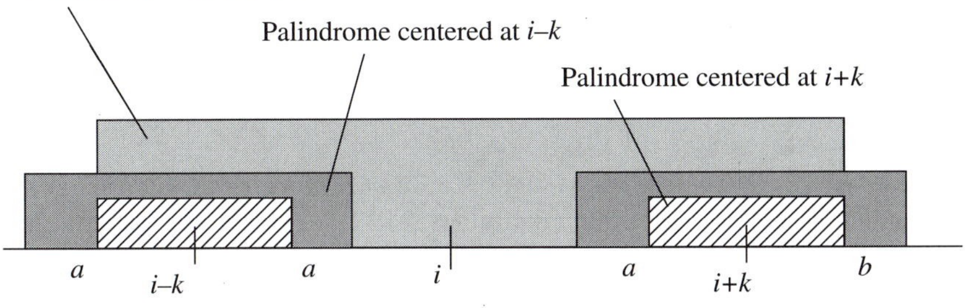

Figure 8.5: Case (b) of proof of Lemma 8.11.

- Case (a): $Rad[i-k] < Rad[i]-k$. The palindrome of radius $Rad[i+k]$ centered at i-k is completely contained in the longest palindrome centered in i. Position $i-k$ is symmetrical to $i+k$ with respect to i. Hence, by symmetry (with respect to position i), the longest palindrome centered in $i+k$ has the same radius as the palindrome centered at $i-k$. This implies the conclusion in this case.
- Case (b): $Rad[i-k] > Rad[i]-k$. The situation is illustrated in Figure 8.5. The maximal palindromes centered at i, $i+k$ and $i-k$ are presented. Symbols a, b are distinct due to the maximality of the palindrome centered in i. Hence, $Rad[i+k] = \min(Rad[i-k], Rad[i]-k)$.

This completes the entire proof of the lemma. ♦

In one stage of the algorithm that computes prefix palindromes, we can update $Rad[i+k]$ for all consecutive positions $k = 1, 2,...$ such that $Rad[i-k] \neq Rad[i]-k$. If the last such k is k', we can then later consider the next value of i as $i+k'$, and start the next stage. This is similar to shifts applied in string-matching algorithms in which values result from a precise consideration on invariants. We obtain the following algorithm for on-line recognition of prefix even palindromes and computation of the table of radii. All positions i satisfying $Rad[i] = i-1$ (occurrences of prefix palindromes) are output.

```
Algorithm Manacher;
{ on-line computation of prefix even palindromes,}
{ and of table Rad; text starts with a unique left end marker}
begin
   i := 2; Rad [1] := 0; j := 0; { j = Rad [i] }
   while i ≤ ⌊n/2⌋ do begin
      while text [i-j] = text [i+1+j] do j := j+1;
      if j = i then write(i); Rad [i] := j;
      k := 1;
      while Rad [i-k] ≠ Rad [i]-k do begin
         Rad [i+k] := min(Rad [i-k], Rad [i]-k); k := k+1;
      end;
```

```
            { inv(i, j): text[i-j] ≠ text[i+1+j] }
            j := max(j-k, 0);
            i := i+k;
        end;
    end.
```

The solution presented for the computation of prefix even palindromes adjusts easily to the table of radii of odd palindromes. They can also be computed in linear time. The same holds true for longest palindromes occurring in the text. Several other problems can be solved in a straightforward way using the table *Rad*.

Theorem 8.12 The longest symmetrical factor and the longest (or shortest) prefix palindrome of a text can be computed in linear time. If *text* has a prefix palindrome, and if s is the length of the smallest prefix palindrome, then s can be computed in time $O(s)$.

We now consider another question regarding regularities in texts. Let P^* be the set of words that are compositions of even palindromes, and let PAL^* denote the set of words composed of any type of palindrome (even or odd). Recall that one-letter words are not palindromes according to our definition (their symmetry is too trivial to be considered as an "interesting" symmetry).

Our aim now is to test whether a word is an *even palstar*, i.e., a member of P^*, or a *palstar*, i.e., a member of PAL^*. We begin with the simpler case of even palstars.

Let *first*(*text*) be a function which value is the first position i in text such that $text[1...i]$ is an *even* palindrome; it is zero if there is no such prefix palindrome. The following algorithm tests even palstars in a natural way. It finds the first prefix even palindrome and cuts it. Afterward, the same process is repeated as often as possible. If we are left with an empty string, then the initial word is an even palstar.

```
function PSTAR(text) : boolean; { is text an even palstar ? }
begin
    s := 0;
    while s < n do begin { cut text[s+1..n] }
        if first(text[s+1..n])=0 then return false;
        s := s+first(text[s+1..n])
    end;
    return true;
end;
```

Theorem 8.13 Even palstars can be tested on-line in linear time.

Proof It is obvious that the complexity is linear, even on-line. In fact, Manacher algorithm computes the function *first* on-line. An easy modification of the algorithm gives an on-line algorithm operating within the same complexity.

However, a more difficult issue is the correctness of function *PSTAR*. Suppose that *text* is an even palstar. It then seems reasonable to expect that its decomposition into even palindromes does not necessarily start with the shortest prefix even palindrome. Fortunately, and perhaps surprisingly, it so happens that we have always had a good decomposition (starting with the smallest prefix palindrome) if *text* is an even palstar. So the greedy strategy of function *PSTAR* is correct. To prove this fact, we need some notation related to decompositions. It is defined only for texts that are even palindromes. Let

$$parse(text) = \min\{s : text[1\ldots s] \in P \text{ and } text[s+1\ldots n] \in P^*\}.$$

Now the correctness of the algorithm results directly from the following fact about even non-empty palstars text.

Claim $parse(text) = first(text)$.

Proof of the claim It follows from the definitions that $first(text) \leq parse(text)$, hence, it is sufficient to prove that the reverse inequality holds. The proof is by contradiction. Assume that *text* is an even palstar such that $first(text) < parse(text)$. Consider two cases,

- case (a): $parse(text)/2 < first(text) < parse(text)$,
- case (b): $2 \leq first(text) \leq parse(text)/2$.

The proof of the claim according to these cases is given in Figure 8.6. This completes the proof of the correctness and the entire proof of the theorem. ♦

If we try to extend the previous algorithm to all palstars, we are led to consider functions *first*1 and *parse*1, analogous to *first* and *parse*, respectively, as follows:

$$parse1(text) = \min\{s : text[1\ldots s] \in PAL \text{ and } text[s+1\ldots n] \in \mathrm{PAL}^*\},$$
$$first1(text) = \min\{s : text[1\ldots s] \in PAL\}.$$

Unfortunately, when *text* is a palstar, the equation $parse1(text) = first1(text)$ is not always true. A counterexample is the text $text = bbabb$. We have $parse1(text) = 5$ and $first1(text) = 2$. If $text = abbabba$, then $parse1(text) = 7$ and $first1(text)=4$. For $text = aabab$, we have $parse1(text) = first1(text)$.

Observe that for the first text, we have $parse1(text) = 2.first1(text)+1$; for the second text, we have $parse1(text) = 2.first1(text)-1$; and for the third text, we have $parse1(text) = first1(text)$. It so happens that it is a general rule that only these cases are possible.

Lemma 8.14 Let *text* be a non-empty palstar, then

$$parse1(text) \in \{first1(text), 2.first1(text)-1, 2.first1(text)+1\}.$$

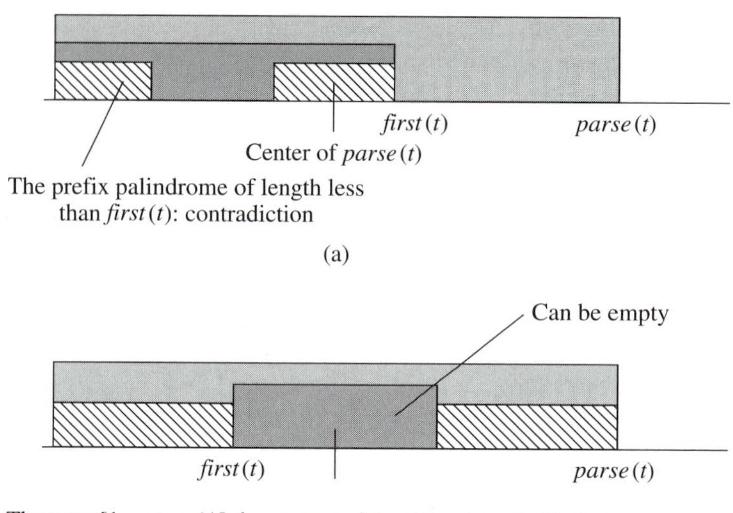
first (t)
parse (t)
Center of parse (t)
The prefix palindrome of length less than first (t): contradiction
(a)
Can be empty
first (t)
parse (t)
The text t[1...parse (t)] decomposes into at least two palindromes; it contradicts the definition of parse (t)
(b)

Figure 8.6: The proof of the claim (by contradiction).

Proof The proof is similar to the proof of the preceding lemma. In fact, the two special cases (2.*first*1(*text*)–1, 2.*first*1(*text*)+1) are caused by the irregularity implied at critical points by considering odd and even palindromes together. Let f = *first*1(*text*), and p = *parse*1(*text*). The proof of the impossibility of the situation ($f < p < 2.f-1$) is essentially presented in the case (a) of Figure 8.6. The proof of the impossibility of two other cases ($p = 2.f$) and ($p > 2.f+1$) is similar. ♦

Assume that we have computed the tables $F[i]$ = *first*1(*text*[i+1...**n**]) and $PAL[i]$=(*text*[i+1...n] is a palindrome) for i = (0, 1, ..., n). Then, the following algorithm recognizes palstars.

```
function PALSTAR(text) : boolean; { palstar recognition }
begin
    palstar[n] := true; { the empty word is a palstar }
    for i := n–1 downto 0 do begin
        f := F[i];
        if f = 0 then palstar[i] := false
        else if PAL[i] then palstar[i] := true
        else palstar[i] := (palstar[i+f] or palstar[i+2f–1]
                                              or palstar[i+2f+1]);
    end;
    return palstar[0];
end.
```

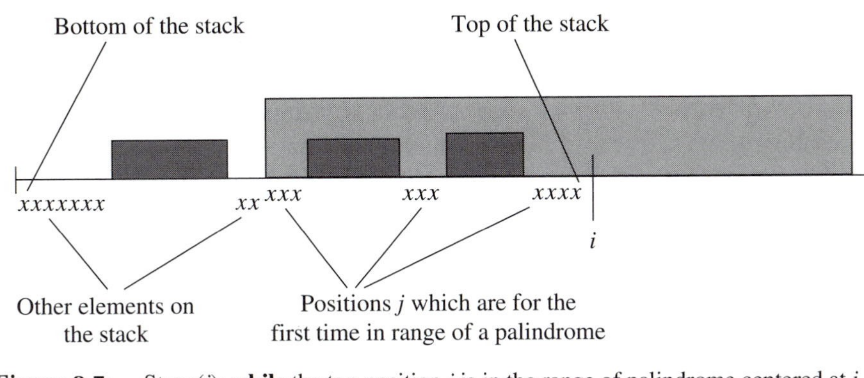

Figure 8.7: Stage(i): **while** the top position j is in the range of palindrome centered at i **do begin** *pop*(*stack*); $F1[j] := 2(j-i)+1$ **end**; *push*(i).

Theorem 8.15 Palstars can be tested in linear time.

Proof Assuming the correctness of function *PALSTAR*, in order to prove that it works in linear time, it is sufficient to show how to compute tables F and *PAL* in linear time. The computation of *PAL* is trivial if the table *Rad* is known. This latter can be computed in linear time. The computation of table F is more difficult. For simplicity we restrict ourselves to odd palindromes, and compute the table

$$F1[j] = \min\{s : text[1\ldots s] \text{ is an odd palindrome, or } s = 0\}.$$

Assume *Rad* is the table of radii of odd palindromes. The radius of the palindrome of size $2k+1$ equals k. We say that j is in the range of an odd palindrome centered at i iff $i-j+1 \le Rad[i]$ (see Figure 8.7).

A stack of positions is used. It is convenient to have some special position at the bottom. Initially the first position 1 is pushed onto the stack, and i is set to 2. One stage begins with the situation depicted in the Figure 8.7. All xs are on the stack (entries of $F1$ waiting for their values). The whole algorithm is:

for i := 2 **to** n **do** stage(i);
for all remaining elements j on the stack **do**
{j is not in a range of any palindrome} $F1[j] := 0$.

The treatment of even palindromes is similar. In that case, $F[i]$ is computed as the minimum value of even or odd palindromes starting at position i. This completes the proof. ♦

Remark In fact the algorithm *PALSTAR* can be transformed into an on-line linear-time algorithm. But this is outside the scope of this book.

It is perhaps surprising that testing whether a text is a composition of a fixed number of palindromes seems more difficult than testing for palstars. Recall that here P denotes the set of even palindromes. It is very easy to recognize compositions of exactly two words from P. The word *text* is such a composition iff for some internal position i $text[1\ldots i]$ and $text[i+1\ldots n]$ are even palindromes. This can be checked in linear time if table *Rad* is already computed. But this approach does not directly produce a linear-time algorithm for P^3. Fortunately, there is another combinatorial property of texts that is useful for that. Its proof is omitted.

Lemma 8.16 If $x \in P$, then there is a decomposition of x into words $x[1\ldots s]$ and $x[s\ldots n]$ such that both are members of P, and that either the first word is the longest prefix palindrome of x, or the second one is the longest suffix palindrome of x.

One can compute the tables of all the longest palindromes starting or ending at positions of the word by a linear-time algorithm very similar to that of table F. Assume now that we have such tables, and also the table *Rad*. One can then check if any suffix or prefix of text is a member of P^2 in constant time because of Lemma 8.16 (only two positions can be checked using preprocessed tables). Now we are ready to recognize elements of P^3 in linear time. For each position i we just check in constant time if $text[1\ldots i] \in P^2$ and $text[i+1\ldots n] \in P$. Similarly, we can test elements of P^4. For each position i we check in constant time if $text[1\ldots i] \in P^2$ and $text[i+1\ldots n] \in P^2$. We have just sketched the proof of the following statement.

Theorem 8.17 The compositions of two, three, and four palindromes can be tested in linear time.

As far as we know, there is presently no algorithm to test compositions of exactly five palindromes in linear time.

We have defined palindromes as nontrivial symmetric words (words of size at least two). One can say that for a fixed k, palindromes of a size smaller than k are also uninteresting. This leads to the definition of PAL_k as palindromes of size at least k. Generalized palstars (compositions of words from PAL_k) can also be defined. For a fixed k, there are linear-time algorithms for such palstars. The structure of algorithms, and the combinatorics of such palindromes and palstars are analogous to what has been presented in the section.

Bibliographic notes

The algorithm KMR of the first section is from Karp, Miller, and Rosenberg [KMR 72]. Another approach is presented in [Cr 81]. It is based on a modification of Hopcroft's partitioning algorithm [Ho 71] (see [AHU 74]), and the algorithm computes vectors NUM_r for all sensible values of r. It yields an optimal computation of all repetitions in a word. The same result has been shown by Apostolico and Preparata [AP 83] as an application of suffix trees, and also by Main and Lorentz [ML 84].

The first $O(n.\log n)$ algorithm for searching a square is by Main and Lorentz [ML 79] (see also [ML 84]). The procedure *righttest* of Section 8.2 is a slight improvement on the original algorithm proposed by these authors. The linear-time algorithm of Section 8.2 is from Crochemore [Cr 83] (see also [Cr 86]). A different method of achieving linear time has been proposed by Main and Lorentz [ML 85].

The algorithm for prefix palindrome in Section 8.3 is from Manacher [Ma 75]. The material in the rest of the section is mainly from Galil and Seiferas [GS 78]. The reader can refer to [GS 78] for the proof of Lemma 8.16, or an on-line linear-time algorithm for *PALSTAR*.

Selected references

[Cr 86] Crochemore, M., "Transducers and repetitions," *Theoret. Comput. Sci.* 45 (1986): 63–86.

[GS 78] Galil, Z., & Seiferas, J., "A linear-time on-line recognition algorithm for Palstars," *J. ACM* 25 (1978): 102–111.

[KMR 72] Karp, R.M., Miller, R.E., & Rosenberg, A.L., "Rapid identification of repeated patterns in strings, arrays and trees," in (*Proc. 4th ACM Symposium on Theory of Computing*, Association for Computing Machinery, New York, 1972): 125–136.

[ML 84] Main, M.G., & Lorentz, R.J., "An O(n log n) algorithm for finding all repetitions in a string," *J. Algorithms* (1984): 422–432.

[Ma 75] Manacher, G., "A new linear-time on-line algorithm for finding the smallest initial palindrome of the string," *J. ACM* 22 (1975): 346–351.

9

Almost optimal parallel algorithms

By an efficient parallel algorithm we mean an NC-algorithm (see Chapter 2). Particularly efficient are optimal and almost optimal parallel algorithms. By optimal we mean an algorithm in which the total number of elementary operations is linear, while an almost optimal algorithm is one that is optimal within a polylogarithmic factor. From the point of view of any application, the difference between optimal and almost optimal algorithms is extremely thin. In this chapter, we present almost optimal algorithms for two important problems: string matching, and suffix tree construction. In the first section we present an almost optimal parallel string-matching algorithm ($O(\log^2 n)$ time, $O(n)$ processors) that is based on a parallel version of the Karp-Miller-Rosenberg algorithm of Chapter 8, and is called *Parallel-KMR*. In fact, it is possible to design a string-matching algorithm using only $n/\log^2 n$ processors (which is strictly optimal), but such an algorithm is much more complicated to describe, and its structure is quite different than that of KMR algorithm. KMR algorithm is conceptually very powerful. The computation of regularities in Section 9.3 is derived from it. And, in the Section 9.4, we present an almost optimal parallel algorithm for suffix tree construction that can be also treated as an advanced version of KMR algorithm.

Methodologically we apply two different approaches to the construction of efficient parallel algorithms:

(1) design of a parallel version of a known sequential algorithm,

(2) construction of a new algorithm with a good parallel structure.

Method (1) works well in the case of almost optimal parallel string-matching algorithms, and square finding. The known KMR algorithm, and the Main-Lorentz algorithm (for squares) already have a good parallel algorithmic structure. However, method (1) works poorly in the case of suffix tree construction, and Huffman coding, for example. Generally, many known sequential algorithms look inherently sequential, and are difficult to parallelize. All algorithms of Chapter 5 for the problem of suffix tree construction, and the algorithm of Chapter 10 for Huffman coding are inherently sequential. For these problems, we need to develop some new approaches and construct different algorithms. In fact, the sequential version of the parallel algorithm for suffix tree construction yields a new, simple, and quite efficient sequential algorithm for this problem (working in $O(n.\log n)$ time). This illustrates an application of parallel computations to efficient sequential computations for textual problems.

Our basic model of parallel computations is a parallel random access machine without write conflicts (PRAM). But the PRAM model with write conflicts (concurrent writes) is also discussed (see Chapter 2). The PRAM model is best suited to work with tree structured objects or tree-like (recursive) structured computations. As an introduction we show such a type of computation. One of the basic parallel operations is the (so-called) *prefix computation*.

Given a vector x of n values compute all prefix products: $y[1] = x[1]$, $y[2] = x[1]\otimes x[2]$, $y[3] = x[1]\otimes x[2]\otimes x[3]$, ... Denote by *prefprod*(x) the function that returns the vector y as value. We assume that $\otimes$ is an associative operation computable on a RAM in $O(1)$ time. We also assume for simplicity that n is a power of two. The typical instances of $\otimes$ are arithmetic operations +, *min* and *max*. The parallel implementation of *prefprod* works as follows. We assign a processor to each entry of the initial vector of length n.

```
function prefprod(x); { the size of x is a power of two }
begin
    n := size(x);
    if n = 1 then return x else begin
        x1 := first half of x; x2 := second half of x;
        for each i ∈ {1,2} do in parallel yi := prefprod(xi);
        midval := y1[n/2];
        for each 1 ≤ j ≤ n/2 do in parallel y2[j] := midval⊗y2[j];
        return concatenation of vectors y1, y2;
    end;
end.
```

Lemma 9.1 Parallel prefix computation can be accomplished in $O(\log n)$ time with $n/\log n$ processors.

Proof The algorithm above for computing *prefprod*(x) takes $O(\log n)$ time and uses n processors. The reduction of the number of processors by a factor $\log n$ is technical.

We partition the vector x into segments of length $\log n$. A processor is assigned to each segment. All such processors simultaneously compute all prefix computations locally for their segments. Each processor does so using a sequential process. We then compress the vector by taking a representative (say the first element) from each segment. A vector x' of size $n/\log n$ is obtained. The function *prefprod* is applied to x' (now $n/\log n$ processors suffice because of the size of x'). Finally, all processors assigned to segments update values for all entries in their own segments using a (globally) correct value of the segment representative. This again takes $O(\log n)$ time, and uses only $n/\log n$ processors. This completes the proof. ♦

Prefix computation will be used in this paper, for example, to compute the set of maximal (in the sense of set inclusion) subintervals of an $O(n)$ set of subintervals of $\{1, 2, \ldots, n\}$.

Another basic parallel method for constructing efficient parallel algorithms is the (so-called) *doubling technique*. Generally speaking, it consists of computing at each step objects whose size is twice as large as before, knowing the previously computed objects. The word "doubling" is often misleading because in many such algorithms the size of objects grows with a ratio $c > 1$, not necessarily with ratio $c = 2$. The typical use of this technique is evident in the proof of the following lemma (see [GR 88]). Suppose we have a vector of size n with some of the positions marked. Denote by *minright*$[i]$ (resp. *maxleft*$[i]$) the nearest marked position to the right (resp. to the left) of position i. Computing vectors *minright* and *maxleft* takes logarithmic parallel time.

Lemma 9.2 From a vector of size n with marked positions, vectors *minright*, and *maxleft* can be computed in $O(\log n)$ time with $n/\log n$ processors.

The doubling technique is also the crucial feature of the structure of the Karp-Miller-Rosenberg algorithm in a parallel setting. At one stage, the algorithm computes the names for all words of size k. At the next stage, using these names, it computes names of words having a size twice as large. We now explain how the algorithm KMR of Chapter 8 can be parallelized.

To make a parallel version of KMR algorithm, it is sufficient to design an efficient parallel version of one stage of the computation, and this essentially reduces to the parallel computation of RENUMBER(x). If this procedure is implemented in $T(n)$ parallel time with n processors, we then have a parallel version of KMR algorithm working in $T(n).\log n$ time also with n processors. This is due to the doubling technique, and the fact that there are only $\log n$ stages. Essentially, the same problems that are computed by a sequential algorithm can be computed by its parallel version in $T(n).\log n$ time.

The time complexity of computing RENUMBER(x) depends heavily on the model of parallel computation used. It is $T(n) = \log n$ without concurrent writes, and it is $T(n) = O(1)$ with concurrent writes. In the latter case, one needs a memory larger than the total number of operations used by what is called a *bulletin board* (auxiliary table with n^2 entries; or, by using some arithmetic tricks, with $n^{1+\varepsilon}$ entries). This looks slightly artificial, though entries of auxiliary memory do not have to be initialized. The details related to the distribution of processors are also very technical in the case of concurrent writes models. Therefore, we present most of the algorithms using the PRAM model without concurrent writes. This generally increases the time by a logarithmic factor. This logarithmic factor also allows more time for solving technical problems related to the assignment of processors to elements to be processed. We will soon indicate how to remove this logarithmic factor when concurrent writes are used. The main difference lies in the implementation of the procedure RENUMBER, and the computation of equivalent classes (classes of objects with the same name).

9.1 Building the dictionary of basic factors in parallel

Given the string *text*, we say that two positions are k-equivalent iff the factors of length k starting at these positions are equal. Such an equivalence is best represented by assigning at each position i a name or a number to the factor of length k starting at this position. The name is denoted by $NUM_k[i]$. This is the *numbering* technique presented in Chapter 8 (see Section 8.1). We shall compute names of all factors of a given length k for k running through 1, 2, 4, ... We consider only those factors of which the length is a power of two. Such factors are called *basic factors*. The *name of a factor* is denoted by its rank in the lexicographic ordering of factors of a given length. We also call these names *k-names*. Factors of length k, and their corresponding k-names can be considered to be the same object. For each k-name r we also require (for further applications) a link $POS[r, k]$ to any one position at which an occurrence of the k-name r starts. We consider only those factors starting at positions [1.. n–1]. The n-th position contains the special end marker #. The end marker has the highest rank in the alphabet. We can generally assume w. l. o. g. that n–1 is a power of two, adding sufficient markers otherwise.

The tables *NUM* and *POS* are together called the *dictionary of basic factors*. This dictionary is the basic data structure used in several applications.

Figure 9.1 displays the data structure for *text* = *abaabbaa#*. Six additional #s are appended to guarantee that each factor of length 8 starting in [1...8] is well-defined. The figure presents tables *NUM* and *pos*. In particular, the entries of *POS*[*, 4] give the lexicographically sorted sequence of factors of length 4. This is the sequence of factors of length 4 starting at positions 3, 7, 1, 4, 8, 2, 6, 5. The ordered sequence is:

aabb, aa##, abaa, abba, a###, baab, baa#, bbaa.

In the case of arrays, basic factors are $k \times k$ subarrays, where k is a power of two. In this situation, $NUM[(i, j), k]$ is the name of the $k \times k$ subarray of a given array *text* having its upper left corner at position (i, j). We will primarily discuss the con-

	i	= 1 2 3 4 5 6 7 8
	$text$	= a b a a b b a a #######
$k = 1$	$NUM[i,k]$ =	1 2 1 1 2 2 1 1
$k = 2$	$NUM[i,k]$ =	2 4 1 2 5 4 1 3
$k = 4$	$NUM[i,k]$ =	3 6 1 4 8 7 2 5
	r	= 1 2 3 4 5 6 7 8
$k = 1$	$POS[r,k]$ =	1 2 undefined
$k = 2$	$POS[r,k]$ =	3 1 8 2 5 undefined
$k = 4$	$POS[r,k]$ =	3 7 1 4 8 2 6 5

Figure 9.1: The dictionary of basic factors: tables of k-names, and of their positions. The k-name at position i is the factor $text[i \ldots i+k-1]$; its name is its rank according to lexicographic ordering of all factors of length k (order of symbols is $a < b < \#$). Integers k are powers of two. The tables can be stored in $O(n.\log n)$ memory.

struction of dictionaries of basic factors for strings. The construction in the two-dimensional case is a simple extension of that used for one-dimensional data.

The central machinery of KMR algorithm is the procedure RENUMBER that is defined now. Let x be a vector or an array of total size n containing elements of some linearly ordered set. We recall the definition of procedure RENUMBER (see Chapter 8). The function RENUMBER assigns to x a vector x'. Let $val(x)$ be the set of values of all entries of x (the alphabet of x). The vector x' satisfies:

(a) if $x[i] = x[j]$ then $x'[i] = x'[j]$, and

(b) $val(x')$ is a subset of $[1 \ldots n]$.

As a side effect, the procedure also computes the table POS: if q is in $val(x')$, then $POS[q]$ is any position i such that $x'[i] = q$. Conditions (a) and (b) are obviously satisfied if $x'[i]$ is defined as the rank of $x[i]$ inside the ordered list of values if $x[j]$'s. But we do not generally assume such a definition because it is often more difficult to compute.

Lemma 9.3 Let x be a vector of length n.

(1) RENUMBER(x) can be computed in $O(\log n)$ time with n processors on an exclusive-write PRAM.

(2) Assume that $val(x)$ consists of integers or pairs of integers in the range $[1 \ldots n]$. Then, RENUMBER(x) can be computed in $O(1)$ time with n processors on a concurrent-write PRAM. In this case, the size of auxiliary memory is $O(n^{1+\varepsilon})$.

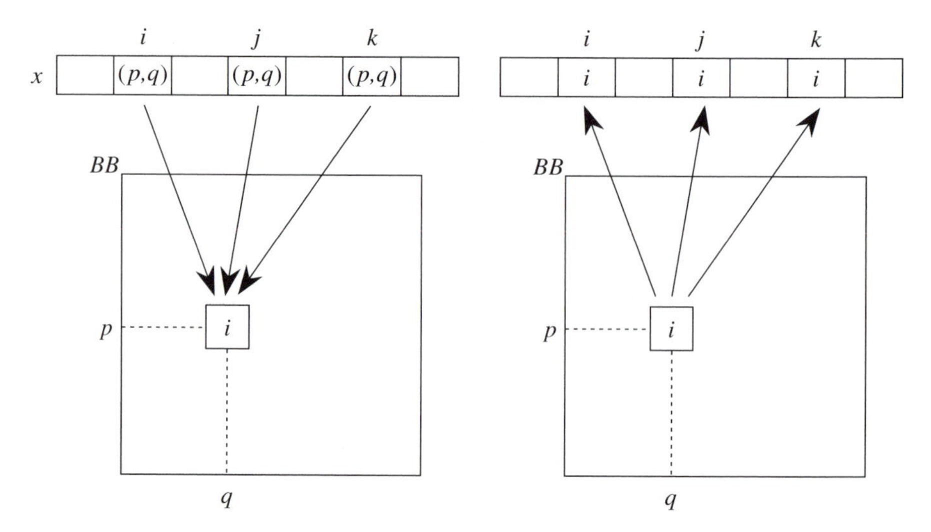

Figure 9.2: Use of a bulletin board. $x[i] = x[j] = x[k] = (p, q)$. The values of x at positions i, j, k are then changed, in two parallel steps (with concurrent writes), to the same value i.

Proof

(1) The main part of the procedure is the sorting (see Chapter 8). Sorting by a parallel merge sort algorithm is known to take $O(\log n)$ time with n processors. Hence, the whole procedure RENUMBER has the same complexity as sorting n elements. This completes the proof of point (1).

(2) Assume that $val(x)$ consists of pairs of integers in the range $[1...n]$. In fact, RENUMBER is used primarily in such cases. We consider an $n \times n$ auxiliary table BB. This table is called the bulletin board. The processor at position i writes its name into the entry (p, q) of table BB, where $(p, q) = x[i]$. We have many concurrent writes here because many processors attempt to write their own position into the same entry of the bulletin board. We assume that one of them (for example, the one with smallest position) succeeds and writes its position i into the entry $BB[p, q]$. Then, for each position j, the processor sets $x'[j]$ to the value of $BB[p, q]$, where (p, q) is the value of $x[j]$. The computation is graphically illustrated in Figure 9.2. The time is constant (two steps). By doing so, we use quadratic memory. But by applying some arithmetic tricks it can be reduced to $O(n^{1+\varepsilon})$. This completes the proof. ♦

Theorem 9.4 The dictionary of basic factors (resp. basic subarrays) of a string (resp. array) of size n can be computed with n processors in $O(\log^2 n)$ time on an exclusive-write PRAM, and in $O(\log n)$ time on a concurrent-write PRAM (in the latter case auxiliary memory space of size $O(n^{1+\varepsilon})$ is used).

Proof We describe the algorithm only for strings. The crucial aspect is the property already used in Chapter 8:

(*) $NUM[i, 2k] = NUM[j, 2k]$ iff ($NUM[i, k] = NUM[j, k]$ and $NUM[i+k, k] = NUM[j+k, k]$).

The dictionary is computed by the algorithm *Parallel-KMR* below. The correctness of the algorithm results from fact (*). The number of iterations is logarithmic, and the dominating operation is the procedure RENUMBER. The thesis now follows from the Lemma 9.3. In the two-dimensional case, there is a fact analogous to (*), which makes the algorithm work similarly. The size of the $n \times n$ array is $N = n^2$. This completes the proof. ♦

```
Algorithm Parallel-KMR;
{ a parallel version of the Karp-Miller-Rosenberg algorithm. }
{ Computation of the dictionary of basic factors of text; }
{ text ends with #, |text| = n, n-1 is a power of two }
begin
   x := text;
   x := RENUMBER(x); { recall that RENUMBER computes POS }
   for i in 1..n do in parallel begin
        NUM [i, 1] := x[i]; POS[1, i] := POS[i];
   end;
   k := 1;
   while k < n-1 do begin
        for i in 1..n-2k+1 do in parallel x[i] := (x[i], x[i+k]);
        delete the last 2k-1 entries of x;
        x := RENUMBER(x);
        for i in 1..n-2k+1 do in parallel begin
             NUM[i, 2k] := x[i]; POS[2k, i] := POS[i];
        end;
        k := 2k;
   end;
end.
```

9.2 Parallel construction of string-matching automata

We first present a simple application of the dictionary of basic factors to string matching. We then describe an application to the computation of failure tables *Bord* and string-matching automata (see Chapter 7).

The table *PREF* is a close "relative" of the failure table *Bord* commonly used in efficient sequential string-matching algorithms and automata (see Chapter 3). The table *PREF* is defined by

$$PREF[i] = \max\{j : text[i \ldots i+j-1] \text{ is a prefix of } text\},$$

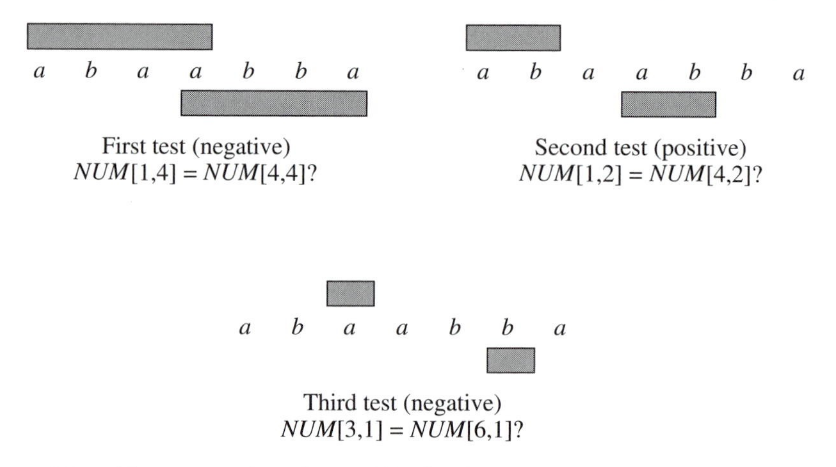

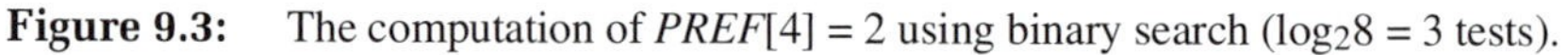

Figure 9.3: The computation of $PREF[4] = 2$ using binary search ($\log_2 8 = 3$ tests).

while the table *Bord* is defined by

$$Bord[i] = \max\{i\text{: } 0 \leq j < i \text{ and } text[1 \ldots j] \text{ is a suffix of } text[1 \ldots i]\}.$$

The table *PREF*, and the related table *SUF*, are crucial tables in the Main-Lorentz square-finding algorithm (see Chapter 8, Section 2), for which a parallel version is presented in Section 9.3.

Theorem 9.5 The table *PREF* can be computed in O(log$n$) time with n processors (without using concurrent writes), if the dictionary of basic factors is already computed.

Proof Essentially it is sufficient to show how to compute table *PREF* within our complexity bounds. One processor is assigned to each position i of *text*, and it computes $PREF[i]$ using a kind of binary search, as indicated in Figure 9.3. This completes the proof. ♦

Corollary 9.6 String matching can be solved in time $O(\log^2 n)$ time with n processors in the exclusive-write PRAM model.

Proof To find occurrences of *pat* in *text*, consider the word *pat#text*. Compute table *PREF* for this word. Then, occurrences of *pat* in *text* appear at positions i of *text* where $PREF[i] = |pat|$. The overall takes $O(\log^2 n)$ time with n processors, as a consequence of Theorems 9.4 and 9.5. ♦

Lemma 9.7 The failure table *Bord* for a pattern *pat* of length n can be computed in $O(\log^2 n)$ time with n processors without using concurrent writes (in $O(\log n)$ time if the dictionary of basic factors is already computed).

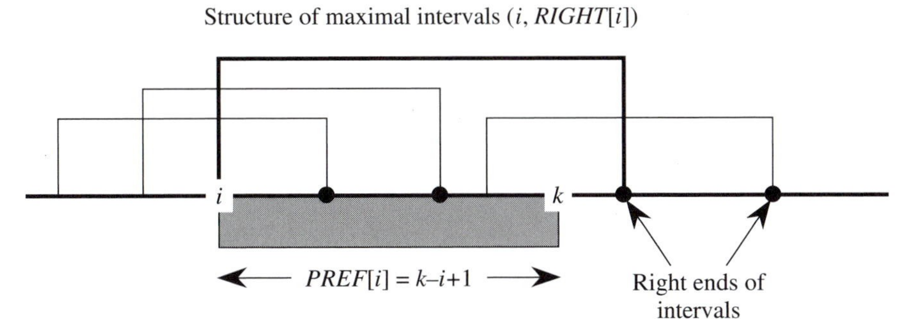

Figure 9.4: Computation of *Bord*[*k*] in the case $i = RIGHT^{-1}[minright[k]] \leq k$.

Proof We can assume that the table *PREF* is already computed (Theorem 9.5). Let us consider the pairs (*i*, *i*+*PREF*[*i*]–1). These pairs correspond to intervals of the form [*i*...*j*]. The first step is to compute all such intervals that are maximal in the sense of set inclusion order. It can be done with a parallel prefix computation. For each *k*, we compute the value

$$maxval(k) = \max(PREF[1], PREF[2]+1, PREF[3]+2, \ldots, PREF[k-1]+k-2).$$

Then, we "turn off" all positions *k* such that $maxval(k) \geq PREF[k]+k-1$. We are left with maximal subintervals (*i*, *RIGHT*[*i*]). Let us (in one parallel step) mark all right ends of these intervals, and compute the table $RIGHT^{-1}[j]$ for all right ends *j* of these intervals. For the other values *j*, $RIGHT^{-1}[j]$ is undefined. Again, using a parallel prefix computation, for each position *k*, we can compute *minright*[*k*], the first marked position to the right of *k*.

Then, in one parallel step, we set

Bord[*k*] := 0 if *minright*[*k*] is undefined, and
Bord[*k*] := $\max(0, k-RIGHT^{-1}[minright[k]]+1)$ otherwise (see Figure 9.4).

This completes the proof. ♦

One may observe that if the table *PREF* is given then even $n/O(\log n)$ processors are sufficient to compute table *Bord*, because our main operations are parallel prefix computations.

Corollary 9.8 The periods of all the prefixes of a word can be computed in $O(\log^2 n)$ time with *n* processors without using concurrent writes (in $O(\log n)$ time if the dictionary of basic factors is already computed).

Proof The period of prefix *pat*[1...*i*] is *i*–*Bord*[*i*] (see Chapter 2). ♦

We now present another consequence of Lemma 9.7 related to the previous result. It is shown in Chapter 7 (see Section 1) how to build the string-matching automaton for *pat*, *SMA*(*pat*). It is the minimal deterministic automaton that accepts the language A^*pat. The construction of such automata is the basis of several string-matching algorithms. Its sequential construction is straightforward, and can be related to the failure function *Bord*. Using this fact, we develop a parallel algorithm to compute *SMA*(*pat*).

Theorem 9.9 Assume that the basic dictionary of *pat* is computed, and the alphabet has $O(1)$ size. Then, we can compute the string-matching automaton *SMA*(*pat*) in $O(\log n)$ time with n processors on an exclusive-write PRAM. The same result holds for a finite set of patterns.

Proof We prove only the one-pattern case. The case with many patterns can be handled in essentially the same way, though trees are to be used instead of one-dimensional tables (see Chapter 7). *Parallel-KMR* algorithm works for trees as well.

We can assume that the failure table *Bord* for *pat* is already known. Our algorithm is essentially a parallel version of the construction of *SMA*(*pat*) shown in Section 7.1. The string-matching automaton has $\{0, 1, \ldots, n\}$ as set of states. The initial state is 0 and n is the only accepting state.

Define first the transition function δ for occurrences of symbols of *pat* only: let $\delta[i, a] = i+1$ for $a = pat[i+1]$ and $\delta[0, a] = 0$ for $a \neq pat[1]$. For each symbol a of the alphabet, define a modified failure table as follows:

for $i < n$, $P_a[i] =$ if $(pat[i+1] = a)$ then i else $Bord[i]$,
and $P_a[n] = Bord[n]$.

Let $P_a^*[i] = P_a^k[i]$, where k is such that $P_a^k[i] = P_a^{k+1}[i]$.

Tables P_a are treated here as functions that can be iterated indefinitely. We can easily compute all tables P_a^* in $O(\log n)$ time with n processors using the doubling technique ($\log n$ executions of $P_a[i] = P_a^2[i]$). Then, the transition table of the automaton is constructed as follows:

for each letter a, and position i such that $\delta[i, a]$ is not already defined
do in parallel $\delta[i, a] := \delta[P_a^*[i], a]$.

This completes the proof. ♦

9.3 Parallel computation of regularities

We continue the applications of KMR algorithm by using the problem of searching for squares in strings. Recall that a square is a non-empty word of the form *ww*. Section 8.2 introduces an algorithm for finding a square factor within a word in sequential linear time. But the algorithm is based on a compact representation of factors of the word. Here we consider the other methods applicable to the same problem and also presented in Section 8.2.

A simple application of failure functions yields a quadratic sequential algorithm. It leads to a parallel NC-algorithm. To do so, we can compute a failure function $Bord_i$ for each suffix $pat[i...n]$ of the word pat. We have already noted that there is a square in pat, prefix of $pat[i...n]$, iff $Bord_i[j] \geq (j-i+1)/2$ for some $j > i$. Computing a linear number of failure functions leads to a quadratic sequential algorithm, and to a parallel NC-algorithm. However, with such an approach, the parallel computation requires a quadratic number of processors. We show how the divide-and-conquer method used in the sequential case saves time and space in parallel computation.

The main portion of the sequential algorithm *SQUARE* used for finding a square in a word (see Section 8.2) is the operation called *test*. This operation applies to squarefree words u and v, and tests whether the word uv contains a square (the square begins in u and ends in v). This operation is a combination of two smaller tests *righttest* and *lefttest*. The first (resp. second) operation tests whether uv contains a square for which the center is in v (resp. u).

The operation *test* can be executed with the two auxiliary tables *PREF* and SUF_u. Recall that $SUF_u[k]$ is the size of longest suffix of $v[1...k]$ which is also a suffix of u. This table SUF_u can be computed in the same way as *PREF* (computing *PREF* for $u^R\&v^R$, for example).

Lemma 9.10 Tables *PREF* and SUF_u being computed, functions *righttest*(u, v), *lefttest*(u, v) and *test*(u, v) can be computed in $O(\log n)$ time with $n/\log n$ processors. The running time is constant using a concurrent-write PRAM having n processors.

Proof Given tables *PREF* and SUF_u, the computation of *righttest*(u, v) reduces to the comparison between k and $PREF[k]+SUF_u[k]$. A square of length $2k$ is found if the latter quantity is greater than or equal to k, as shown in Figure 8.2. The evaluation of *righttest* is done by inspecting in parallel all positions k on v. Hence, in $O(\log n)$ time (and constant time with the concurrent-writes model), the time to collect the Boolean value, we can compute *righttest*. By grouping the values of k in intervals of length $\log n$ we can even reduce the number of processors to $n/\log n$. The same holds for *lefttest*, and also for *test* which is a combination of the two others. This completes the proof. ♦

The function *test* yields a recursive parallel algorithm for testing occurrences of squares.

```
function SQUARE(text): boolean;
{ returns true if text contains a square }
{ W.l.o.g. n = |text| is a power of two }
begin
   if n > 1 then begin
```

```
        for i ∈ {1, n/2+1} do in parallel
            if SQUARE(text [i..i+n/2-1]) then return true;
        { if the algorithm has not yet stopped then ... }
        if (test(text [1..n/2], text [n/2+1..n]) then return true;
    end;
    return false; { if the "true" has not been yet returned }
end;
```

Theorem 9.11 The algorithm *SQUARE* tests the squarefreeness of a word of length n in $O(\log^2 n)$ parallel time using n processors in the PRAM model without concurrent writes.

Proof The complexity of the algorithm *SQUARE* is essentially due to the computation of basic factors in $O(\log^2 n)$ (Theorem 9.4). Each recursive step during the execution of the algorithm takes $O(\log n)$ time with n processors as shown by Lemma 9.10. The number of step is $\log n$. This produces the result. ♦

The parallel squarefreeness test above is the parallel version of the corresponding algorithm *SQUARE* of Chapter 8. Section 8.2 contains another serial algorithm that solves the same problem in linear time on fixed alphabets. We now present the parallel version of the latter algorithm.

Recall that the second algorithm is based on the *f*-factorization of *text*. It is a sequence of non-empty words $(v_1, v_2, \ldots, v_m)$ such that $text = v_1v_2\ldots v_m$ and defined by

- $v_1 = text[1]$, and
- for $k > 1$, v_k is defined as follows. If $|v_1v_2\ldots v_{k-1}| = i$ then v_k is the longest prefix u of $text[i+1\ldots n]$ that occurs at least twice in $text[1\ldots i]u$. If there is no such u then $v_k = text[i+1]$. Recall also that $pos(v_k)$ is the position of the first occurrence of v_k in *text* (it is the length of the shortest word y such that yv_k is a prefix of *text*).

Theorem 9.12 The *f*-factorization of *text* of size n, and the associated table *pos*, can be computed in $O(\log n)$ time with n processors of a CRCW PRAM.

Proof

(a) First construct the suffix tree $T(text)$ by the algorithm of [AILSV 88]. The leaves of $T(text)$ correspond to suffixes of text, and the path from the root to the leaf numbered i spells the suffix starting at position i in *text*.

(b) For each node v of $T(text)$ compute the minimal value of the leaf in the subtree rooted at v. This can be executed using a tree-contraction algorithm, for example the one in [GR 88].

(c) For each leaf i, compute the first node v (bottom-up) on the path from i to the root with a value $j < i$. Denote by $size(i)$ the length of the string spelled by the path from the root to the node v. The value j equals $pos(i)$. If there is no such node then $pos(i) = i$ and $size(i) = 1$. The computation of tables $pos(i)$ and $size(i)$ can be executed efficiently in parallel as follows:

- Let $Up[v, k]$ be the node lying on the path from v to the root whose distance from v is 2^k. If there is no such node then we set $Up[v, k]$ to *root*.
- For each node v of the tree, compute the table $MinUp[v, k]$ for $k = 1, 2, \ldots, \log n$. The value of $MinUp[v, k]$ is the node with the smallest value on the path from v to $Up[v, k]$. Both tables can be easily computed in $O(\log n)$ time with $O(n)$ processors.
- Next, the values of $pos(i)$ can be computed by assigning one processor to each leaf i. The assigned processor makes a kind of binary search to find the first $j < i$ on the path from i to the root. This takes logarithmic time.

(d) Let $Next(i) = i + size(i)$. Compute in parallel all powers of $Next$. The factorization can then be deduced because the position of the element v_{i+1} of $(v_1, v_2, \ldots, v_m)$ is $Next^i(1)$.

The table *pos* has been already computed at point (c). This completes the proof. ♦

The parallel version of the function *linear-SQUARE* of Section 8.2 is shown below. Now the key point is that the computation of $righttest(v_1v_2\ldots v_{k-2}, v_{k-1}v_k)$, implied by the test of conditions (2) and (3), can be executed in logarithmic time using only $O(|v_{k-1}v_k|)$ processors. All the tests can be performed independently.

```
function Parallel-test(text): boolean;
begin
    compute in parallel
    the f-factorization (v1, v2, .., vm) of text;
    for each k ∈ {1, 2, .., m} do in parallel
        if (1) or (2) or (3) hold then { see Lemma 8.9 }
            return true;
    return false;
end;
```

Theorem 9.13 The squarefreeness test for a given text *text* of size n can be computed in $O(\log n)$ time using $O(n)$ processors of a CRCW PRAM.

Proof Each condition can be checked in the algorithm *Parallel-test* in $O(\log n)$ time using $O(|v_{k-1}v_k|)$ processors, due to Lemma 9.10. Thus, the number of processors needed is at most the total sum of all $|v_{k-1}v_k|$, which is obviously $O(n)$. This completes the proof. ♦

The next application of KMR algorithm is to algorithmic questions related to palindromes. We consider only even length palindromes: let *PAL* denote the set of such non-empty even palindromes (words of the form ww^R). Recall that *Rad*[*i*] is the radius of the maximal even palindrome centered at position *i* in the text *text*,

$$Rad[i] = \max\{k : text[i-k+1 \ldots i] = (text[i+1 \ldots i+k])^R\}.$$

If *k* is the maximal integer satisfying $text[i-k+1 \ldots i] = (text[i+1 \ldots i+k])^R$, then we say that *i* is the center of the maximal palindrome $text[i-k+1 \ldots i+k]$. We identify palindromes with their corresponding subintervals of $[1 \ldots n]$.

Lemma 9.14 Assume that the dictionaries of basic factors for *text* and its reverse $text^R$ are computed. Then the table *Rad* of maximal radii of palindromes can be computed in $O(\log n)$ time with *n* processors.

Proof The proof is essentially the same as the one for the computation of table *PREF*: a variant of parallel binary search is used. ♦

Denote by *Firstcentre*[*i*] (resp. *Lastcentre*[*i*]) the first (resp. last) center, to the right of *i* (including *i*), of palindromes containing position *i* (see Figure 9.7).

Lemma 9.15 If the table *Rad* is computed then the table *Firstcentre* (resp. *Lastcentre*) can be computed in $O(\log n)$ time with *n* processors.

Proof We first describe an $O(\log^2 n)$ time algorithm. It uses the divide-and-conquer approach. A notion of half-palindrome is introduced and shown in Figure 9.5.

The value *Firstcentre*[*k*] is the first (to the right of *k* including *k*) right end of an half-palindrome containing *k*.

The structure of the algorithm is recursive:

> The word is divided into two parts $x1$ and $x2$ of equal sizes. In $x1$ we disregard the half-palindromes with right end in $x2$. Then, we compute in parallel the table *Firstcentre* for $x1$ and $x2$ independently. The computed table, at this stage, yields the correct final value for positions in $x2$. However, the part of the table for $x1$ may be incorrect due to the fact that we disregarded the $x1$ half-palindromes ending in $x2$. To compensate for this, we apply procedure UPDATE($x1$, $x2$).

The table is updated looking only at previously disregarded half-palindromes. The structure of these half-palindromes is shown in Figure 9.6.

Maximal palindrome centered at i Its corresponding half-palindrome

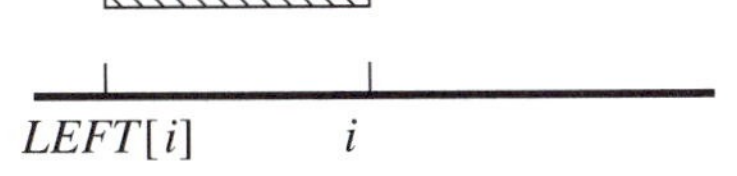

$LEFT[i]$ i

Figure 9.5: Half-palindromes.

The half-palindromes can be treated as subintervals $[i \ldots j]$. We introduce a (partial) ordering on half-palindromes as follows $[i \ldots j] \leq [k \ldots l]$ iff $(i \leq k$ and $j \leq l)$.

```
procedure UPDATE(x1, x2);
{ only half-palindromes starting in x1 and ending in x2 are considered }
begin
    remove all half-palindromes but minimal ones;
{ this can be done in logn time by parallel prefix computation }
{ we are left with the situation presented in Figure 9.7 }
    we compute for each position k the first (to the left)
    starting position s[k] of an half-palindrome;
    for each position k in x1 do in parallel
        { see Figure 9.7 }
        Firstcentre[k] := min(Firstcentre[k], right end of half-
                                    palindrome starting at s[k]);
end;
```

At the beginning of the whole algorithm $Firstcentre[k]$ is set to $+\infty$. One can see that the procedure UPDATE takes $O(\log n)$ time using $|x1|+|x2|$ processors. The depth of the recursion is logarithmic, therefore, the whole algorithm takes $O(\log^2 n)$ time.

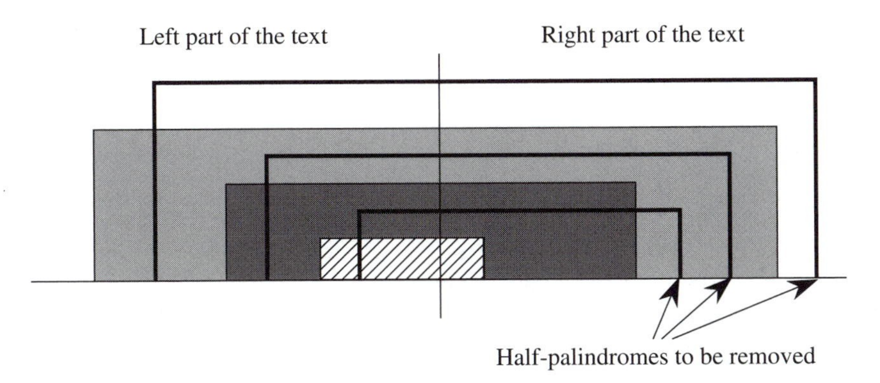

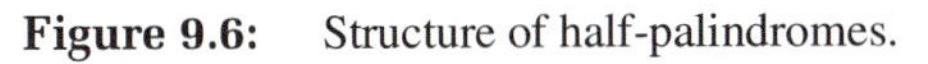

Figure 9.6: Structure of half-palindromes.

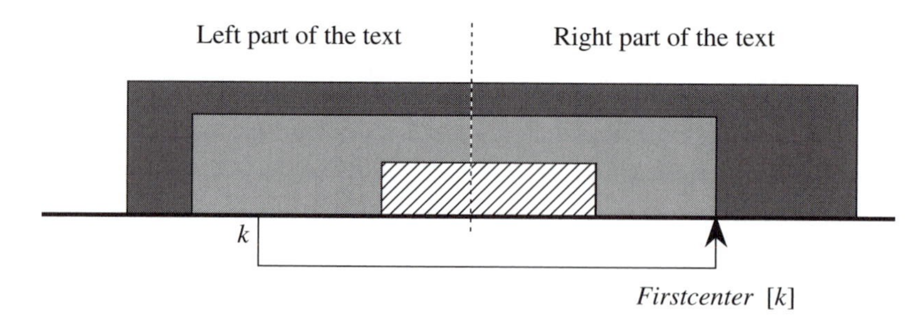

Figure 9.7: Table *Firstcenter*.

We now briefly describe how to obtain $O(\log n)$ time with the same number of processors. We look at the global recursive structure of the previous algorithm. The initial level of the recursion is the zero level, the last one has depth $\log n-1$. The crucial observation is that we can simultaneously execute UPDATE at all levels of the recursion. However, at the q-th level of the recursion, instead of minimizing the value of $Firstcentre[k]$, we store the values computed at this level in a newly introduced table $Firstcentre_q$: for each position k considered at this level the procedure sets $Firstcentre_q[k]$ to the right end of half-palindrome starting at $s[k]$ (see UPDATE). Hence, the procedure UPDATE is slightly modified; at level q it computes the table $Firstcentre_q$. Assume that all entries of all tables are initially set to infinity.

After computing all tables $Firstcentre_q$ we assign a processor to each position k. It computes sequentially (for its position) the minimum of $Firstcentre_q[k]$ over recursion levels $k = 0, 1,..., \log n-1$. This globally takes $O(\log n)$ time with n processors. But we compute $O(\log n)$ tables $Firstcentre_q$, each of size n, so we have to be certain that n processors suffice for computing all these data in $O(\log n)$ time. The computation of $Firstcentre_q$, at a given level of recursion, can be executed using simultaneous applications of parallel prefix computations (to compute the first marked element to the left of each position k) for each pair $(x1, x2)$ of factors (at the q-th level we have 2^q such pairs). The prefix computations at a given level together employ $O(n)$ elementary sequential operations. Hence the total number of operations for all levels is $n\ O(\log n)$. It is simple to compute $Firstcentre_q$ for a fixed q using n processors. This requires $O(n.\log n)$ processors for all levels q, however, we have only n processors.

At this moment, we have an algorithm working in $O(\log n)$ time in which the total number of elementary operations is $O(n\log n)$. We can apply Brent's lemma: if $M(n)/P(n) = T(n)$, then the algorithm performing the total number of $M(n)$ operations and working in parallel time $T(n)$ can be implemented to work in $O(T(n))$ time with $P(n)$ processors (see, for example, [GR 88]).

In our case $M(n) = O(n.\log n)$ and $T(n) = O(\log n)$. Brent's lemma is a general principle. Its application depends on the detailed structure of the algorithm as to whether it is possible to redistribute processors efficiently. However, in our case we deal with computations of a very simple structure: multiple applications of parallel prefix computations. Hence n processors suffice to make all computations in $O(\log n)$ time. This completes the proof. ♦

Theorem 9.16 Even palstars can be tested in $O(\log n)$ time with n processors on the exclusive-write PRAM model, if the dictionary of basic factors is already computed.

Proof Let *first* be a table in which the i-th value is the first position j in *text* such that $text[i...j]$ is an even non-empty palindrome; it is zero if there is no such prefix even palindrome. Define $first[n] = n$. This table can be easily computed using the values of $Firstcentre[i]$. Then, the sequential algorithm tests even palstars in the following natural way: it finds the first prefix even palindrome and cuts it; such a process is then repeated as often as possible; if we conclude with an empty string, we then return "true": the initial word is an even palstar. The correctness of the algorithm is then analogous to that of Theorem 8.13.

We make a parallel version of the algorithm. Compute table $first^*[i] = first^k[i]$, where k is such that $first^k[i] = first^{k+1}[i]$, using a doubling technique. Now the text is an even palstar iff $first^*[1] = n$. This can be tested in one step.

This ends the proof. ♦

Unfortunately the natural sequential algorithm described above for even palstars does not work for arbitrary palstars. But, in this more general case, a similar table *first* can be defined and computed in essentially the same manner as for even palindromes. Palstar recognition then reduces to the following reachability problem: is there a path according to *first* from position 1 to n? It is known that the positions are nodes of a directed graph in which the maximal outdegree is three. It is rather easy to solve this reachability problem in linear sequential time, but we do not know how to solve it using an almost optimal parallel algorithm. In fact, the case of even palstars can also be viewed as a reachability problem: its simplicity is related to the fact that in this case the outdegree of each vertex in the corresponding graph is just one.

Lemma 9.17 The table *Lastcentre* can be computed in $O(\log n)$ time with n processors on the exclusive-write model.

Proof We compute the maximal (with respect to inclusion) half-palindromes. We mark their left most positions, and then, the table *Lastcentre* is computed according to Figure 9.8. For position k, we only have to find the first marked position to the left of k (including k). This completes the proof. ♦

Theorem 9.18 Compositions of k palindromes, for $k = 2, 3$ or 4, can be tested in $O(\log n)$ time with $n/\log n$ processors in the exclusive-write PRAM model, if the dictionary of basic factors is computed.

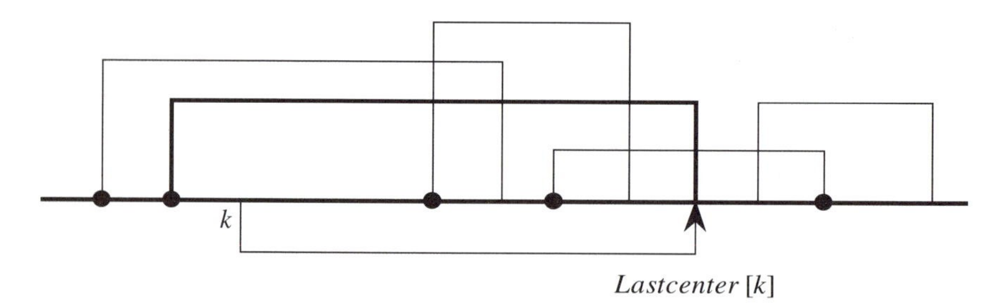

Figure 9.8: Table *Lastcenter*.

Proof The parallel algorithm is a simple parallel version of the sequential algorithm in [GS 78] that applies to $k = 2, 3, 4$. The key point is that if a text x is a composition uv of even palindromes then there is a composition $u'v'$ such that u' is a maximal prefix palindrome or v' is a maximal suffix palindrome of x. The maximal prefix (suffix) palindrome for each position of the text can be computed efficiently using the table *Lastcentre* (in the case of suffix palindromes the table is computed for the reverse of the string).

Given the tables *Rad* and *Lastcentre* (also for the reversed string) the question "is the text $x[1\ldots i]$ a composition of two palindromes?" can be answered in constant time using one processor. The computation of logical "or," or questions of this type, can be executed using a parallel prefix computation. The time is $O(\log n)$, and $O(n/\log n)$ processors suffice (see Lemma 9.1). This completes the proof. ♦

Several combinatorial algorithms on graphs and strings consider strings on an ordered alphabet. A basic algorithm often used in this context is the computation of maximal suffixes, or of minimal non-empty suffixes, according to the alphabetic ordering. These algorithms are strongly related to the computation of the Lyndon factorization of a word that is recalled now. A Lyndon word is a non-empty word that is minimal among its non-empty suffixes. Chen, Fox, and Lyndon have proven that any word x can be uniquely factorized as $l_1 l_2 .. l_h$ such that $h \geq 0$, the l_is are Lyndon words, and $l_1 \geq l_2 \geq \ldots \geq l_h$. It is known also that l_h is the minimal non-empty suffix of x. In the next parallel algorithm, we will use the following characterization of the Lyndon word factorization of x.

Lemma 9.19 The sequence of non-empty words $(l_1, l_2, \ldots, l_h)$ is the Lyndon factorization of the word x iff $x = l_1 l_2 .. l_h$, and the sequence $(l_1 l_2 \ldots l_h, l_2 l_3 \ldots l_h, \ldots, l_h)$ is the longest sequence of suffixes of x in decreasing alphabetical order.

Proof We only prove the "if" part and leave the "only if" part to the reader.

Let $s_1 = l_1 l_2 \ldots l$, $s_2 = l_2 l_3 \ldots l_h$, $\ldots$, $s_h = l_h$, $s_{h+1} = \varepsilon$. As a consequence of the maximality of the sequence, one may note that, for each $i = 1, \ldots, h$, any suffix w longer than s_{i+1} satisfies $w > s_i$. To prove that $(l_1, l_2, \ldots, l_h)$ is the Lyndon factorization of x, we have only to show that each element of the sequence is a Lyndon word. It is the case for l_h since, again due to the maximality condition, l_h is the minimal non-empty suffix of x, and therefore is less than any of its non-empty suffixes. Assume, ab absurdo, that v is both a non-empty proper prefix and suffix of l_i, and let w be such that $vw = s_i$. Since vs_{i+1} is a suffix of x longer than s_{i+1}, we have $vs_{i+1} > s_i$. The latter expression implies $s_i > w$, which is a contradiction. This proves that no non-empty proper suffix v of l_i is a prefix of l_i. But again, since $vs_{i+1} > s_i$, we must have $v > s_i$, and also $v > l_i$. Therefore, l_i is a Lyndon word. This ends the proof. ♦

Theorem 9.20 The maximal suffix of text of length n and its Lyndon factorization can be computed in $O(\log^2 n)$ time with n processors on the exclusive-write PRAM.

Proof We first apply the KMR algorithm of Section 3 to the word x#...# using the exclusive-write model. Assume here (contrary to the previous sections) that the special character # is the minimal symbol. Essentially, it makes no difference for KMR algorithm. We then get the ordered sequence of basic factors of x given by table *POS*. After padding enough dummy symbols to the right, each suffix of x becomes also a basic factor. Hence, for each position i, we can get $Rank[i]$: the rank of the i-th suffix $x[i]\ldots x[n]$ in the sequence of all suffixes. One may note that the padded character # has no effect on the alphabetical ordering since it is smaller than any other character.

Let us define $L[i]$ to be the position j such that both $j \le i$ and $Rank[j] = \min\{Rank[j'] : 1 \le j' \le i\}$. By the above characterization of the Lyndon word factorization of x, $L[i]$ yields the starting position of the Lyndon factor containing i. Values $L[i]$ can be computed by prefix computation. Define an auxiliary table $G[i] = L[i]-1$. Now, it is sufficient to compute the sequence $G[n]$, $G^2[n]$, $G^3[n]$, ..., $G^h[n] = 0$. In fact, such a sequence gives the required list of positions, starting with position n. It is quite simple to mark all positions of $[0..\ n]$ contained on this list in $O(\log^2 n)$ parallel time. These positions break the word down into its Lyndon factorization. This completes the proof. ♦

9.4 Parallel computation of suffix trees

Efficient parallel computation of the suffix tree of a text is a delicate problem. The dictionary of basic factors leads to a rather simple solution, but one that is not optimal. As already noted in Chapter 8, the equivalences on factors of a text are strongly related to its suffix tree. And the naming technique used by KMR algorithm to compute the dictionary is precisely the technique for handling the equivalence classes.

To build the suffix tree of a text, a coarse approximation of it is first built. Afterward the tree is refined step by step. This approximation more closely resembles the suffix tree that is reached at the end of the process. During the construction, we have to answer quickly the queries of the form: are the k-names (i.e., the factors of length k) starting at positions i and j identical? This is necessary for computing the

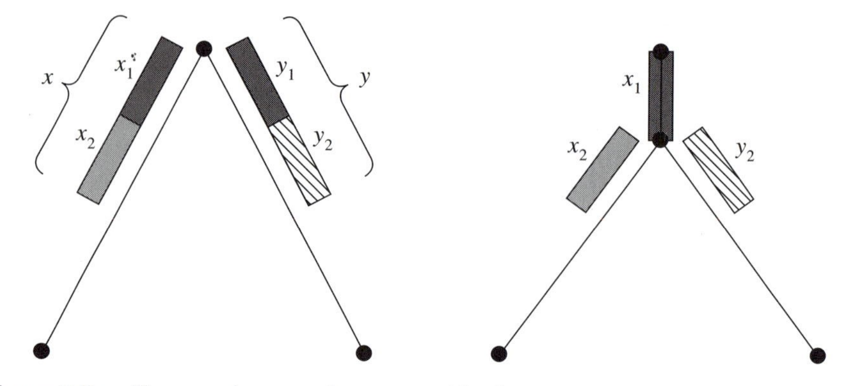

Figure 9.9: If $x \neq y$ and $x_1 = y_1$, then $x_2 \neq y_2$. After insertion of a new node, *invar*(2*k*) is locally satisfied (on the left), *invar*(*k*) holds locally (on the right).

k-equivalence relations defined on the nodes of intermediate trees. Roughly speaking, such relations help to refine the tree and get better approximations of the final suffix tree.

Assume n–1 is a power of two. The n-th letter of *text* is a special symbol. A suitable number of special symbols is padded to the end of *text* in order to have well-defined factors of length n–1 starting at positions 1, 2, ..., n–1. We say that two words x and y are k-equivalent if they have the same prefix of length 2^k.

We build a series of a logarithmic number of trees T_{n-1}, $T_{(n-1)/2}$, ..., T_1: each successive tree is an approximation of the suffix tree. The key invariant of the construction is:

> *invar*(k): for each internal node v of T_k there are no two distinct outgoing edges for which the labels have the same prefix of length k. The label of the path from the root to the leaf i equals *text*[$i...i+n$]. There is no internal node of outdegree one.

Note that the parameter k is always a power of two. This gives the logarithmic number of iterations.

Remark If *invar*(1) holds, then we have the tree T_1 that is essentially the suffix tree T(*text*). Just a trivial modification may be needed to delete all #s padded for technical reasons, except one.

The core of the construction is the procedure REFINE(k) that transforms T_{2k} into T_k. The procedure maintains the invariant if *invar*(2k) is satisfied for T_{2k}, then *invar*(k) holds for T_k after running REFINE(k) on T_{2k}. The correctness (preservation of invariant) of the construction is based on the trivial fact expressed graphically in Figure 9.9.

The procedure REFINE(k) consists of two stages:

(1) insertion of new nodes, one per each non-singleton k-equivalence class;

(2) deletion of nodes of outdegree one.

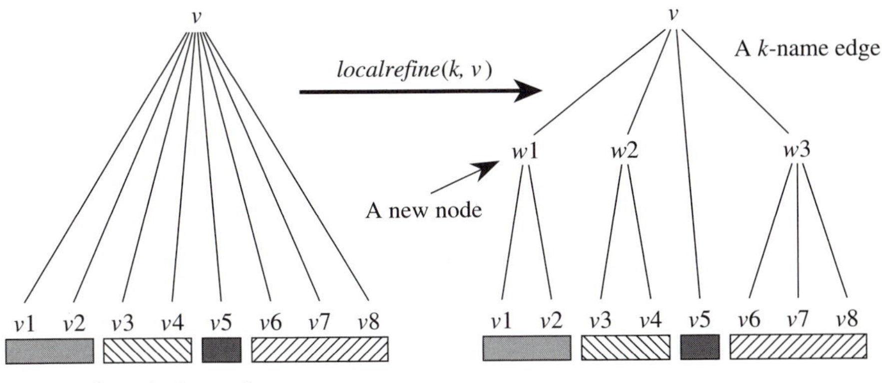

Figure 9.10: Local refinement. The sons (of node v) for which the edge labels have the same k-prefixes are k-equivalent.

In the first stage the operation *localrefine*(k, v) is applied in parallel to all internal nodes v of the current tree. This local operation is graphically presented in the Figure 9.10. The k-equivalence classes, labels of edges outgoing a given node, are computed. For each non-singleton class, we insert a new (internal) node.

To achieve a more complete parallelization, each operation *localrefine*(k, v) is performed with as many processors as the number of sons of v (one processor per son). Hence, each local refinement is executed in logarithmic time, as well as the entire REFINE operation. Since we apply REFINE $\log n$ times, the global time complexity is $O(\log^2 n)$. The number of processors is linear as is the sum of outdegrees of all internal nodes.

The algorithm is informally presented on the example text *abaabbaa#* (see Figure 9.11). We start with the tree $T(8)$ of all factors of length 8 starting at positions 1, 2, ..., 8. The tree is almost a suffix tree: the only condition that is violated is that some internal node v (here, in fact, the root) has two distinct outgoing edges in which the labels have a common non-empty prefix. We attempt to satisfy the condition by successive refinements: the prefixes violating the condition become smaller and smaller, divided by two at each stage, until they become empty. This is illustrated by Figures 9.11, 9.12, 9.13, 9.14, and 9.15.

The informal description of the construction of the suffix tree $T(text)$ is summarized by the algorithm below.

```
Algorithm
{ parallel suffix tree construction; n-1 is a power of two }
  procedure REFINE(k);
  begin
     for each internal node v of T do in parallel begin
        localrefine(k, v); delete all nodes of outdegree one;
     end;
  end;
```

```
{ main algorithm }
begin
    let T be the tree of height 1 whose leaves are 1..n-1, and
    the label of the i-th edge is [i, i+n];
    k := n-1;
    repeat { T satisfies invar(k) }
        k := k/2; REFINE(k); { T satisfies invar(k) }
    until k = 1;
    delete extra #'s;
    return T;
end.
```

Remark Observe that the previous strategy followed to build the suffix tree $T(text)$ can be implemented in the sequential RAM model (one processor). Then, because the computation of equivalence classes can be executed by a sequential linear time algorithm (with the help of radix sorting), we get a new $O(n.\log n)$ time sim-

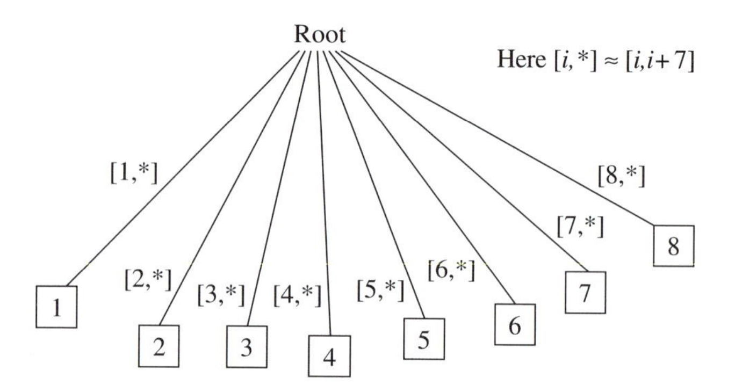

Figure 9.11: The tree T_8 for $text = abaabbaa\#$. The 8-equivalence is the 4-equivalence, hence $T_8 = T_4$. But the 2-equivalence classes are {6, 2}, {3, 7}, {4, 1}, {8}, {5}. We apply REFINE(2) to get T_2.

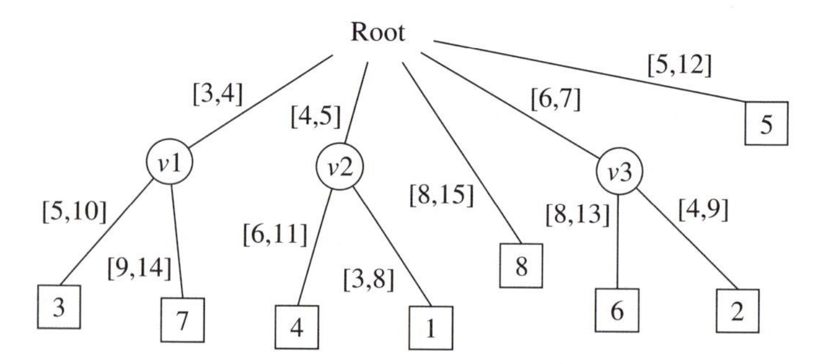

Figure 9.12: Tree T_2. Now the 1-equivalence classes are {3}, {7}, {4}, {1}, {$v1$, v2, 8}, {6, 2}, {$v3$, 5}. Three new nodes have to be created to obtain T_1.

ple algorithm for the computation of suffix trees. The suffix tree constructions of Chapter 5 are inherently sequential, because the text is scanned from left to right or in the reverse direction. The nature of the present algorithm is "more parallel," and has almost no reason to be considered for use in sequential computations.

Theorem 9.21 The suffix tree of text of length n can be constructed in $O(\log^2 n)$ parallel time with n processors on a PRAM with exclusive writes. (With a more elaborate construction, even read conflicts can be avoided.)

Proof The complexity bounds follows directly from the construction, and from the algorithm computing the dictionary of basic factors. At every stage, it is possible to

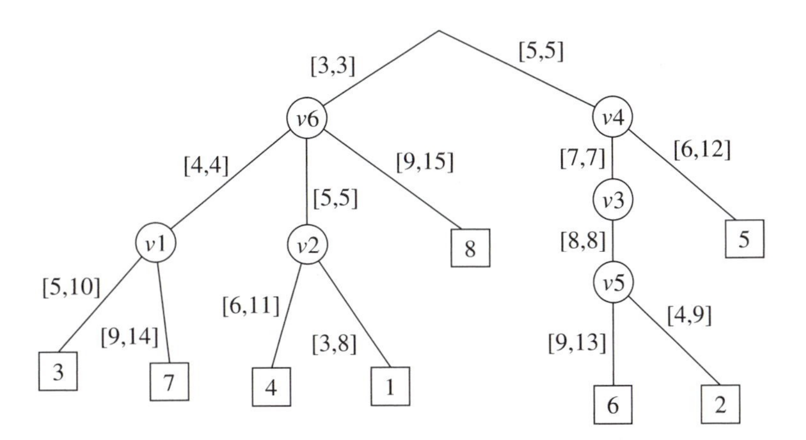

Figure 9.13: The tree after the first stage of REFINE(1): insertion of new nodes $v4$, $v5$, $v6$.

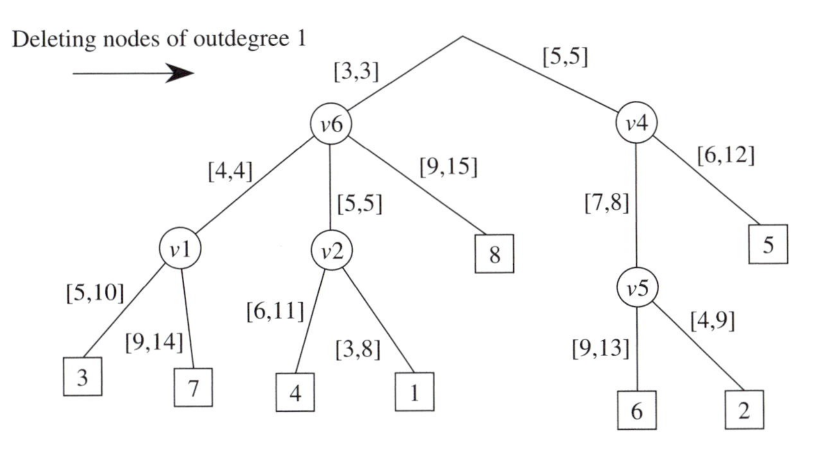

Figure 9.14: Tree T_1 after the second stage of REFINE(1): deletion of nodes of outdegree one.

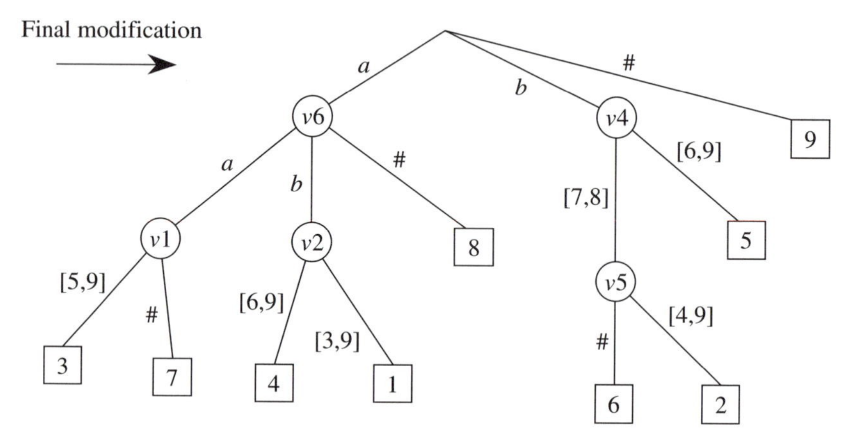

Figure 9.15: The suffix tree $T(abaabbaa\#)$. It results from T_1 by a "cosmetic" modification, elimination of extra #.

calculate for each node how many new sons need to be created for this node. Then, in logarithmic time, we can assign new processors to the newly created nodes. All calculations are simple because we can calculate the number of new nodes traversing the tree in pre-order, for example. Such traversal of the tree is easily parallelized efficiently. This completes the proof. ♦

In the construction above, the running time can be reduced by the logarithmic factor if concurrent writes are allowed. Basically, the method is analogous to the computation of the dictionary of basic factors. The bulletin board technique is used similarly. However, it becomes more difficult to assign processors to newly created nodes.

Theorem 9.22 Suffix trees can be built in $O(\log n)$ parallel time with n processors in the PRAM with write conflicts.

9.5* Parallel computation of dawgs

In this section we extend the results of the previous section to the computation of dawgs. We show how to efficiently compute in parallel the directed acyclic word graph of all factors of a text, and also its compressed version—the minimal factor automaton. The construction of dawgs presented here consists essentially of a parallel transformation of suffix trees. The basic procedure is the computation of equivalent classes of subtrees. We encode the suffix tree with a string in such a way that the subtrees correspond to subwords of the coding string.

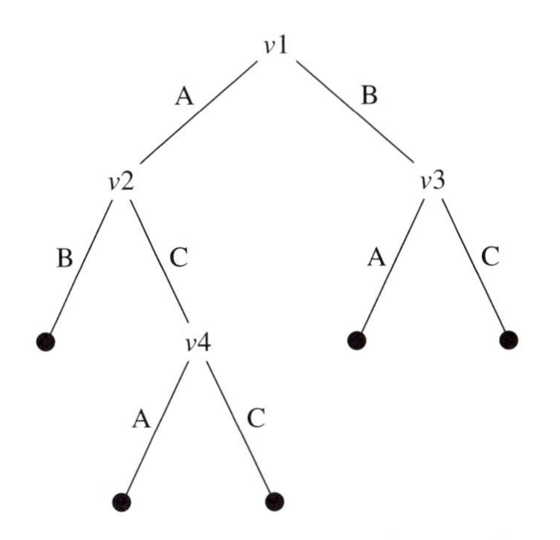

Figure 9.16: *code*($v1$) = "ABBCAACCCABAACCB"; *code*($v3$) = *code*($v4$). The code x of the tree T is constructed in parallel by the Euler tour method. The labels of edges outgoing a given node are sorted. The nodes $v4$ and $v3$ are roots of isomorphic subtrees because $x[5\ldots8] = x[12\ldots15]$.

Lemma 9.23 Let T be a rooted ordered tree in which the edges are labeled with constant-size letters. Then, we can compute the isomorphic classes of subtrees of T with $O(n)$ processors in $O(\log^2 n)$ time in the exclusive write PRAM model, and in $O(\log n)$ time in the concurrent write PRAM model.

Proof We use an Euler tour technique. We refer the reader to [TV 85] and [GR 88] for a more detailed exposition of this technique. The edge with label X is replaced by two edges: the forward edge with label X, and the backward edge with label X'. The tree becomes an Eulerian directed graph.

An Euler tour x is computed such that the edges outgoing a given node of the tree appear in their original order. Such a tour can be computed in time $O(\log n)$ with n processors [TV 85]. The main advantage to using this technique is the transformation of tree problems into problems on strings. Figure 9.16 demonstrates an Euler tour technique for a sample tree.

Let *code*(v) denote the part of the traversal sequence x corresponding to the subtree rooted at v. It is possible to compute in $O(\log n)$ time with n processors the integers i and j such that *code*(v) = $x[i\ldots j]$ for each node v. The subtrees rooted at $v1$ and $v2$ are isomorphic iff *code*($v1$) = code($v2$).

We build the dictionary of basic factors for x. Then, each subtree gets a constant-size name corresponding to its code. These names are given directly by the dictionary if the length is a power of two. If not, the composite name is created (decomposing the code of the subtree into two overlapping strings having length a power of two). Afterward, two subtrees are isomorphic iff they have the same names, and the

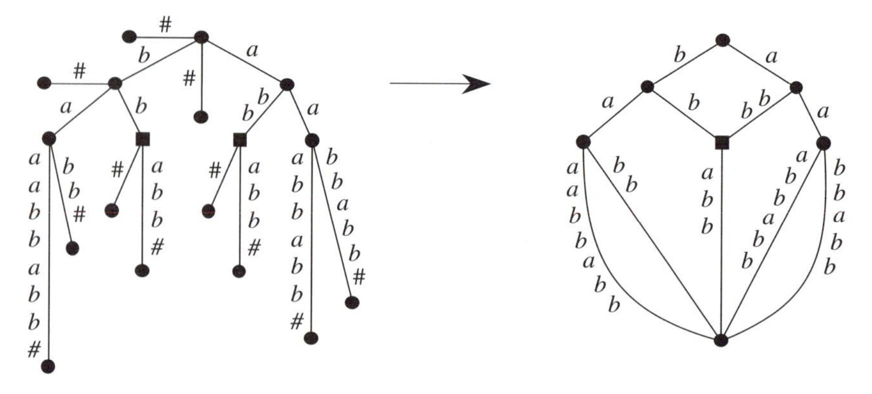

Figure 9.17 Identification of isomorphic classes of nodes of the suffix tree. In the algorithm, labels of edges are constant-size names corresponding to themselves.

crucial point here is that the names are of $O(1)$ size. The procedure RENUMBER can be applied for computing the equivalence classes of subtrees with the same name. This completes the proof. ♦

Theorem 9.24 *DAWG(text)* can be constructed in $O(\log^2 n)$ time in the exclusive write PRAM model, and in $O(\log n)$ time in the concurrent-write PRAM model, using $O(n)$ processors.

Proof We demonstrate the algorithm on the example string *text* = *baaabbabb*. We assume that the suffix tree of text is already constructed. For technical reasons, the end marker # is added to the *text*. This end marker is later ignored in the final dawg (however, it is necessary at this stage).

By the algorithm given in the proof of the preceding lemma, the equivalence classes of isomorphic subtrees are computed. The roots of isomorphic subtrees are identified, and we get an approximate version G' of *DAWG(text)* (see Figure 9.17). The difference between the actual structure and the dawg is related to lengths of strings that are labels of edges. In the dawg each edge is labeled by a single symbol. The "naive" approach could be to decompose each edge labeled by a string of length k into k edges and then merge some of them; but the intermediate graph could have a quadratic number of nodes. We apply such an approach with the following modification. By the weight of an edge we mean the length of its label. For each node v, we compute the heaviest (with the biggest weight) incoming edge. Denote this edge by *inedge*(v). Then, in parallel, for each node v, we perform a local transformation *local-action*(v). It is crucial that all these local transformations are independent and can be performed simultaneously for all v.

The transformation *local-action*(v) consists of decomposing the edge *inedge*(v) into k edges, where k is the length of the label z of this edge. The label of i-th edge is i-th letter of z. New nodes $k-1$ are introduced: $(v, 1)$, $(v, 2)$, ..., $(v, k-1)$. The node $(v,$

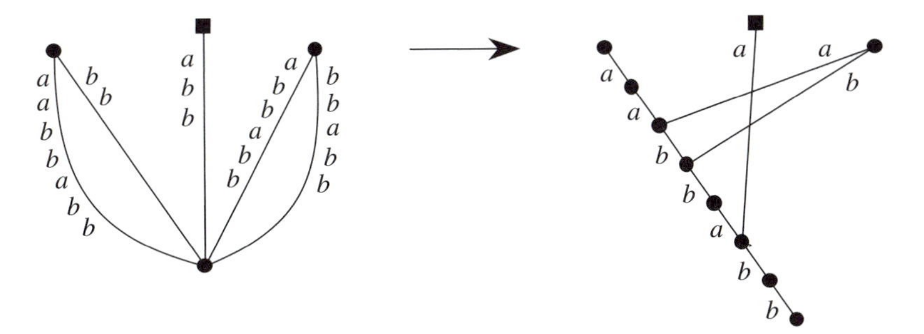

Figure 9.18: The local transformation. New nodes are introduced on the heaviest incoming edge.

i) is at distance i from v. For each other edge incoming v, $e = (w, v)$ we perform the following additional action in parallel. Suppose that the label of e has length p, and that its first letter is a. If $p > 1$, we remove the edge e, and create an edge from w to $(v, p-1)$ for which the label is the letter a. This is graphically illustrated in Figure 9.18.

Then we execute the statement: **for** all non-root nodes v **in parallel do** *local-action*(v). The resulting graph for our example string *text* is presented in the Figure 9.19. Generally, this graph is the required *DAWG*(*text*). This completes the proof. ♦

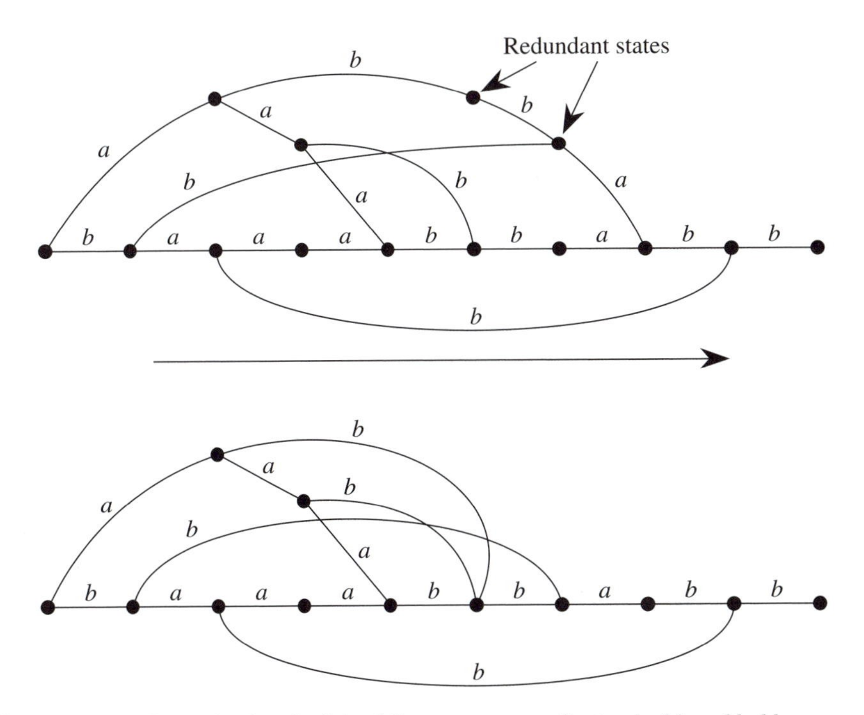

Figure 9.19: Dawg (top) and minimal factor automaton (bottom) of *baaabbabb*.

The dawg *DAWG*(*text*) can be further compacted. In fact, the structure can be reduced in the sense of minimization of automata. What remains is to merge equivalent nodes according to the paths starting from them. The reduction is possible if we do not mark nodes associated with suffixes of the text. In other words, the structure then accepts the set of all factors of *text* without any distinction between them. We call the resulting structure the minimal factor automaton of *text*, and denote it by *F*(*text*). Figure 9.19 illustrates the notion on the word *text* = *baaabbabb*. We first show how to determine the nodes of *DAWG*(*text*) associated with suffixes of *text*. The overall result yields the minimal automaton accepting suffixes of *text*, *S*(*text*). Then, we give the construction of *F*(*text*) derived from that of *DAWG*(*text*).

Theorem 9.25 The minimal suffix automaton of a text of length of length n can be constructed in $O(\log^2 n)$ time in the exclusive-write PRAM model, and in $O(\log n)$ time in the concurrent-write PRAM model, using $O(n)$ processors.

Proof We admit that the minimal suffix automaton *S*(*text*) is just *DAWG*(*text*) in which nodes associated with suffixes of *text* are marked as accepting states. We can assume that we know the values of nodes of the suffix tree (labels of the path from the root to nodes). In the process of identification of the roots of isomorphic subtrees, we can assign names to the nodes of the dawg the name of node v is the word corresponding to the longest path from the root to v. Then, a node is an accepting state of *S*(*text*) iff its name is the suffix of the input word *text*. This can be checked for each node v in constant time with the dictionary of basic factors. This completes the proof. ♦

Theorem 9.26 The minimal factor automaton of a text of length of length n can be constructed in $O(\log^2 n)$ time in the exclusive-write PRAM model, and in $O(\log n)$ time in the concurrent-write PRAM model, using $O(n)$ processors.

Sketch of the proof Construct the suffix tree $T' = T(w)$. Let $G = DAWG(text)$ and $G' = DAWG(\#text)$. Let w be the reverse of #*text*. There is a one-to-one correspondence between nodes of G' and T', if we reverse all labels of nodes of T'. The original label of the node v of T' is the word corresponding to the path from the root to v. However, we modify these labels by reversing these words. We identify all nodes of G with their equivalents in T' by an efficient parallel algorithm based on the dictionary of basic factors.

The tree T' is the tree of so-called suffix-links of G'. It provides useful information about G' and G, since G is a subgraph of G'. Let V' be the set of all nodes (including root) reachable in G from the root, after removing the edge labeled #; the subgraph G' induced by V' equals G. Now the graph $G = DAWG(text)$ can be treated as a finite deterministic automaton accepting all factors of text (the set *Fac*(*text*)). All nodes are accepting states. Two nodes of G are equivalent iff they are equivalent in the sense of this automaton. It happens that equivalence classes are very simple, each equivalence class consists of at most two nodes. It is simple to compute these equiv-

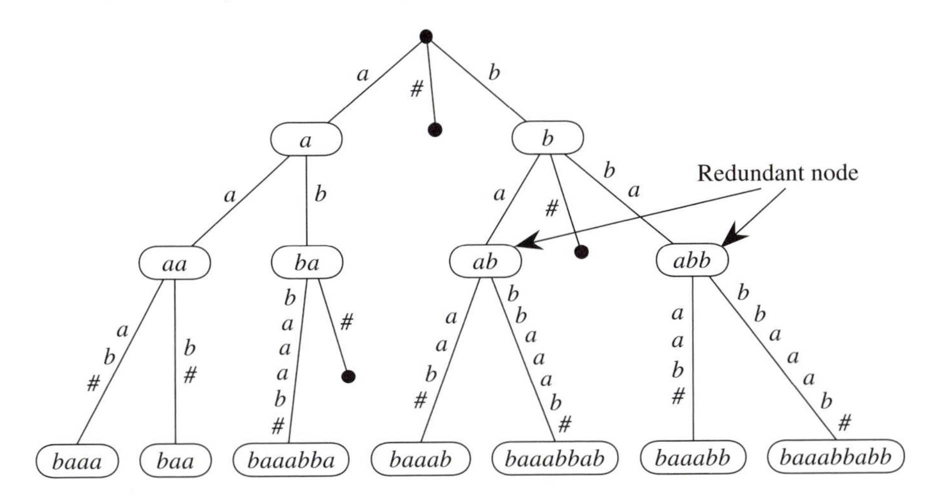

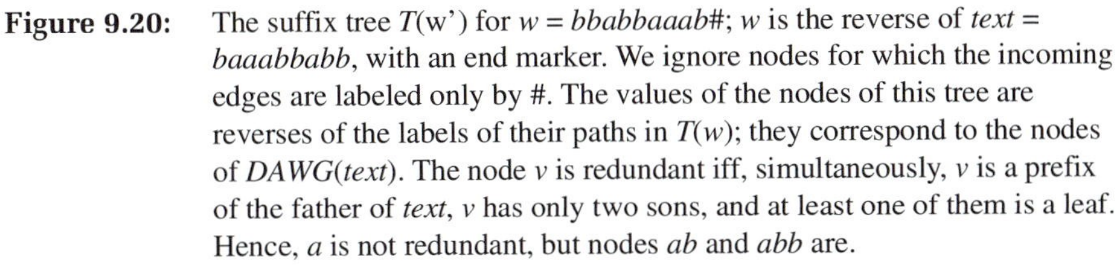
Figure 9.20: The suffix tree T(w') for $w = bbabbaaab\#$; w is the reverse of *text* = *baaabbabb*, with an end marker. We ignore nodes for which the incoming edges are labeled only by #. The values of the nodes of this tree are reverses of the labels of their paths in $T(w)$; they correspond to the nodes of $DAWG(text)$. The node v is redundant iff, simultaneously, v is a prefix of the father of *text*, v has only two sons, and at least one of them is a leaf. Hence, a is not redundant, but nodes ab and abb are.

alence classes in parallel using the tree T'. We say that the node (state) v is redundant if it is contained in a two-element equivalence class, and its partner v' (in the equivalence class) has a longer value. For a redundant node v, denote by $\pi(v)$ the node v'. Essentially, the problem of the computation of the minimal factor automaton reduces to the computation of redundant nodes, and of the function π. Let us identify nodes with their values. Let z be the father of *text* in T'.

Claim The node v of G is redundant iff v is a prefix of z (not necessarily proper), v has only two sons in T', and one of its sons is a leaf. Let b be the letter preceding the rightmost occurrence of z in *text*. Assume that v is a redundant node, and that v' is a son of v such that the label of the edge (v, v') does not start with the letter b. Then $\pi(v) = v'$.

We leave the technical proof of the claim to the reader (as a nontrivial exercise). Section 6.4 may be helpful for that purpose. In the example of Figure 9.20, $z = abb$, its prefixes are a, ab, and abb. Only ab and abb are redundant; $\pi(ab) = baaab$ and $\pi(abb) = baaabb$.

The claim above allows us to compute efficiently in parallel all redundant nodes, and the function π. After completion, we execute the statement

```
for each redundant node v do in parallel
    identify v with π(v), redirect all edges coming
    from non-redundant nodes to v onto π(v);
```

The computation of redundant nodes is illustrated in Figure 9.20. Once we know redundant nodes, the entire minimization process can be executed efficiently in parallel. This completes the proof. ♦

The algorithm also gives another sequential linear-time construction of dawgs, since classes of isomorphic subtrees can be computed in linear sequential time traversing the tree in a bottom-up manner and using bucket sort. It is possible to use a different efficient parallel algorithm for dawg construction that transforms the suffix tree of the reverse input word into the dawg of the word. In this case, the correspondence between such suffix trees and dawgs, as described in Chapter 6, can be used. In the previous presentation, we parallelize the transformation of suffix trees into dawgs, which relies on subtrees isomorphism. It is also possible to parallelize the other transformation, but this becomes much more technical.

The parallel algorithm for dawg construction can be further extended, and applied to the computation (in parallel) of complete inverted files. This can be understood as the construction of the suffix dawg of a finite set of texts.

Bibliographic notes

Elementary techniques for the construction of efficient parallel algorithms on a PRAM can be found in [GR 88]. It is also a source of a detailed description of optimal pattern-matching algorithm for strings (see Chapter 14). The method developed in *Parallel-KMR* algorithm leads to a less efficient algorithm, but is more general and simpler to explain. Moreover, the running times we get with KMR algorithm are usually within a logarithmic factor of best known algorithms. The dictionary of basic factors is from [CR 91c]. Cole [Co 87] has designed a parallel merge sort that can be used for one implementation of the procedure RENUMBER of Section 9.1.

The material of Sections 9.2 and 9.3 is from [CR 91c]. Apostolico has given an optimal algorithm to detect squares in texts [Ap 92].

Efficient parallel algorithms for the construction of suffix trees have been discovered independently by Landau, Schieber, Vishkin, and by Apostolico and Iliopoulos. The combined version appears in [AILSV 88]. This paper also contains the trick for the reduction of bulletin board space from n^2 to $n^{1+\varepsilon}$.

The construction of suffix and minimal factor automata is from [CR 90], and is based on the close relationship between suffix trees and dawgs given in [CS 84] (see Chapter 6). A more efficient algorithm to test the squarefreeness of a text, achieving an optimal speed-up, has been designed by Apostolico et al. [ABG 92] (see also [Ap 92]), but the methods are much more complicated than the ones presented in this chapter.

Technical properties used to derive parallel detections of palindromes are essentially from [GS 78].

The properties required to prove Theorem 9.26 can be found in [BBEHCS 85] (see also Section 6.4).

Selected references

[AILSV 88] Apostolico, A., Iliopoulos, C., Landau, G.M., Schieber, B., & Vishkin, U., "Parallel construction of a suffix tree with applications," *Algorithmica* 3 (1988): 347–365.

[CR 91c] Crochemore, M., & Rytter, W., "Usefulness of the Karp-Miller-Rosenberg algorithm in parallel computations on strings and arrays," *Theoret. Comput. Sci* 88 (1991): 59–82.

[GR 88] Gibbons, A., & Rytter, W., *Efficient Parallel Algorithms*, Cambridge University Press, Cambridge, U.K., 1988.

[Já 92] Jájá, J., *An Introduction to Parallel Algorithms*, Addison-Wesley, Reading, Mass., 1992.

10

Text compression techniques

The aim of data compression is to provide representations of data in a reduced form. The information carried by data is left unchanged by the compression processes considered in this chapter. There is no loss of information. In other words, the compression processes that we discuss are reversible.

The main interest of data compression is its practical nature. Methods are used both to reduce the memory space required to store files on hard disks or other similar devices, and to accelerate the transmission of data in telecommunications. This feature remains important particularly due to the rapid increase of mass memory, because the amount of data increases accordingly (to store images produced by satellites or scanners, for example). The same argument applies to transmission of data, even if the capacity of existing media is constantly improved.

We describe data compression methods based on substitutions. The methods are general, which means that they apply to data about which little is known. Semantic data compression techniques are not considered. Therefore, compression ratios must be appreciated on that condition. Standard methods usually save about 50% memory space.

Data compression methods attempt to eliminate redundancies, regularities, and repetitions in order to compress the data. It is not surprising then that algorithms have features in common with others described in preceding chapters.

After Section 10.1 on elementary notions about the compression problem, we consider the classical Huffman statistical coding (Sections 10.2 and 10.3). It is implemented by the UNIX (system V) command "pack." It admits an adaptive version well suited for telecommunications, and implemented by the "compact" command of UNIX (BSD 4.2). Section 10.4 deals with the general problem of factor encoding, and contains the efficient Ziv-Lempel algorithm. The "compress" command of UNIX (BSD 4.2) is based on a variant of this latter algorithm.

10.1 Substitutions

The input of a data compression algorithm is a text. It is denoted by s, for *source*. It should be considered as a string on the alphabet $\{0, 1\}$. The output of the algorithm is also a word of $\{0, 1\}^*$ denoted by c, for *encoded text*. Data compression methods based on substitutions are often described with the aid of an intermediate alphabet A on which the source s translates into a text *text*. The method is then defined by the mean of two morphisms g and h from A^* to $\{0, 1\}^*$. The text *text* is an inverse image of s by the morphism g, which means that its letters are coded with bits. The encoded text c is the direct image of *text* by the morphism h. The set $\{(g(a), h(a)): a \in A\}$ is called the *dictionary* of the coding method. When the morphism g is known or implicit, the description of the dictionary is given simply by the set $\{a, h(a): a \in A\}$.

We only consider data compression methods that have no loss of information. This implies that a *decoding function* f exists such that $s=f(c)$. Again, f is often defined through a decoding function h' such that $text = h'(c)$, and then f itself is the composition of h' and g. The lossless information constraint implies that the morphism h is one-to-one, and that h' is its inverse morphism. This means that the set $\{h(a): a \in A\}$ is a uniquely decipherable code.

The pair of morphisms, (g, h), leads to a classification of data compression methods with substitutions. We get four principal classes according to whether g is uniform (i.e., when all images of letters are words of the same length) or not, and whether the dictionary is fixed or computed during the compression process. Most elementary methods do not use any dictionary. Strings of a given length are sometimes called *blocks* in this context, while factors of variable lengths are called *variables*. A method is then qualified as *block-to-variable* if the morphism g is uniform, or *variable-to-variable* if neither g nor h are assumed to be uniform.

The efficiency of a compression method that encodes a text s into a text c is measured through a *compression ratio*. It can be $|s|/|c|$, or its inverse $|c|/|s|$. It is sometimes more logical to compute the amount of space saved by the compression, $(|s|-|c|)/|c|$.

	Block-to-Variable	**Variable-to-Variable**
	differential encoding	repetition encoding
fixed dictionary	statistical encoding(Huffman)	factor encoding
evolutive dictionary	sequential statistical encoding (Faller and Gallager)	Ziv and Lempel's algorithm

```
1100001  1100001  1100001  1100001  1100001  1100001
   a        a        a        a        a        a
                    encoded by
              &        a        6
           0100110  1100001  000110
```

Figure 10.1: Repetition coding (with ASCII code).

The remainder of this section is devoted to the description of two basic methods. They appear on the first line of the previous table *repetition encoding* and *differential encoding*.

The aim of repetition encoding is to efficiently encode repetitions. Let *text* be a text on the alphabet A. Let us assume that *text* contains a certain quantity of repetitions of the form $aa...a$ for some character a ($a \in A$).Within *text*, a sequence of n letters a, can be replaced by $\&an$, where "&" is a new character ($\& \notin A$). This corresponds to the usual mathematical notation a^n. When the repetitions of only one letter are encoded in such a manner, the letter itself does not need to appear, so that the repetition is encoded by $\&n$. This is commonly considered for "space deletion."

The string $\&an$ that encodes a repetition of n consecutive occurrences of a is itself encoded on the binary alphabet $\{0, 1\}$. In practice, letters are often represented by their ASCII code. Therefore, the code word of a letter belongs to $\{0, 1\}^k$ with k=7 or 8. Generally there is no problem in choosing the character "&" (in Figure 10.1, the real ASCII symbol & is used). Both symbols & and a appear in the coded text c under their ASCII form. The integer n of the string $\&an$ should also by encoded on the binary alphabet. Note that it is not sufficient to translate n by its binary representation, because we would be unable to localize it at decoding time inside the stream of bits. A simple way to cope with this is to encode n by the string $0^l\text{bin}(n)$, where $\text{bin}(n)$ is the binary representation of n, and l is the length of it. This works well because the binary representation of n starts with a "1" (because n>0). There are even more sophisticated integer representations, but none really suitable for the present situation. Conversely, a simpler solution is to encode n on the same number of bits as letters. If this number is k=8 for example, the translation of any repetition of length less than 256 has length 3. Thus it is useless to encode a^2 and a^3. A repetition of more than 256 identical letters is cut into smaller repetitions. These limitations reduce the efficiency of the method.

The second very useful elementary compression technique is differential encoding, also called relative encoding. We explain it using an example. Assume that the source text s is a series of dates

1981, 1982, 1980, 1982, 1985, 1984, 1984, 1981, ...

These dates appear in binary form in s, so at least 11 bits are necessary for each of them. But, the sequence can be represented by the other sequence

(1981, 1, –2, 2, 3, –1, 0, –3, ...),

assuming that the integer 1 in place of the second date is the difference between the second and the first dates. An integer in the second sequence, except the first one, is the difference between two consecutive dates. The decoding process is obvious, and processes the sequence from left to right, which is well adapted for texts. Again, the integers of the second sequence appear in binary in the coded text c. In the example, all but the first can be represented with only 3 bits each. This is how the compression is realized on suitable data.

More generally, differential encoding takes a sequence $(u_1, u_2, \ldots, u_n)$ of data, and represents it by the sequence of differences $(u_1, u_2-u_1, \ldots, u_n-u_{n-1})$ where "–" is an appropriate operation.

Differential encoding is commonly used to create archives of successive versions of a text. The initial text is fully memorized on the tape. And, for each following version, only the differences from the preceding one are kept on the tape. Several variations on this idea are possible. For example, $(u_1, u_2, \ldots, u_n)$ can be translated by $(u_1, u_2-u_1, \ldots, u_n-u_1)$, considering that the differences are all computed relatively to the first element of the sequence. This element can also change during the process according to some rule.

Very often several compression methods are combined to realize a whole compression process. A good example of this strategy is given by the application to facsimile (FAX) machines. Pages to be transmitted are made of lines, each of 1728 bits. A differential encoding is first applied on consecutive lines. Therefore, if the nth line is

0101001 0101010 1001001 1101001 1011101 1000000...

and the $n+1^{\text{th}}$ line is

0101000 0101011 0111001 1100101 1011101 0000000...

the following line is to be sent

0000001 0000001 1110000 0001100 0000000 1000000...

Of course, no compression at all would be achieved if the line were sent as it is. There are two solutions to encoding the line of differences. The first solution encodes runs of "1" occurring in the line both by their length and their relative position in the line. Therefore, we get the sequence

(7, 1), (7, 4), (8, 2), (10, 1), ...

in which the representation on the binary alphabet gives the coded text. The second solution makes use of statistical Huffman codes to encode successive runs of 0s and runs of 1s. These codes are defined in the next section.

A good compression ratio is generally obtained when the transmitted image contains written text. But it is clear that "dense" pages lead to a small compression ratio, and the that best ratios are reached with blank (or black) pages.

10.2 Static Huffman coding

The most common technique for compressing a text is redefining the code words associated with the letters of the alphabet. This is achieved by *block-to-variable* compression methods. According to the pair of morphisms (g, h) introduced in Section 10.1, this means that g represents the usual code attributed to the characters (ASCII code, for example). More generally, the source is factorized into blocks of a given length. Once this length is chosen, the method reduces to the computation of a morphism h that minimizes the size of the encoded text $h(text)$. The key to finding h is to consider the frequency of letters, and to encode frequent letters by short code words. Methods based on this criterion are called *statistical compression methods*.

Computing h requires finding the set $C = \{h(a): a \in A\}$, which must be a uniquely decipherable code in order to permit the decoding of the compressed text. Moreover, to get an efficient decoding algorithm, C is chosen as an instantaneous code, i.e., a *prefix code*, which means that no word of C is prefix of another word of C. It is quite surprising that this does not reduce the power of the method. This is due to the fact that any code has an equivalent prefix code with the same distribution of code word lengths. The property, stated in the following theorem, is a consequence of what is known as the Kraft-McMillan inequalities related to code word lengths, which are recalled first.

Kraft's inequality There is a prefix code with word lengths $l_1, l_1, \ldots, l_k$ on the alphabet $\{0, 1\}$ iff

$$(*) \sum_{i=1}^{k} 2^{-l_i} \leq 1.$$

McMillan's inequality There is a uniquely decipherable code with word lengths $l_1, l_1, \ldots, l_k$ on the alphabet $\{0, 1\}$ iff the inequality (*) holds.

Theorem 10.1 A uniquely decipherable code with prescribed word lengths exists iff a prefix code with the same word lengths exists.

Huffman's method computes the code C according to a given distribution of frequencies of the letters. This method is both optimal and practical. The entire Huffman's compression algorithm proceeds in three steps. In the first step, the numbers of occurrences of letters (blocks) are computed. Let n_a be the number of times

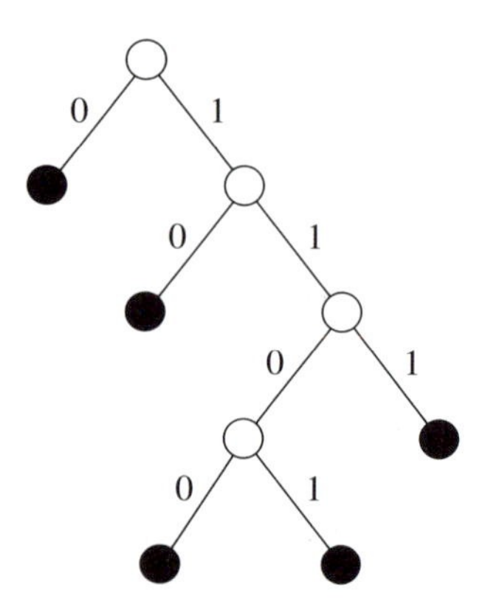

Figure 10.2: The trie of the prefix code {0, 10, 1100, 1101, 111}. Black nodes correspond to code words.

letter a occurs in *text*. In the second step, the set $\{n_a: a \in A\}$ is used to compute a prefix code C. Finally, in the third step, the text is encoded with the prefix code C found previously. Note that the prefix code should be appended to the coded text because the decoder needs it. It is commonly put inside a *header* (of the compressed file) which contains additional information on the file. Instead of computing the exact numbers of occurrences of letters in the source, a prefix code can be computed on the base of a standard probability distribution of letters. In this situation, only the third step is applied to encode a text, which gives a very simple and fast encoding procedure. Obviously, however, the coding is no longer optimal for a given text.

The core of Huffman's algorithm is the computation of a prefix code, over the alphabet {0, 1}, corresponding to the distribution of letters $\{n_a: a \in A\}$. This part of the algorithm builds the word trie T of the desired prefix code. The "prefix" property of the code ensures that there is a one-to-one correspondence between code words and leaves of T (see Figure 10.2). Each code word (and leaf of the trie) corresponds to some number n_a, and yields the encoding $h(a)$ of the letter a.

The size of the coded text is

$$|h(text)| = \sum_{\alpha \in A} n_a.|h(a)|.$$

On the trie of the code C the equality translates into

$$|h(text)| = \sum_{fa \in T} n_a.level(f_a),$$

where f_a is the leaf of T associated with letter a, and $level(f_a)$ is its distance to the root of T. The problem of computing a prefix code $C = \{h(a): a \in A\}$ such that $|h(text)|$ is minimum becomes a problem on trees:

- find a minimum weighted binary tree T in which the leaves $\{f_a: a \in A\}$ have initial weights $\{n_a: a \in A\}$,

where the weight of T, denoted by $W(T)$, is understood as the quantity $\Sigma n_a.level(f_a)$. The algorithm following builds a minimum weighted tree in a bottom-up fashion, grouping together two subtrees under a new node. In other words, at each stage the algorithm creates a new node that is made a father of two existing nodes. The weight of a node is the weight of the subtree below it. There are several possible trees of minimum weight. All trees that can be created by the Huffman algorithm are called *Huffman trees*.

Example Let *text* = *abracadabra*. The number of occurrences of letters are

$$n_a = 5, n_b = 2, n_c = 1, n_d = 1, n_r = 2.$$

The tree of a possible prefix code built by Huffman's algorithm is shown in Figure 10.3. The prefix code is:

$$h(a) = 0, h(b) = 10, h(c) = 1100, h(d) = 1101, h(r) = 111.$$

The coded text is then

0 10 111 0 1100 0 1101 0 10 111 0

which is a word of length 23. If the initial code words of letters have length 5, we get the compression ratio 55/23 ≈ 2.4. However, if the prefix code (or its trie) has to be memorized, this additionally takes at least 9 bits, which reduces the compression ratio to 55/32 ≈ 1.7. If the initial coding of letters also has to be stored (with at least

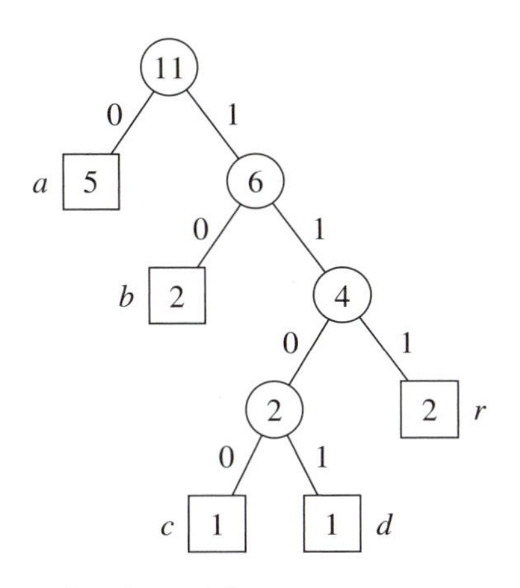

Figure 10.3: Huffman tree for *abracadabra*.

25 bits), the compression ratio is even worse: $55/57 \approx 0.96$. In the latter case, the encoding leads to an expansion of the source text. ♦

As noted in the previous example, the header of the coded text must often contain sufficient information on the coding to allow a later decoding of the text. The information necessary is the prefix code computed by Huffman's algorithm, and the initial code words of letters. Altogether, this takes approximately $2|A|+k|A|$ bits (if k is the length of initial code words), because the structure of the trie can be encoded with $2|A|-1$ bits.

```
Algorithm Huffman { minimum weighted tree construction }
begin
    for a in A do create a one-node tree fa with weight w(fa) = na;
    L := queue of trees fa in increasing order of weights;
    N := empty queue; { for internal nodes of the final tree }
    while |L|+|N| > 1 do begin
        let u and v be the elements of L ∪ N with smallest weights;
        { u and v are found from heads of lists }
        delete u and v from (heads of) the queues;
        create a tree x with a new node as root,
                                    and u and v as left and right subtrees;
        w(x) := w(u)+w(v);
        add the tree x to the end of queue N;
    end;
    return the remaining tree in L ∪ N;
end.
```

Theorem 10.2 Huffman algorithm produces a minimum weighted tree in $O(|A|.\log|A|)$ time.

Proof The correctness is based on several observations. First, a Huffman tree is a binary complete tree, in the sense that all internal nodes have exactly two sons. Otherwise, the weight of the tree could be decreased by replacing a one-son node by its own son. Second, there is a Huffman tree T for $\{n_a: a \in A\}$ in which two leaves at the maximum level are siblings, and have minimum weights among all leaves (and all nodes, as well). Possibly exchanging leaves in a Huffman tree gives the conclusion. Third, let x be the father of two leaves f_b and f_c in a Huffman tree T. Consider a weighted tree T' for $\{n_a: a \in A\}-\{n_b, n_c\}+\{n_b+ n_c\}$, assuming that x is a leaf of weight $n_b+ n_c$. Then,

$$W(T) = W(T')+n_b+ n_c.$$

Thus, T is a Huffman tree iff T' is. This is the way the tree is built by the algorithm, joining two trees of minimal weights into a new tree.

The sorting phase of the algorithm takes $O(|A|.\log|A|)$ time with any efficient sorting method. The running time of the instructions inside the "while" loop is constant because the minimal weighted nodes in $L \cup N$ are at the beginning of the lists. Therefore, the running time of the "while" loop is proportional to the number of created nodes. Since exactly $|A|-1$ nodes are created, this takes $O(|A|)$ time. ♦

The performance of Huffman codes is related to a measure of information of the source text, called the *entropy of the alphabet.* Let p_a be $n_a/|text|$. This quantity can be viewed as the probability that letter a occurs at a given position in the text. This probability is assumed to be independent of the position. Then, the entropy of the alphabet according to the p_as is defined as

$$H(A) = -\sum_{a \in A} p_a.\log p_a.$$

The entropy is expressed in bits (log is a base-two log). It is a lower bound of the average length of the code words $h(a)$,

$$m(A) = \sum_{a \in A} p_a.|h(a)|.$$

Moreover, Huffman codes give the best possible approximation of the entropy (among methods based on a recoding of the alphabet). This is summarized in the following theorem whose proof relies on the Kraft-McMillan inequalities.

Theorem 10.3 The average length of any uniquely decipherable code on the alphabet A is at least $H(A)$. The average length $m(A)$ of a Huffman code on A satisfies $H(A) \leq m(A) \leq H(A)+1$.

The average length of Huffman codes is exactly the entropy $H(A)$ when, for each letter a of A, $p_a = 2^{-|h(a)|}$ (note that the sum of all p_as is 1). The ratio $H(X)/m(X)$ measures the efficiency of the code. In English, the entropy of letters according to a common probability distribution on letters is close to 4 bits. And the efficiency of a Huffman code is more than 99%. This means that if the English source text is an 8-bit ASCII file, the Huffman compression method is likely to divide its size by two. The Morse code (developed for telegraphic transmissions), which also takes into account probabilities of letters, is not a Huffman code, and has an efficiency close to 66%. This not as good as any Huffman code, but Morse code incorporates redundancies in order to make transmissions safer.

In practice, Huffman codes are built for ASCII letters of the source text, and also for digrams (factors of length 2) occurring in the text instead of letters. In the latter case, the source is factorized into blocks of length 16 bits (or 14 bits). On larger texts, the length of blocks chosen can be higher to capture some dependencies between consecutive letters, but the size of the alphabet grows exponentially with the length of blocks.

Huffman's algorithm generalizes to the construction of prefix codes on alphabets of size m larger than two. The trie of the code is then an almost m-ary tree. Internal nodes have m sons, except maybe one node which is a father of less than m leaves.

The main default of the entire Huffman compression algorithm is that the source text must be read twice: the first time to compute the frequencies of letters, and the second time to encode the text. Only one reading of the text is possible if one uses known average frequencies of letters. But then, the compression is not necessarily optimal on a given text. The next section presents a solution for avoiding two readings of the source text.

There is another statistical encoding that produces a prefix code. It is known as the Shannon-Fano coding. It builds a weighted tree, as Huffman's method does, but the process works top-down. The tree and all its subtrees are balanced according to their weights (sum of occurrences of characters associated with leaves). The result of the method is not necessarily an optimal weighted tree, so its performance is generally within that of Huffman coding.

10.3 Dynamic Huffman coding

We now describe an adaptive version of Huffman's method. With this algorithm, the source text is read only once. Moreover, the memory space required by the algorithm is proportional to the size of the current trie, that is, to the size of the alphabet. The encoding of letters of the source text is realized while the text is read. In some situations, the compression ratio is even better than the ratio of Huffman's method.

Assume that za (z a word, a a letter) is a prefix of *text*. We consider the alphabet A of letters occurring in z, plus the extra symbol # that stands for all letters not occurring in z but possibly appearing in *text*. Let us denote by T_z any Huffman tree built on the alphabet $A \cup \{\#\}$ with the following weights:

$$n_a = \text{number of occurrences of } a \text{ in } z,$$
$$n_\# = 0.$$

We denote by $h_z(a)$ the code word corresponding to a, and determined by the tree T_z. Note that the tree has only one leaf of zero weight, namely, the leaf associated with #.

The encoding of letters proceeds as follows. In the current situation, the prefix z of *text* has already been coded, and we have the tree T_z together with the corresponding encoding function h_z. The next letter a is then encoded by $h_z(a)$ (according the tree T_z). Afterward, the tree is transformed into T_{za}. At decoding time the algorithm reproduces the same tree at the same time. However, the letter a may not occur in z, in which case it cannot be translated as any other letter. Here, the letter a is encoded by $h_z(\#)g(a)$, that is, the code word of the special symbol # according the tree T_z, followed by the initial code of letter a. This step is also reproduced without any problem at decoding time.

The reason why this method is practically effective is due to an efficient procedure updating the tree. The procedure UPDATE used in the algorithm following can be implemented in time proportional to the height of the tree. This means that the compression and decompression algorithms work in real time for any fixed alphabet, as is the case in practice. Moreover, the compression ratio obtained by the method is close to that of Huffman's compression algorithm.

```
Algorithm { adaptive Huffman coding }
begin
    T := Tε;
    while not end of text do begin
        a := next letter of text;          { h is implied by T }
        if a is not a new letter then write(h(a))
        else write(h(#)g(a)); { g(a) = initial code word for a }
        T := UPDATE(T);
    end;
end.
```

The key point of the adaptive compression algorithm is a fast implementation of the procedure UPDATE. It is based on a characterization of Huffman trees, known as the siblings property. This property does not hold in general for minimum weighted trees.

Theorem 10.4 (siblings property) Let T be a complete binary weighted tree (with p leaves) in which leaves have positive weights, and the weight of any internal node is the sum of weights of its sons. Then, T is a Huffman tree iff its nodes can be arranged in a sequence $(x_1, x_2, \ldots, x_{2n-1})$ such that:

1. the sequence of weights $(w(x_1), w(x_2), \ldots, w(x_{2n-1}))$ is in increasing order, and
2. for any i $(1 \le i<n)$, the consecutive nodes x_{2i-1} and x_{2i} are siblings (they have the same father).

Proof If T is a tree built by Huffman algorithm, the ordering on nodes is simply given by the order in which nodes are deleted from queues during the run of the algorithm. The "if" part of the proof is by induction on the number p of leaves. Consider the two nodes x_1 and x_2 of the list. It is rather obvious that they are leaves because weights are strictly positive integers. The leaves x_1 and x_2 can be chosen first by the Huffman algorithm because they have minimum weights. Let x be their father. The rest of the construction is executed as if we had only p–1 leaves, x_1 and x_2 being

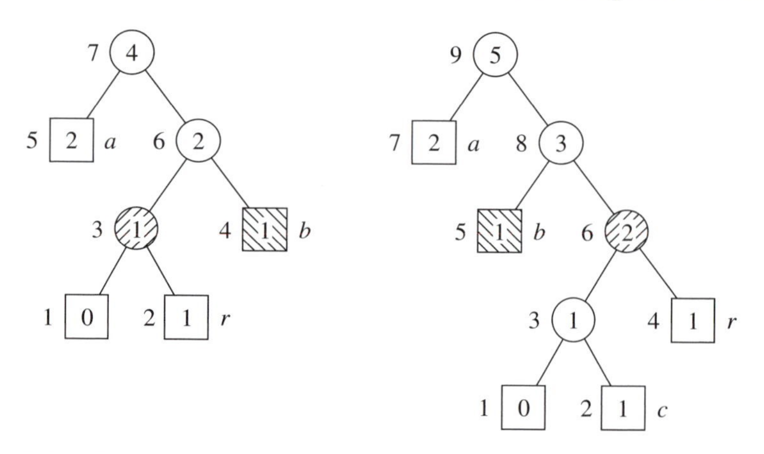

Figure 10.4: Transformation of T_{abra} into T_{abrac}. Marked nodes have been exchanged. Numbers beside nodes give an ordering satisfying the siblings property.

substituted by x. By induction, the existence of the ordering proves that the tree in which x is a leaf is a Huffman tree. Thus, the initial tree in which x is the father of x_1 and x_2 is also a Huffman tree (see the proof of Theorem 10.2). ♦

The characterization of Huffman trees by the siblings property remains true if only one leaf has a null weight. During the sequential encoding, the transformation of the current tree T_z into T_{za} starts by incrementing the weight of the leaf x_i that corresponds to a. If point 1 of the siblings property is no longer satisfied, node x_i is exchanged with the node x_j for which j is the greatest integer such that $w(x_j) < w(x_i)$. If necessary, the same operation is repeated on the father of x_i, and so on. The exchange of nodes is, in fact, the exchange of the corresponding subtrees (see Figure 10.4) The tree structure is not affected by exchanges because weights strictly increase from leaves to the root (except maybe in the 3-node tree containing the leaf associated with #), so that a node cannot be exchanged with any of its ancestors. This proves that the procedure UPDATE can be implemented in time proportional to the height of the tree. Thus, we have proved the following.

Lemma 10.5 Procedure UPDATE can be implemented to work in $O(|A|)$ time.

Example Figure 10.5 shows the sequential encoding of *abracadabra*. Letters are assumed to be initially encoded on 5 bits ($a \rightarrow 00000$, $b \rightarrow 00001$, $c \rightarrow 00010$, ..., $z \rightarrow 11010$). The entire translation of *abracadabra* is:

```
00000 000001 0010001 0 10000010 0 110000011 0 110 110 0
  a      b       r   a     c    a     d     a  b   r  a
```

We get a word of length 45. These 45 bits are to be compared with the 57 bits obtained by the original Huffman algorithm. For this example, the dynamic algorithm gives a better compression ratio, say 55/45 = 1.22. ♦

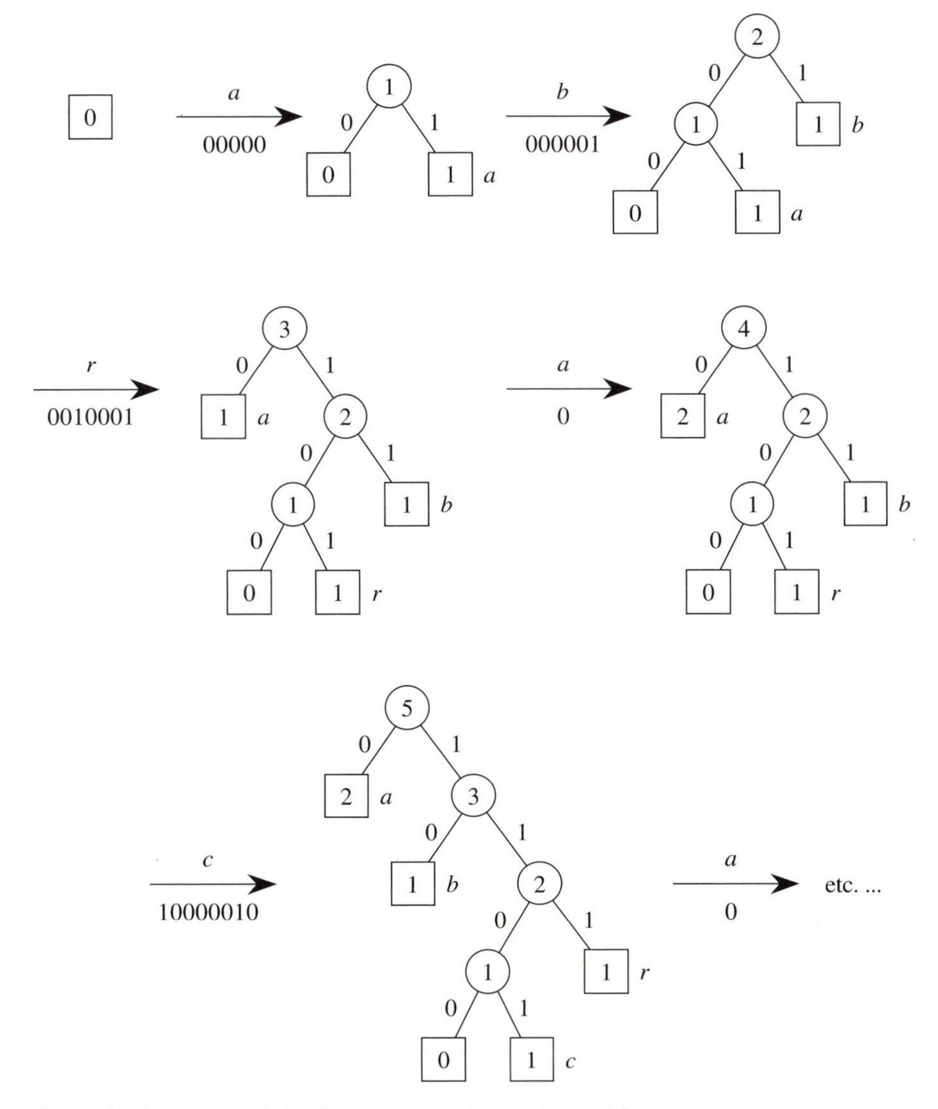

Figure 10.5: Dynamic Huffman compression of *abracadabra*.

The precise analysis of the adaptive Huffman compression algorithm has led to an improved compression algorithm. The key for the improvement is to choose a specific ordering for the nodes of the Huffman tree. Indeed, one may note that in the ordering given by the siblings property, two nodes of same weight are exchangeable. The improvement is based on a specific ordering that corresponds to a width-first tree-traversal of the tree from the leaves to the root. Moreover, at each level in the tree, nodes are ordered from left to right, and leaves of a given weight precede internal nodes of the same weight. The algorithm derived from this is efficient for texts of just a few thousand characters. The encoding of larger texts save almost one bit per character compared with Huffman's algorithm.

10.4 Factor encoding

Data compression methods with substitutions gain their power when the substitution applies to variable-length factors rather than blocks. The substitution is defined by a dictionary:

$$D = \{(f, c): f \in F, c \in C\},$$

where F is a (finite) set of factors of the source text s, and C is the set of their corresponding code words. The set F acts as the alphabet A of Section 10.1. The source text is a concatenation of elements of F.

Example Let *text* be a text composed of ordinary ASCII characters encoded on 8 bits. For C, one may choose the 8-bit words that correspond to no letter of *text*. Then, F can be a set of factors occurring frequently inside the text *text*. Replacing factors of F in *text* by letters of C compresses the text. This results in increasing the alphabet on which *text* is written. ♦

In the general case of factor encoding, a data compression scheme must solve the three following points:

- find the set F of factors that are to be encoded,
- design an algorithm to factorize the source text according to F,
- compute a code C in one-to-one correspondence with F.

When the text is given, the computation of such an optimal encoding is a NP-complete problem. The proof can be executed inside the following model of encoding. Let A be the alphabet of the *text*. The encoding of *text* is a word of the form $d\#c$ where d ($\in A^*$) is supposed to represent the dictionary, # is a new letter (not in A), and c is the compressed text. The part c of the encoded string is written on the alphabet $A \cup \{1, 2, \ldots, n\} \times \{1, 2, \ldots, n\}$. A pair (i, j) occurring in c means a reference to the factor of length j that occurs at position i in d.

Example On $A = \{a, b, c\}$, let $text = aababbabbabbc$. Its encoding can be $bbabb\#aa(1, 4)a(0, 5)c$. The explicit dictionary is then

$$D = \{(babb, (1, 4)), (bbabb, (0, 5))\} \cup A \times A. ♦$$

Within the above model, the length of $d\#c$ is the number of occurrences of both letters and pairs of integers that appear in it. For example, this length is 12 in the previous example. The search for a shortest word $d\#c$ that encodes a given *text* reduces to the SCS problem—the Shortest Common Superstring problem for a finite set of words—which is a classical NP-complete problem.

When the set F of factors is known, the main problem is to factorize the source s efficiently according to the elements of F, that is, to find factors $f_1, f_2, \ldots, f_k \in F$ such that

$$s = f_1 f_2 \ldots f_k.$$

The problem arises from the fact that F is not necesarily a unique decipherable code, so several factorizations are often possible. It is important that the integer k be as small as possible, and the factorization is said to be an *optimal factorization* when k is minimal.

The simplest strategy for factorizing s is to use a greedy algorithm. The factorization is computed sequentially. Therefore, the first factor f_1 is naturally chosen as the longest prefix of s that belongs to F. And the decomposition of the remainder of the text is done iteratively in the same way.

Remark 1 If F is a set of letters or digrams ($F \subseteq A \cup A \times A$), the greedy algorithm computes an optimal factorization. The condition may seem quite restrictive, but in French, for example, the most frequent factors ("er", "en") have length 2.

Remark 2 If F is a factor-closed set (all factors of words of F are in F), the greedy algorithm also computes an optimal factorization.

Another factorization strategy, called *semi-greedy* here, leads to optimal factorizations under broader conditions. Moreover, its time complexity is similar to the previous strategy.

Semi-greedy factorization strategy of *s*

- let $m = \max\{|uv|: u, v \in F$, and uv is a prefix of $s\}$;
- let f_1 be an element of F such that $f_1 v$ is prefix of s, and $|f_1 v| = m$, for some $v \in F$;
- let $s = f_1 s'$;
- iterate the same process on s'.

Example Let $F = \{a, b, c, ab, ba, bb, bab, bba, babb, bbab\}$, and $s = aababbabbabbc$. The greedy algorithm produces the factorization

$$s = a\ ab\ ab\ babb\ ab\ b\ c$$

which has 7 factors. The semi-greedy algorithm gives

$$s = a\ a\ ba\ bbab\ babb\ c$$

which is an optimal factorization. Note that F is prefix-closed (prefixes of words of F are in F) after adding the empty word. ♦

The interest in the semi-greedy factorization algorithm is due to the following lemma in which the proof is left as an exercise. As we shall see later in this section, the hypothesis of the set of factors F originates naturally for some compression algorithms.

Lemma 10.6 If the set F is prefix-closed, the semi-greedy factorization strategy produces an optimal factorization.

When the set F is finite, the semi-greedy algorithm may be realized with the help of a string-matching automaton (see Chapter 7). This leads to a linear-time algorithm to factorize a text.

Finally, if the set F is known, and if the factorization algorithm has been designed, the next step to consider when performing a whole compression process is to determine the code C. As for the statistical encodings of Sections 10.2 and 10.3, the choice of code words associated with factors of F can take their frequencies into account. This choice can be made once for all, or in a dynamic way during the factorization of s. Here is an example of a possible strategy: the elements of F are put into a list and encoded by their position in the list. A move-to-front procedure realized during the encoding phase tends to attribute small positions, and thus short encodings, to frequent factors. The idea of encoding a factor by its position in a list is applied in the next compression method.

Factor encoding becomes even more powerful with an adaptive feature. It is realized by the Ziv-Lempel method (ZL method, for short). The dictionary is built during the compression process. The code words of factors are their positions in the dictionary. Therefore, we regard the dictionary D as a sequence of words $(f_0, f_1, \ldots)$. The algorithm encodes positions in the most efficient way according to the present state of the dictionary, using the function lg defined by:

$lg(1) = 1$ and,
$lg(n) = \lceil \log_2(n) \rceil$, for $n > 1$.

```
Algorithm ZL; { Ziv-Lempel compression algorithm }
{ encodes the source s on the binary alphabet }
begin
   D := {ε}; x := s#;
   while x ≠ ε do begin
      fk := longest word of D such that x = fk a y, for some a ∈ A;
      a := letter following fk in x;
      write k on lg|D| bits;
      write the initial code word of a on lg|A| bits;
      add fk a at the end of D;
      x := y;
   end;
end.
```

Example Let $A = \{a, b, \#\}$. Assume that the initial code words of letters are 00 for a, 01 for b, and 10 for #. Let $s = aababbabbabb\#$. Then, the decomposition of s is

s = a ab abb abba b b#

that leads to

c = 000 101 1001 1100 00001 10110.

After that, the dictionary D contains seven words:

$D = (\varepsilon, a, ab, abb, abba, b, b\#)$. ♦

Intuitively, ZL algorithm compresses the source text because it is able to discover some repeated factors within it. The second occurrence of such a factor is encoded only by a pointer onto its first position, which can save a large amount of space.

From the implementation point of view, the dictionary D in ZL algorithm can be stored as a word trie. The addition of a new word in the dictionary takes a constant amount of time and space.

There is a large number of possible variations on ZL algorithm. The study of this algorithm has been stimulated by its high performance in practical applications. Note, for example, that the dictionary built by ZL algorithm is prefix-closed, so the semi-greedy factorization strategy may help to reduce the number of factors in the decomposition.

The model of encoding valid for that kind of compression method is more general than the preceding one. The encoding c of the source text s is a word on the alphabet $A \cup \{1, 2, \ldots, n\} \times \{1, 2, \ldots, n\}$. A pair (i, j) occurring in c references a factor of s itself: i is the position of the factor, and j is its length.

Example Again let $s = aababbabbabb\#$. It can be encoded by the word

$c = a\ (0, 1)b\ (1, 2)b\ (3, 3)a\ b\ (11, 1)\#$,

of length 10, which corresponds to the factorization found by ZL algorithm. ♦

The number of factors of the decomposition of the source text reduces if we consider a decomposition of the text similar to the f-factorization of Section 8.2. The factorization of s is a sequence of words $(f_1, f_2, \ldots, f_m)$ such that $s = f_1f_2\ldots f_m$, and that is iteratively defined by (see Figure 10.6):

f_i is the shortest prefix of $f_if_{i+1}\ldots f_m$ which does not occurs before in s.

Example The factorization of $s = aababbabbabb\#$ is

$(a, ab, abb, abbabb\#)$

which accounts to encode s into

$c = a\ (0, 1)b\ (1, 2)b\ (3, 6)\#$,

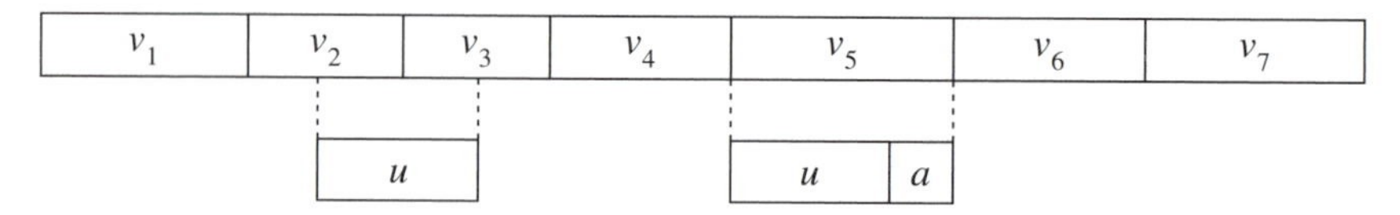

Figure 10.6: Efficient factorization of the source. v_5 is the shortest factor not occurring on its left.

word of length 7 only, compared with the previous factorization. ♦

Theorem 10.7 The *f*-factorization of a string *s* can be computed in linear time.

Proof Use the directed acyclic word graph *DAWG*(*s*), or the suffix tree *T*(*s*). ♦

The number of factors arising in the *f*-factorization of a text, in a certain sense, measures the complexity of the text. The complexity of sequences is related to the notion of entropy of a set of strings as stated below. Assume that the possible set of texts of length n (on the alphabet A) is of size $|A|^{hn}$ (for all length n). Then h (≤ 1) is called the entropy of the set of texts. For example, if the probability of appearance of a letter in a text does not depend on its position, then h is the entropy $H(A)$ considered in Section 10.2.

Theorem 10.8 The number m of elements of the *f*-factorization of long enough texts is upper bounded by $hn/\log n$, for almost all texts.

We end this section by reporting some experiments on compression algorithms. Table 10.1 gives the results. Rows are indexed by algorithms, and columns by types of text files. "Uniform" is a text on a 20-letter alphabet generated with a uniform distribution of letters. "Repeated alphabet" is a repetition of the word *abc*...*zABC*...*Z*. Compression of the five files has been executed using the Huffman method, Ziv-Lempel algorithm (more precisely, COMPRESS command of UNIX), and the compression algorithm, called FACT, based on the *f*-factorization. Huffman method is the most efficient only for the file "Uniform," which is not surprising. For other files, the results for COMPRESS and FACT are similar.

Table 10.1: Sizes of some compressed files (best scores in bold).

Source	French text	C program	Lisp program	Uniform	Repeated alphabet
Initial length	62816	684497	75925	70000	530000
Huffman	53.27%	62.10 %	63.66 %	**55.58** %	72.65 %
COMPRESS	**41.46** %	34.16 %	40.52 %	63.60 %	2.13 %
FACT	47.43 %	**31.86** %	**35.49** %	73.74 %	**0.09** %

10.5* Parallel Huffman coding

The sequential algorithm for Huffman coding is quite simple, but unfortunately it appears to be inherently sequential. Its parallel counterpart is much more complicated, and requires a new approach. The global structure of Huffman trees must be explored in depth. In this section, we give a polylogarithmic parallel-time algorithm to compute a Huffman code. The number of processors is $M(n)$, where $M(n)$ is the number of processors needed for a (min, +) multiplication of two n by n real matrices in logarithmic parallel time. We assume, for simplicity, that the alphabet is binary.

A binary tree T is said to be *left-justified* iff it satisfies the following properties:

1. the depths of the leaves are in non-increasing order from left to right,
2. if a node v has only one son, then it is a left son,
3. let u be a left brother of v, and assume that the height of the subtree rooted at v is at least l. Then the tree rooted at u is full at level l, which means that u has 2^l descendants at distance l.

Most important for the present problem is the following property of left-adjusted trees.

Basic property Let T be a left-justified binary tree. Then, it has a structure as illustrated in Figure 10.7. All hanging subtrees have height at most $\log n$.

Lemma 10.9 Assume that the weights $p_1, p_2, \ldots, p_n$ are pairwise distinct and in increasing order. Then, there is Huffman tree for $(p_1, p_2, \ldots, p_n)$ that is left-justified.

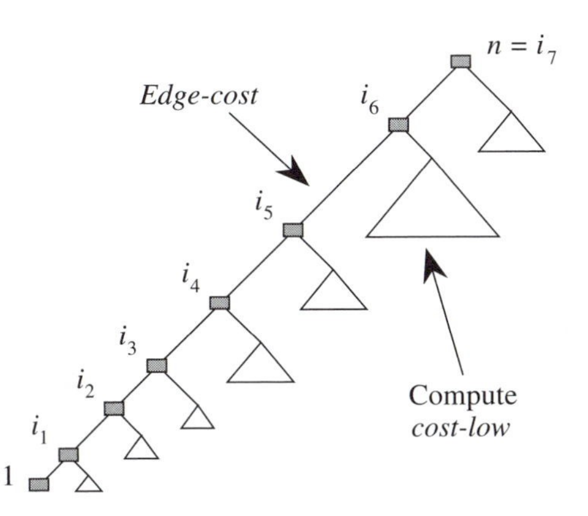

Figure 10.7: A left-justified Huffman tree. The hanging subtrees are of logarithmic height.

Proof We first show the following claim:

> For each tree T we can find a tree T' satisfying (1), (2), (3), in which the leaves are a permutation of leaves of T, and such that the depths of corresponding leaves in the trees T and T' are the same.

This claim can be proven by induction with respect to the height h of the tree T. Let $T1$ be derived from T by the following transformation: cut all leaves of T at maximal level; the fathers of these leaves become new leaves, called special leaves in $T1$. The tree $T1$ has height h–1. It can be transformed into a tree $T1$' satisfying (1), by applying the inductive assumption. The leaves of height h–1 of $T1$' form a segment of consecutive leaves from left to right. It contains all special leaves. We can permute the leaves at height h–1 in such a way that special leaves are at the left. Then, we insert the deleted leaves back as sons of their original fathers (special leaves in $T1$ and $T1$'). The resulting tree T' satisfies the claim.

Let us consider a Huffman tree T. It can be transformed into a form satisfying conditions (1) to (3), with the weight of the tree left unchanged. Hence, after the transformation, the tree is also of minimal weight. Therefore, we can consider that our tree T is optimal *and* is left-justified. It is sufficient to prove that leaves are in increasing order of their weights p_i, from left to right. But this is straightforward since the deepest leaf has the smallest weight. Hence, the tree T satisfies all requirements. This completes the proof. ♦

Theorem 10.10 The weight of a Huffman tree can be computed in $O(\log^2 n)$ time with n^3 processors. The corresponding tree can be constructed within the same complexity bounds.

Proof Let $weight[i, j] = p_{i+1}+p_{i+2}+\ldots+p_j$. Let $cost[i, j]$ be the weight of a Huffman tree for $(p_{i+1}, p_{i+2}, \ldots, p_j)$, and in which the leaves are keys $K_{i+1}, K_{i+2}, \ldots, K_j$. Then, for $i<j-1$, we have:

$$(*)\ cost[i, j] = \min\{cost[i, k]+cost[k, j]+weight[i, j]: i < k < j\}.$$

Let us use the structure of T illustrated in Figure 10.7. All hanging subtrees are shallow; they are of height at most $\log n$. Therefore, we first compute the weights of such shallow subtrees.

Let $cost\text{–}low[i, j]$ be the weight of the Huffman tree of logarithmic height, in which the leaves are keys $K_{i+1}, K_{i+2}, \ldots, K_j$.

The table *cost–low* can be easily computed in parallel by applying (*). We initialize $cost\text{–}low[i, i+1]$ to p_i, and $cost\text{–}low[i, j]$ to ∞, for all other entries. Then, we repeat $\log n$ times the same parallel-do statement:

for each $i, j, i < j-1$ **do in parallel**
$cost\text{–}low[i, j] := \min\{cost\text{–}low[i, k]+cost\text{–}low[k, j]: i < k < j\} + weight[i, j]\}$.

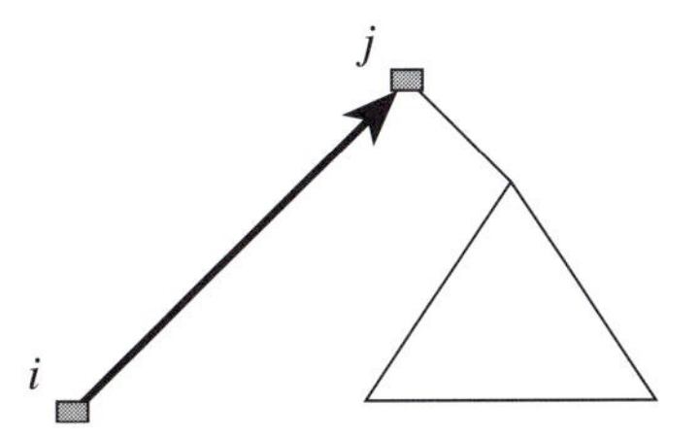

Figure 10.8: $edge\text{–}cost[i, j] = cost\text{–}low[i, j] + weight[1, j]$.

We need n processors for each operation "min" concerning a fixed pair (i, j). Since there are n^2 pairs (i, j), globally we use a cubic number of processors to compute cost–lows. Now, we have to find the weight of an optimal decomposition of the entire tree T into a left most branch, and hanging subtrees. The consecutive points of this branch correspond to points $(1, i)$. Consider the edge from $(1, i)$ to $(1, j)$, and identify it with the edge (i, j). The contribution of this edge to the total weight is illustrated in Figure 10.8.

We assign to the edge (i, j) the cost given by the formula:

$$edge\text{–}cost(i, j) = cost\text{–}low[i, j] + weight[1, j].$$

It is easy to deduce the following fact:

the total weight of T is the sum of costs of edges corresponding to the left most branch.

Once we have computed all cost-lows we can assign the weights to the edges correspondingly to the formula, and we have an acyclic directed graph with weighted edges. The cost of the Huffman tree is reduced to the computation of the minimal cost from 1 to n in this graph. This can be executed by $\log n$ squarings of the weight matrix of the graph. Each squaring corresponds to a (min, +) multiplication of $n \times n$ matrices, and, therefore, can be executed in $\log n$ time with n^3 processors. Hence $\log^2 n$ time is sufficient for the entire process. This completes the proof of the first part of the theorem.

Given costs, the Huffman tree can be constructed within the same complexity bounds. We refer the reader to [AKLMT 89]. This completes the proof of the whole theorem. ♦

In fact the matrices which occur in the algorithm have a special property called the *quadrangle inequality* (see Figure 10.9). This property allows the number of processors to be reduced to a quadratic number.

A matrix C has the quadrangle inequality iff for each $i < j < k < l$ we have:

$$C[i, k] + C[j, l] \leq C[i, l] + C[j, k].$$

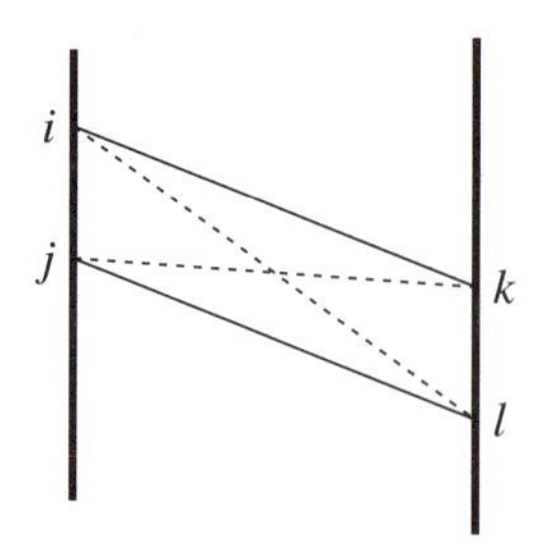

Figure 10.9: The quadrangle inequality: (cost of plain lines) ≤ (cost of dashed lines).

Let us consider matrices that are strictly upper triangular (elements below the main diagonal and on the main diagonal are null). Such matrices correspond to weights of edges in acyclic directed graphs. Denote by © the (min, +) multiplication of matrices:

$$A \copyright B = C \text{ iff } C[i, j] = \min\{A[i, k]+B[k, j]: i < k < j\}.$$

The proof of the following fact is left to the reader.

Lemma 10.11 If the matrices A and B satisfy the quadrangle inequality then A © B also satisfies this property.

For matrices occurring in the Huffman tree algorithm, the (min, +) multiplication is simpler in parallel, and the number of processors is reduced by a linear factor, due to the following lemma.

Lemma 10.12 If the matrices A and B satisfy the quadrangle inequality, then A © B can be computed in $O(\log^2 n)$ time with $O(n^2)$ processors.

Proof Let us fix j, and denote by $CUT[i]$ the smallest integer k for which the value of $A[i, k]+B[k, j]$ is minimal. The computation of A © B reduces to the computation of vectors CUT for each j. It is sufficient to show that, for j fixed, this vector can be computed in $O(\log^2 n)$ time with n processors.

The structure of the algorithm is the following:

let i_{mid} the middle of interval $[1, 2, \ldots, n]$; compute $CUT[i_{mid}]$ in $O(\log n)$ time with n processors assuming that the value of *cut* is in the whole interval $[1, 2, \ldots, n]$.

Afterward it is easy to see, due to the quadrangle inequality (see Figure 10.10), that

$CUT[i] \leq CUT[i_{mid}]$ for all $i \leq i_{mid}$,
$CUT[i] \geq CUT[i_{mid}]$ for all $i \geq i_{mid}$.

Using this information we compute $CUT[i]$ for all $i \leq i_{mid}$, knowing that its value is above $CUT[i_{mid}]$. Simultaneously we compute $CUT[i]$ for all $i > i_{mid}$, knowing that its value is not above $CUT[i_{mid}]$. Let m be the size of the interval in which we expect to

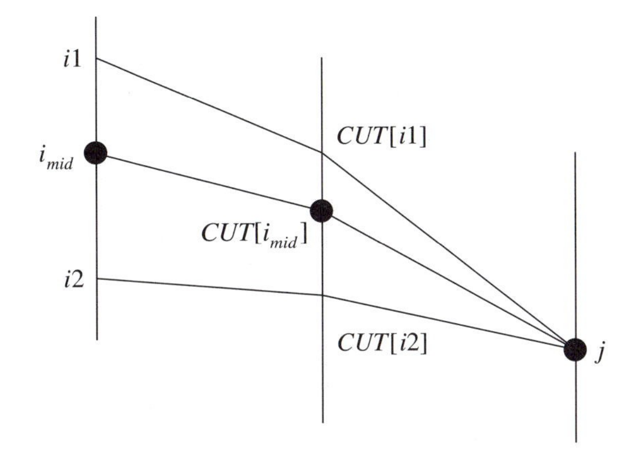

Figure 10.10: Computing cuts using a divide-and-conquer strategy.

find the value of $CUT[i]$ (for i in the interval of size n). We have the following equation for the number $P(n, m)$ of processors:

$$P(n, m) \leq \max(m, P(n, m_1)+P(n, m_2)), \text{ where } m_1+m_2 = m.$$

Obviously $P(n, n) = O(n)$. The depth of the recursion is logarithmic. Each evaluation of a minimum also takes logarithmic time. Hence, we get the required total time. For a fixed j, $P(n, n)$ processors are adequate. Altogether, for all j, we need only a quadratic number of processors. This completes the proof. ♦

The lemma implies that the parallel Huffman coding problem can indeed be solved in polylogarithmic time with only a quadratic number of processors. This is still not optimal, since the sequential algorithm makes only $O(n.\log n)$ operations. We refer the reader to [AKLMT 89] for an optimal algorithm.

Bibliographic notes

An extensive exposition of the theory of codes is given in [BP 85].

Elementary methods of data compression may be found in [He 87]. Practical programming aspects of data compression techniques are presented by Nelson in [Ne 92].

The construction of a minimal weighted tree is from Huffman [Hu 51]. The Shannon-Fano method is presented independently by Shannon (1948) and Fano (1949), see [BCW 90].

In the seventies, Faller [Fa 73] and Gallager [Ga 78] independently designed a dynamic data compression algorithm based on Huffman's method. Practical versions of dynamic Huffman coding have been designed by Cormack and Horspool [CH 84], and Knuth [Kn 85]. The precise analysis of sequential statistical data compression has been done by Vitter in [Vi 87], where an improved version is given.

NP-completeness of various questions on data compression can be found in [St 77]. The idea of the semi-greedy factorization strategy is from Hartman and Rodeh [HR 84]. And the dynamic factor encoding using the move-to-front strategy if by Bentley, Sleator, Tarjan, and Wei [BSTW 86].

In 1977, Ziv and Lempel designed the main algorithm of Section 10.4 [ZL 77] (see also [ZL 88]). The notion of word complexity, and Theorem 10.8, appears in [LZ 76]. The corresponding linear-time computations are by Rodeh, Pratt, and Even [RPE 81] (with suffix trees) and Crochemore [Cr 83] (with suffix dawgs). A large number of variants of Ziv-Lempel algorithm may be found in [BCW 90] and [St 88]. An efficient implementation of a variant of ZL algorithm is by Welch [We 84]. The experimental results of Section 10.4 are from Zipstein [Zi 92]. References and results relating compression ratios and entropy may be found in [HPS 92].

Some data compression methods do not use substitutions. Application of arithmetic coding gives a typical example that often leads to higher efficiency because it can be combined with algorithms that evaluate or approximate the source probabilities. A software version of data compression based on arithmetic coding is by Witten, Neal, and Cleary [WNC 87]. It is not clear to whom application of arithmetic coding to compression should be attributed, see [BCW 90] for historical remarks on this point.

The Huffman coding in parallel setting (see Section 10.5) is from Atallah, Kosaraju, Larmore, Miller, and Teng [AKLMT 89]. The number of processors is reduced by observing that matrices are "concave." In [AKLMT 89], an optimal parallel algorithm for constructing almost optimal prefix codes is also presented.

Selected references

[BCW 90] Bell, T.C., Cleary, J.G., & Witten, I.H., *Text Compression*, Prentice Hall, Englewood Cliffs, NJ, 1990.

[BP 85] Berstel, J., & Perrin, D., *Theory of Codes*, Academic Press, Orlando, FL, 1985.

[He 87] Held, G., *Data Compression—Techniques and Applications, Hardware and Software Considerations*, John Wiley & Sons, New York, NY, 1987. 2nd edition.

[Ne 92] Nelson, M., *The Data Compression Book*, M&T Publishing, Inc., Redwood City, CA, 1992.

[St 88] Storer, J.A., *Data Compression: Methods and Theory*, Computer Science Press, Rockville, MD, 1988.

11

Approximate pattern matching

In practical pattern-matching applications, the exact matching is not always pertinent. It is often more important to find objects that match a given pattern in a reasonably approximate way. In this chapter, approximation is measured mainly by the so-called edit distance: the minimal number of local edit operations needed to transform one object into another. Such operations are well suited for strings, but they are less natural for two-dimensional patterns or for trees. The analogue for DNA sequences is called the *alignment problem* (see Figure 11.1). Algorithms are mainly based on the algorithmic method called *dynamic programming*. We also present problems strongly related to the notion of edit distance, namely, the computation of longest common subsequences, string matching allowing errors, and string matching with don't care symbols. At the end of this chapter, we give the scheme of an algorithm for efficiently computing the edit distance in parallel.

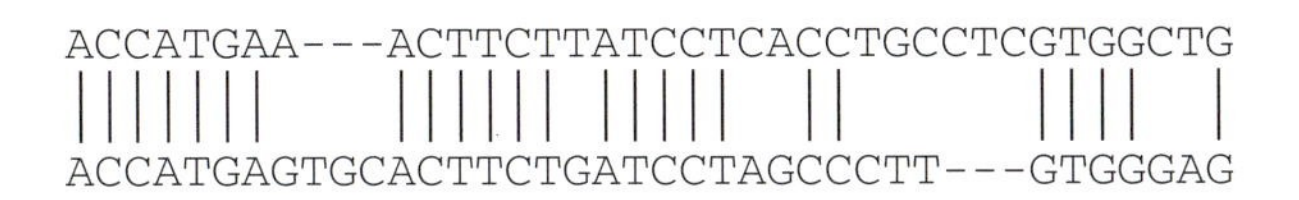

Figure 11.1: Alignment of two DNA sequences showing changes, insertions (- in top line), and deletions (- in bottom line).

11.1 Edit distance—Serial computation

An immediate question arising in applications is how to test the equality of two strings allowing some errors. The errors correspond to differences between the two words. We consider in this section three types of differences between two strings x and y:

change—symbols at corresponding positions are distinct,
insertion—a symbol of y is missing in x at a corresponding position,
deletion—a symbol of x is missing in y at a corresponding position.

It is perfectly natural to require the minimum number of differences between x and y. We translate this as the smallest possible number of operations (change, deletion, insertion) to transform x into y. This is called the *edit distance* between x and y, and denoted by $edit(x, y)$. It is clear that it is a distance between words, in the mathematical sense. This means that the following properties are satisfied:

- $edit(x, y) = 0$ iff $x = y$,
- $edit(x, y) = edit(y, x)$ (symmetry),
- $edit(x, y) \leq edit(x, z) + edit(z, y)$ (triangle inequality).

The symmetry of *edit* comes from the duality between deletion and insertion: a deletion of the letter a of x in order to get y corresponds to an insertion of a into y to get x.

Below is an example of transformation of $x = wojtk$ into $y = wjeek$.

```
w o j t   k
  ↓             deletion
      ↓         change
        ↓       insertion
w   j e e k
```

This shows that $edit(wojtk, wjeek) \leq 3$, because it uses three operations. In fact, this is the minimum number of edit operations to transform *wojtk* into *wjeek*.

From now on, we consider that words x and y are fixed. The length of x is m, and the length of y is n, and we assume that $n \geq m$. We define the table *EDIT* by

$$EDIT[i, j] = edit(x[1\ldots i], y[1\ldots j]),$$

with $0 \leq i \leq m$, $0 \leq j \leq n$. The boundary values are defined as follows (for $0 \leq i \leq m$, $0 \leq j \leq n$):

$$EDIT[0, j] = j, EDIT[i, 0] = i.$$

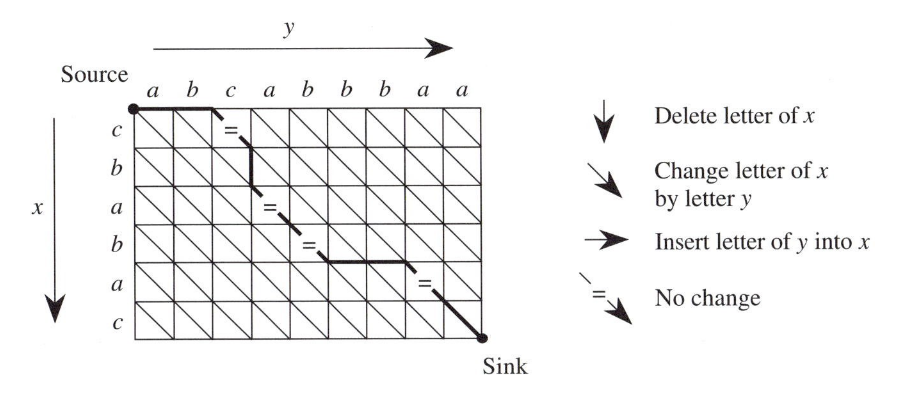

Figure 11.2: The path corresponds to the sequence of edit operations: *insert*(a), *insert*(b), *delete*(b), *insert*(b), *insert*(b), *change*(c, a).

There is a simple formula for computing other elements

$$(*) \quad EDIT[i, j] = \min(EDIT[i-1, j]+1, EDIT[i, j-1]+1, EDIT[i-1, j-1]+ \partial(x[i], y[j])),$$

where $\partial(a, b) = 0$ if $a = b$, and $\partial(a, b) = 1$ otherwise. The formula reflects the three operations, deletion, insertion, and change, in that order.

There is a graph theoretical formulation of the editing problem. We can consider the grid graph, denoted by G, composed of nodes (i, j) (for $0 \leq i \leq m$, $0 \leq j \leq n$). The node $(i-1, j-1)$ is connected to the three nodes $(i-1, j)$, $(i, j-1)$, (i, j) when they are defined (i.e., when $i \leq m$, or $j \leq n$). Each edge of the grid graph has a weight corresponding to the recurrence (*). The edges from $(i-1, j-1)$ to $(i-1, j)$ and $(i, j-1)$ have weight 1, as they correspond to the insertion or deletion of a single symbol. The edge from $(i-1, j-1)$ to (i, j) has weight $\partial(x[i], y[j])$. Figure 11.2 shows an example of grid graph for words *cbabac* and *abcabbbaa*. The edit distance between words x and y equals the length of a least weighted path in this graph from the *source* (0, 0) (left upper corner) to the *sink* (m, n) (right bottom corner).

```
function edit(x, y) { computation of edit distance }
{ |x| = m, |y| = n, EDIT is a matrix of integers }
begin
   for i := 0 to m do EDIT [i, 0] := i;
   for j := 1 to n do EDIT [0, j] := j;
   for i := 1 to m do
       for j := 1 to n do
           EDIT [i, j] = min(EDIT [i-1, j]+1, EDIT [i, j-1]+1,
                             EDIT [i-1, j-1] + ∂(x [i], y[j]));
   return EDIT [m, n]
end.
```

The algorithm above computes the edit distance of strings x and y. It stores and computes all values of the matrix *EDIT*, although only the one entry $EDIT[m, n]$ is required. This serves to save time, and this feature is called the *dynamic programming method*. Another possible algorithm to compute $edit(x, y)$ could be to use the classical Dijkstra's algorithm for shortest paths in the grid graph.

Theorem 11.1 Edit distance can be computed in $O(mn)$ time using $O(m)$ additional memory.

Proof The time complexity of the algorithm above is obvious. The space complexity results from the fact that we do not have to store the entire table. The current and the previous columns (or lines) are sufficient for carrying out computations. ♦

Remark We can assign specific costs to edit operations, depending on the type of operations and on the type of symbols involved. Such a generalized edit distance can still be computed using a formula analogous to (*).

Remark As noted in the proof of the previous theorem, the whole matrix *EDIT* does not need to be stored (only two columns are necessary at a given step) in order to compute only $edit(x, y)$. However, we can keep it in memory if we want to compute a shortest sequence of edit operations transforming x into y. This is essentially the same process as reconstructing a shortest path from the table of sizes of all-pairs shortest paths.

11.2 Longest common subsequence problem

In this section, we consider a problem that illustrates a particular case of the edit distance problem of the previous section. This is the example of computing *longest common subsequences* (see Figure 11.3). We denote by $lcs(x, y)$ the maximal length of subsequences common to x and y. For fixed x and y, we denote by $LCS[i, j]$ the length of a longest common subsequence of $x[1...i]$ and $y[1...j]$.

There is a strong relationship between the longest common subsequence problem and a particular edit distance computation. Let $edit_{di}(x, y)$ be the minimal number of operations *delete* and *insert* necessary to transform x into y. This corresponds to a restricted edit distance where changes are not allowed. A change in the edit distance of Section 11.1 can be replaced by a pair of operations, deletion and insertion. The following lemma shows that the computation of $lcs(x, y)$ is equivalent to the evaluation of $edit_{di}(x, y)$. The statement is not true in general for all edit operations, or for edit operations having distinct costs. However, the restricted edit distance $edit_{di}$ remains to give weight 2 to changes, and weight 1 to deletions and insertions. Recall that x and y have respective length m and n, and let $EDIT_{di}[i, j]$ be $edit_{di}(x[1...i], y[1..j])$.

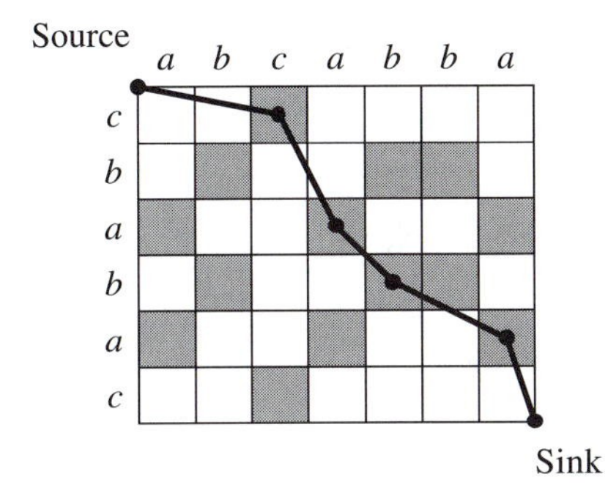

Figure 11.3: The size of the longest subsequence is the maximal number of shaded boxes on a monotonically decreasing path from source to sink. Compare to Figure 11.4.

Lemma 11.2 $2lcs(x, y) = m + n - edit_{di}(x, y)$, and, for $0 \leq i \leq m$ and $0 \leq j \leq n$, $2LCS[i, j] = i + j - EDIT_{di}[i, j]$.

Proof The equation can be easily proved by induction on i and j. This is also apparent on the graphical representation in Figure 11.4. Consider a path from the source to the sink in which diagonal edges are only allowed when the corresponding symbols are equal. The number of diagonal edges in the path is the length of a subsequence common to x and y. Horizontal and vertical edges correspond to a sequence of edit operations (delete, insert) to transform x into y. If we assign cost 2 to the diagonal edges, the length of the path is exactly the sum of the lengths of words, $m+n$. The result is shown in Figure 11.4. ♦

As a consequence of Lemma 11.2, computing $lcs(x, y)$ takes the same amount of time as computing the edit distance of the strings. A longest common subsequence can even be found with linear extra space (see bibliographic references).

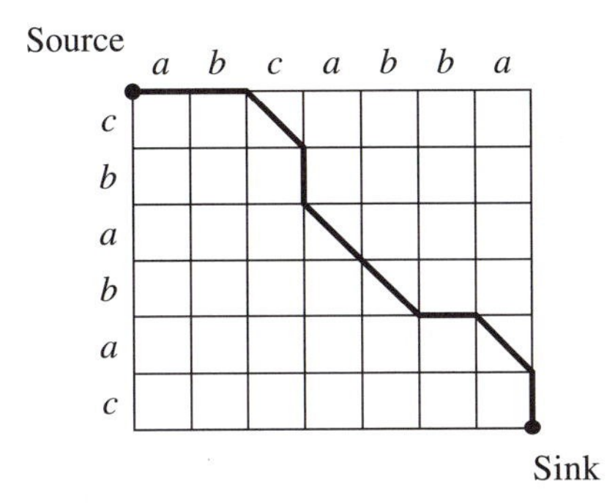

Figure 11.4: Assigning cost 2 to diagonal edges, the length of the path is $m+n$. Diagonal edges correspond to equal letters, other edges to edit operations.

Theorem 11.3 A longest common subsequence of x and y can be computed in $O(mn)$ time using $O(mn)$ additional memory.

Proof Assume that the table $EDIT_{di}$ is computed for zero-one costs of edges. Table *LCS* can be recomputed from Lemma 11.2. A longest common subsequence can be constructed from the table *LCS*. ♦

Let r be the number of shaded boxes in Figure 11.3. More formally, it is the number of pairs (i, j) such that $x[i] = y[j]$. If r is small (which happens often in practice) compared to mn, then there is an algorithm to compute longest common subsequence that is faster than the previous algorithm.

The algorithm is given below. It processes the word x sequentially from left to right. Consider the situation when $x[1\ldots i-1]$ has just been processed. The algorithm maintains a partition of positions on the word y into intervals I_0, I_1, ..., I_k,.... They are defined by

$$I_k = \{j: lcs(x[1\ldots i-1], y[1\ldots j]) = k\}.$$

In other words, positions in a class I_k correspond to prefixes of y having the same maximal length of common subsequence with $x[1\ldots i-1]$.

Consider, for instance, $y = abcabbaba$ and $x = cb\ldots$. Figure 11.5 (top) shows the partition $\{I_0, I_1, I_2\}$ of positions on y. For the next symbol of x, letter a, the figure (bottom) displays the modified partition $\{I_0, I_1, I_2, I_3\}$. The computation reduces to shifting to the right, like bowls on a wire, positions corresponding to occurrences of a inside y.

The algorithm *lcs* below implements this strategy. It makes use of operations on intervals of positions: *CLASS*, *SPLIT*, and *UNION*. They are defined as follows. For a position p on y, $CLASS(p)$ is the index k of the interval I_k to which it belongs. When p is in the interval $[f, f+1, \ldots, g]$, and $p \neq f$, then $SPLIT(I_k, p)$ is the pair of intervals $([f, f+1, \ldots, p-1], [p, p+1, \ldots, g])$. Finally, *UNION* is the union of two intervals; in the algorithm, only unions of disjoint consecutive intervals are performed.

```
function lcs(x, y) : integer; { Hunt-Szymanski algorithm }
{ m = |x| and n = |y| }
begin
   I0 := {0,1,...,n}; for k := 1 to n do Ik = ∅;
   for i := 1 to m do
      for each p position of x[i] inside y in decreasing order do
      begin
         k := CLASS(p);
         if k = CLASS(p-1) then begin
            (Ik, X) := SPLIT(Ik, p);
            Ik+1 := UNION(X, Ik+1);
         end;
      end;
   return CLASS(n);
end;
```

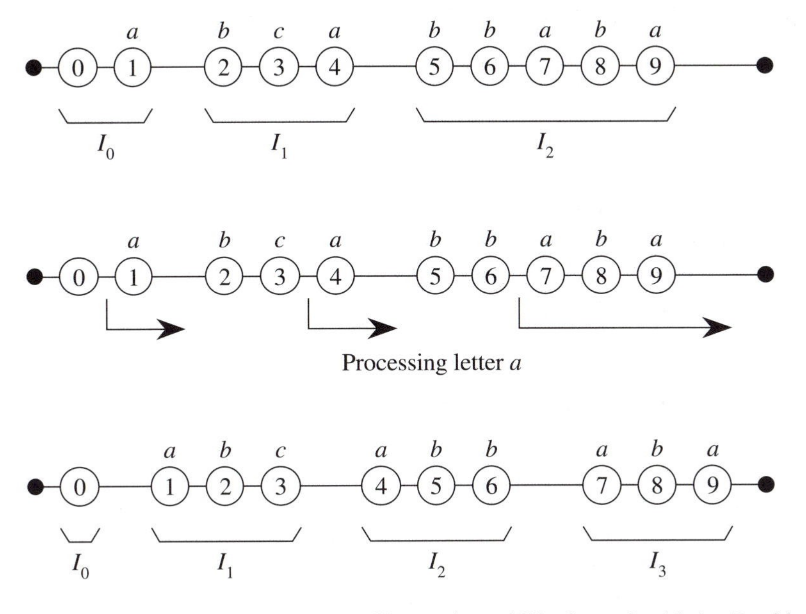

Figure 11.5: Hunt-Szymanski strategy: like an abacus! Word $y = abcabbaba$. Partitions of positions just before processing letter a of $x = cba$ (top), and just after (bottom).

Theorem 11.4 Algorithm *lcs* computes the length of a longest common subsequence of words of length m and n ($m \leq n$) in $O((n+r)\log n)$ time, where $r = |\{(i, j); x[i] = y[j]\}|$.

Proof The correctness of the algorithm is left as an exercise. The time complexity of the algorithm strongly relies on an efficient implementation of intervals I_ks. Using an implementation with B-trees, it can be shown that each operation *CLASS*, *SPLIT*, and *UNION* takes $O(\log n)$ time. Preprocessing lists of occurrences of letters of the word y takes $O(n\log n)$ time. The rest of the algorithm takes $O(r\log n)$ time. The result follows. ♦

According to Theorem 11.4, if r is small, the computation of *lcs* by the last algorithm takes $O(n\log n)$ time, which is faster than with the dynamic programming algorithm. But, r can be of order mn (in the trivial case where $x = a^m$, $y = a^n$, for example), and then the time complexity becomes $O(mn\log n)$, which is larger than with the dynamic programming algorithm.

The problem of computing lcss can be reduced to the computation of the longest increasing subsequence of a given string of elements belonging to a linearly ordered set. Let us write the coordinates of shaded boxes (as in Figure 11.3) from the first to the last row, and from left to right within rows. By doing so, we get a string w. No matrix table is needed to build w. The words x and y themselves suffice. For example, for the words of Figure 11.3 we get the sequence w = ((1,3), (2,2), (2,5), (2,6), (3,1), (3,4), (3,7), (4,2), (4,5), (4,6), (5,1), (5,4), (5,7), (6,3)).

Define the following linear order on pairs of positions

$$(i, j) << (k, l) \text{ iff } ((i = k) \,\&\, (j > l)) \text{ or } ((i < k) \,\&\, (j < l)).$$

Then the longest increasing (according to <<) subsequence of the string w gives the longest common subsequence of the words x and y. There is an elegant algorithm to compute such increasing subsequences running in $O(r\log r)$ time. It presents an alternative to the above algorithm.

11.3 String matching with errors

String matching with errors differs only slightly from the edit distance problem. Here, we are given pattern *pat* and text *text*, and we want to compute min(*edit*(*pat*, y): $y \in Fac(text)$). Simultaneously, we want to find a factor y of *text* realizing the minimum and the position of one of its occurrences in *text*. We consider the table *SE* (that stands for String matching with Errors), of the same type as table *EDIT*:

$$SE[i, j] = \min(edit(pat[1\ldots i], y) : y \in Fac(text[1\ldots j])).$$

The computation of table *SE* can be executed with dynamic programming. It is very similar to the computation of table *EDIT* (see Section 11.1).

Theorem 11.5 The problem of string matching with errors can be solved in $O(mn)$ time.

Proof Surprisingly the algorithm is almost the same as that for computing edit distances. The only difference is that we initialize $SE[0, j]$ to 0, instead of to j for *EDIT*. This is because the empty prefix of *pat* matches an empty factor of *text* (no error). The formula (*) also works for *SE*. Then, $SE[m, n]$ is the distance between *pat* and one of its best matches y in the text. To find an occurrence of y and its position in *text*, we can use the same graph-theoretic approach as we used for the computation of longest common subsequences (see Section 11.2). A suitable path should be recovered using the computed table *SE*. This completes the proof. ♦

One of the most interesting problems related to string matching with errors concerns the case in which the allowed number of errors is bound by a constant k. The number k is usually understood as a small fixed constant. We show that the problem can be solved in $O(n)$ time, or more exactly in $O(kn)$ time, if k is not fixed. For a fixed value of the parameter k, this gives an algorithm having an optimal asymptotic time complexity. Recall that the string edit table *SE* is computed according to the recurrence:

$$(*) \quad SE[i, j] = \min(SE[i-1, j]+1, SE[i, j-1]+1, SE[i-1, j-1]+\partial(x[i], y[j])),$$

for words x and y.

Suppose that we have a fixed bound k on the number of errors. We have to compute only those entries of the table SE that contain values not exceeding k. The basic difficulty, in order to reduce the time complexity of the algorithm, is that the size of the table SE is $O(mn)$, but we require that the complexity is $O(kn)$. Only $O(kn)$ entries of the table must be considered. The basic algorithmic trick is to consider only so-called *d-nodes*, which are entries of the table SE satisfying special conditions. The d-nodes are defined in such a way that altogether we have, for $d = 0, 1, \ldots, k$, $O(kn)$ such nodes.

We consider diagonals of the table SE. Each diagonal is oriented top-down, left-to-right. We define a d-node as the last pair (i, j) on a given diagonal with $SE[i, j] = d$. Note that it is possible that a diagonal has no d-node. The approximate string-matching problem reduces to the computation of d-nodes. And it is clear that there is an occurrence of the pattern with d errors ending at position j in *text* iff (m, j) is a d-node.

Computation of d-nodes is executed in the order $d = 0,1, \ldots, k$. Computation of 0-nodes is equivalent to string matching without errors. Assume that we have already computed the $(d-1)$-nodes. We show how to compute d-nodes. Two auxiliary concepts are necessary: d-special nodes, and maximal subpaths of zero-weight on a given diagonal.

A d-special node is a node reachable from a $(d-1)$-node by a single edge of weight one. The computation of d-special nodes is illustrated in Figure 11.6.

For a node (i, j), define the node $NEXT(i, j) = (i+t, j+t)$ as a lowest node on the same diagonal as (i, j) reachable from (i, j) by a subpath of zero weight. The subpath can be of zero length, and, in this case, $NEXT(i, j) = (i, j)$.

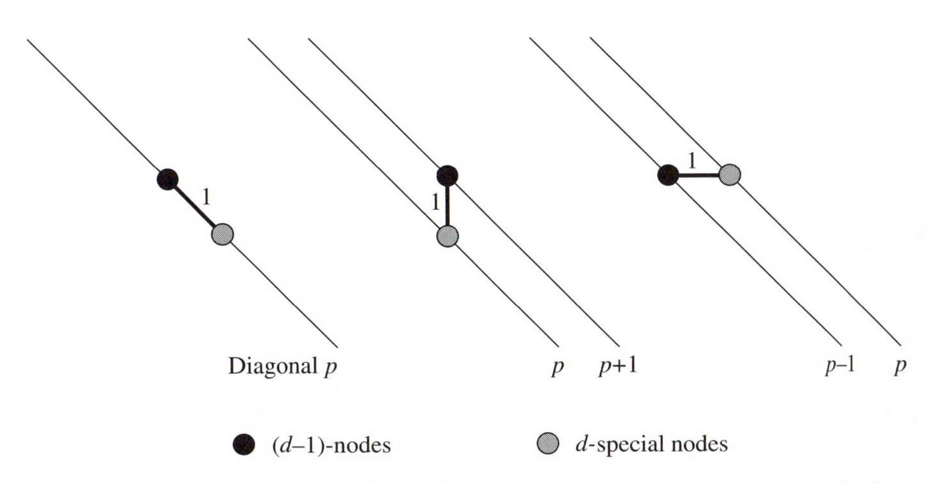

Figure 11.6: The computation of d-special nodes: nodes reachable by one edge of weight 1 from a $(d-1)$-node.

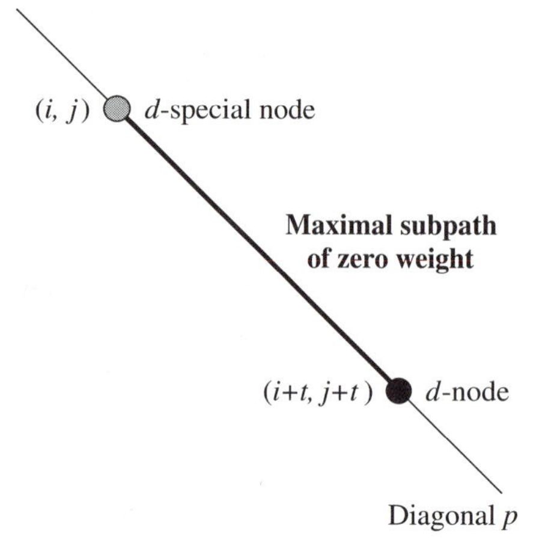

Figure 11.7: The computation of d-nodes from d-special nodes.

Once d-special nodes are computed, the d-nodes can be easily found, as suggested in Figure 11.7. The structure of the whole algorithm is given below.

```
Algorithm Approximate_string_matching with at most k errors;
begin
    compute 0-nodes { exact string-matching }
    for d := 1 to k do begin
        compute d-special nodes; { see Figure 11.6 }
        { computation of d-nodes }
        S := {NEXT(i,j) : (i,j) is a d-special node};
        for each diagonal p do
            select, on diagonal p, lowest node which is in the set S;
        selected nodes form the set of d-nodes
    end;
end.
```

Theorem 11.6 Assume that the alphabet is of a constant size. Approximate string matching with k errors can be achieved in $O(kn)$ time.

Proof It is sufficient to prove that we can run $O(kn)$ calls of the function *NEXT* in $O(kn)$ time. We show that, after a linear-time preprocessing, each value $NEXT(i, j)$ can be computed in constant time.

The equation $NEXT(i, j) = (i+t, j+t)$ means that t is the size of the longest common prefix of $pat[i...m]$ and $text[j...n]$. Assume that we have computed the common suffix tree for *pat* and *text*. The computation of the longest common prefix of two suffixes is equivalent to the computation of their lowest common ancestor (LCA) in the tree. There is an algorithm (mentioned in Section 5.7) that preprocesses any tree in linear time in order to allow further LCA queries in constant time. This completes the proof. ♦

Remark It can be shown that the same algorithm, implemented in the PRAM model, yields an efficient parallel algorithm for the approximate string-matching problem.

11.4 String matching with don't care symbols

In this section, we assume that pattern *pat* and text *text* can contain occurrences of the symbol ∅, called the *don't care symbol*. Several different don't care symbols can be considered instead of only one, but the assumption is that they are all undistinguishable from the point of view of string matching. These symbols match any other symbol of the alphabet. We define an associated notion of *matching* on words as follows. We say that two symbols a, b, *match* if they are equal, or if one of them is a don't care symbol (see Figure 11.8). We write $a \approx b$ in this case. We say that two strings u and v (of same length) match if $u[i] \approx v[i]$ for any position i. String matching with don't care symbols entails the problem of finding a factor of *text* that matches the pattern *pat* according to the present relation ≈.

The string matching with don't care symbols does not use any of the techniques developed for other string-matching questions. This is because the relation ≈ is not transitive ($a \approx \varnothing$ and $\varnothing \approx b$ does not imply $a \approx b$). Moreover, if symbol comparisons (involving only the relation ≈) are the only access to input texts, then there is a quadratic lower bound for the problem, which additionally proves that the problem is quite different from other string-matching problems. The algorithm presented later is an interesting example of a reduction of a textual problem to a problem in arithmetics.

Theorem 11.7 If symbol comparisons are the only access to input texts, then $O(n^2)$ such comparisons are necessary to solve the string matching with don't care symbols.

Proof Consider a pattern of length m, and a text of length $n = 2m$, both consisting entirely of don't care symbols ∅. Occurrences of the pattern start at all positions $0 \ldots m$. If the comparison "$pat[j] \approx text[i]$," for $1 \leq j \leq m$ and $m < i \leq n$, is not done,

$a\ a\ a\ \varnothing\ b\ a\ b\ b\ \varnothing\ \varnothing\ a\ a\ b\ \varnothing$
$a\ \varnothing\ a\ b\ b\ \varnothing\ \varnothing\ b\ b\ a\ a\ a\ b\ a$

Figure 11.8: Matching words with the *don't care symbol* ∅.

then we can replace *pat*[*j*] and *text*[*i*] by two distinct symbols a, b (that are not don't care symbols). The output then remains unchanged, but one of the occurrences is disqualified. Hence, the algorithm is not correct. This proves that all comparisons "$pat[j] \approx text[i]$," for $1 \le j \le m$ and $m < i \le n$, must be executed. This gives $m(m+1)/2 = O(n^2)$ comparisons. This completes the proof. ♦

Contrary to what occurs elsewhere, we temporarily assume that positions in *pat* and *text* are numbered from zero (and not from one). We start with an algorithm that "multiplies" two words in a manner similar to how two binary numbers are multiplied, but ignoring the carry. We define the product operation • as the composition of ≈ and the logical "and" in the following sense. If x, y are two strings, then $z = x \bullet y$ is defined by

$$z[k] = \mathrm{AND}(x[i] \approx y[j]: i+j = k).$$

In other words, it is the logical "and" of all values $x[i] \approx y[j]$ taken over all i, j such that both $i+j = k$ and $x[i]$, $y[j]$ are defined. We can write symbolically $\bullet = (\approx, \text{and})$.

Let p be the reverse of pattern *pat*, and consider $z = p \bullet text$. Let us examine the value of $z[k]$. We have $z[k]$ = true iff ($p[m-1] \approx text[k-m+1]$ and $p[m-2] \approx text[k-m+2]$ and ... and $p[0] \approx text[k]$). Therefore, "$z[k]$ = true" exactly means that there is an occurrence of *pat* ending at position k in *text*. Hence, the string matching with don't care symbols reduces to the computation of products •.

Let us define an operation on logical vectors similar to •. If x, y are two logical vectors, then $z = x \lozenge y$ is defined by

$$z[k] = \mathrm{OR}(x[i] \text{ and } y[j]: i+j = k).$$

For a word x and a symbol a, denote by *logical*(a, x) the logical vector in which the i-th component is *true* iff $x[i] = a$. Define also

$$LOGICAL_{a,\, b}(x, y) = logical(a, x) \lozenge logical(b, y).$$

The following fact is now apparent:

> For two words x, y, the vector $x \bullet y$ equals the negation of logical OR of all vectors $LOGICAL_{a,\, b}(x, y)$ over all distinct symbols a, b that are not don't care symbols.

A consequence of the above fact is that, for a fixed-size alphabet, the complexity of evaluating the product • is of the same order as that of computing the operation ◊.

Now, we show that the computation of the operation ◊ can be reduced to the computation of the ordinary product * of two integers. Let x, y be two logical vectors of size n. Let $k = \log n$. Replace logical values *true* and *false* by ones and zeros, respectively. Next, insert an additional group of k zeros between each two consecutive digits. We obtain two numbers x', y' (in binary representation). Let z' be the integer, the product of the numbers, $z' = x * y$. The vector $z = x \lozenge y$ can be recovered from z' as follows. Take the first digit (starting at position 0), and then each $(k+1)$-th digit of z'; convert ones and zeros into *true* and *false*, respectively. In this way, we have proven the following statement.

Theorem 11.8 The string-matching problem with don't care symbols can be solved in $IM(n\log n)$ time, where $IM(r)$ denotes the complexity of multiplying two integers of size r.

The value $IM(r)$ depends heavily on the model of computations considered. If bit operations are counted, then the best known solution for the problem is given by the Schönhage-Strassen multiplication, which works in time only slightly larger than $O(r\log r)$. Even with the serial model of computation considered throughout the book, uniform RAM model (with logarithmic-sized integers), then, the best known complexity is at least $O(r\log r)$. In fact, no linear-time algorithm for the problem is known. This gives an $O(n\log^2 n)$-time algorithm for the string-matching problem with don't care symbols.

String matching with don't care symbols generates a methodological interest because of its relationship to arithmetics. It should be also interesting to find relationships between some other typical textual problems to arithmetics.

11.5* Edit distance—efficient parallel computation

The parallel computation of the edit distance can be ruled out as a shortest path problem for the corresponding grid graph G (see Section 11.1). For simplicity assume that words x and y are of the same length n. Then, G has $(n+1)^2$ nodes. Let A be the matrix of weights associated to edges of G. We can use (min, +) matrix multiplications to obtain the required value of the edit distance. Assume that the weight from the sink node to itself is zero. Then,

$$edit(x, y) = A^n[0, n].$$

The matrix A^n can be computed using successive squarings (or an adaptation of it, if n is not a power of 2):

repeat $\log n$ **times** $A := A^2$.

Obviously, k^3 processors are sufficient multiplying two $k \times k$ matrices in $O(\log k)$ time on a CREW PRAM. In our case $k = (n+1)^2$, so, this proves that n^6 processors suffice to compute the edit distance. The time of computation is $O(\log^2 n)$. However, there is a more efficient algorithm, due to the special structure of the grid graph G. The grid graph can be decomposed into four grid graphs of the same type (see Figure 11.9). The partition of G leads to a kind of parallel divide-and-conquer computation.

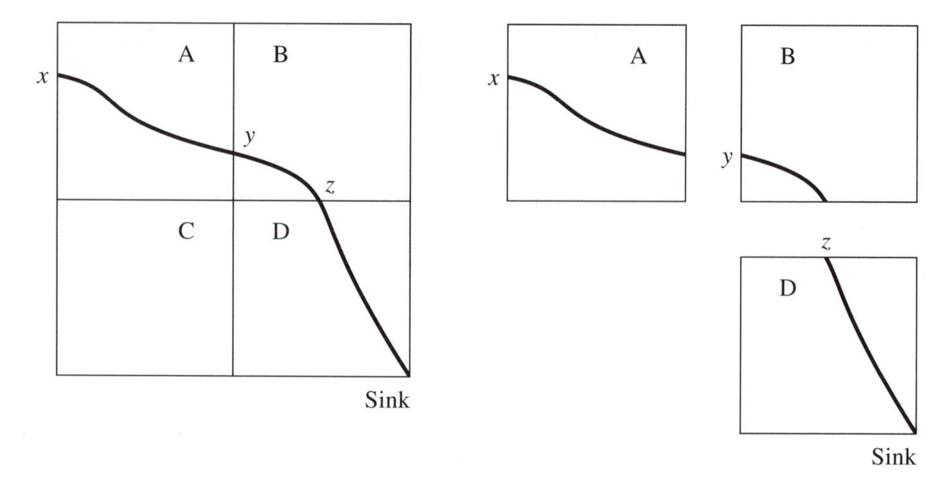

Figure 11.9: Decomposition of the shortest path problem into subproblems.

Theorem 11.9 The edit distance of two words of length n can be computed in $\log^3 n$ time with n^2 processors.

Proof The edit distance problem reduces to the computation of a shortest path from the source (left upper corner) to the sink (right lower corner) on the weighted grid graph G. We compute (recursively) the matrix A of all costs of paths, from positions x on the left or top boundary to any position y on the right or bottom boundary of G.

Let us partition the grid into four identical subgrids, as in Figure 11.9. If we know the matrices of costs between boundary points for all subgrids, then all costs of paths between boundary positions of the whole grid can be computed with $O(M(n))$ processors in $O(\log^2 n)$ time, using a constant number of matrix multiplications. Here $M(n)$ denotes the number of processors needed to multiply, in $\log^2 n$ time, two matrices satisfying the quadrangle inequality. In Chapter 10, we have seen that $M(n) = O(n^2)$. For $n/2 \times n/2$ subgrids we need $M(n/2)$ processors at one level of recursion. Altogether $M(n)$ processors suffice, since we have the inequality $4M(n/2) \leq M(n)$ (because $M(n) = c\, n^2$). This completes the proof of the theorem. ♦

Bibliographic notes

The edit distance computation can be attributed to Needleman and Wunsch [NW 70] and to Wagner and Fischer [WF 74]. Various applications of sequence comparisons are presented in a book edited by Sankoff and Kruskal [SK 83]. Algorithms of Sections 11.1 and 11.2 are widely used for molecular sequence comparisons, for which a large number of variants have been developed (see, for example [GG 89]). Computation of a longest common subsequence (not only its length) in linear space is from

Hirschberg [Hi 75]. The last algorithm of Section 11.2 is from Hunt and Szymanski [HS 77]. It is the base of the "diff" command of UNIX system. An algorithm for the longest increasing subsequence can be found in [Ma 89].

There have been numerous substantial contributions to the problem, among them are those by Hirschberg [Hi 77], Nakatsu, Kambayashi, and Yajima [NKY 82], Hsu and Du [HD 84], Ukkonen [Uk 85b], Apostolico [Ap 86] and [Ap 87], Myers [My 86], Apostolico and Guerra [AG 87], Landau and Vishkin [LV 89], Apostolico, Browne, and Guerra [ABG 92], Ukkonen and Wood [UW 93].

A subquadratic solution (in $O(n^2/\log n)$ time) to the computation of edit distances has been given by Masek and Paterson [MP 80] for fixed-size alphabets.

The first efficient string matching with errors is by Landau and Vishkin [LV 86b]. A parallel algorithm is given in [GG 87].

The computation of lowest common ancestors (LCA) is discussed by Schieber and Vishkin in [SV 88] (see the bibliographic notes at the end of Chapter 5).

The best asymptotic time complexity of the string matching with don't care symbols is achieved by the algorithm of Fischer and Paterson [FP 74].

Practical approximate string matching is discussed by Baeza-Yates and Gonnet [BG 92], and by Wu and Manber [WM 92]. The solutions are close to each others. the second algorithm has been implemented under UNIX as command "agrep".

The parallel algorithm computing the edit distance (see Section 11.5) is by Apostolico, Atallah, Larmore, and McFaddin [AALF 88]. It is also observed independently in [Ry 88] that the reduction of the edit distance problem to the shortest path problem for grid graphs leads to the use of parallel matrix multiplication. For parallel approximate string matching the reader can also refer to [LV 89].

Selected references

[GG 88] Galil, Z., & Giancarlo, R., "Data structures and algorithms for approximate string matching," *J. Complexity* 4 (1988): 33–72.

[LV 89] Landau, G.M., & Vishkin, U., "Fast parallel and serial approximate string matching," *J. Algorithms* 10 (1989): 158–169.

[SK 83] Sankoff, D., & Kruskal, J.B., *Time Warps, String Edits and Macromolecules: The Theory and Practice of Sequence Comparison*, Addison-Wesley, Reading, Mass., 1983.

[Uk 85b] Ukkonen, E., "Algorithms for approximate string matching," *Information and Control* 64 (1985): 100–118.

12

Two-dimensional pattern matching

The chapter is devoted to pattern matching in two-dimensional structures. These objects can be considered as images represented by matrices of pixels (*bit-map* images). Questions related to pattern matching in strings extend naturally to similar questions on images. However, not all algorithms allow simple extensions in this sense. The two-dimensional pattern matching is interesting due to its relationship to image processing. The efficiency of algorithms is even more important in the two-dimensional case because the size of the problem, the number of pixels of images, is very large in practical situations.

We mainly consider rectangular images. The pattern-matching problem is to locate an $m \times m'$ pattern array PAT inside an $n \times n'$ (text) array T. The position of an occurrence of PAT in T is a pair (i, j), such that $PAT = T[i+1 \ldots i+m, j+1 \ldots j+m']$ (see Figure 12.1).

We present two different solutions to two-dimensional pattern matching. The first reduces the problem to multi-pattern matching. The second is based on two-dimensional periodicities and the notion of duels. We then consider non-rectangular patterns in relation to approximate matching. The method of sampling is presented for two-dimensional patterns and appears to be very powerful for almost all patterns. In Section 12.8, we apply the naming technique to several problems concerning regularities in images. Finally, the last three sections prove that the problem admits a linear-time solution, independent of the size of the alphabet.

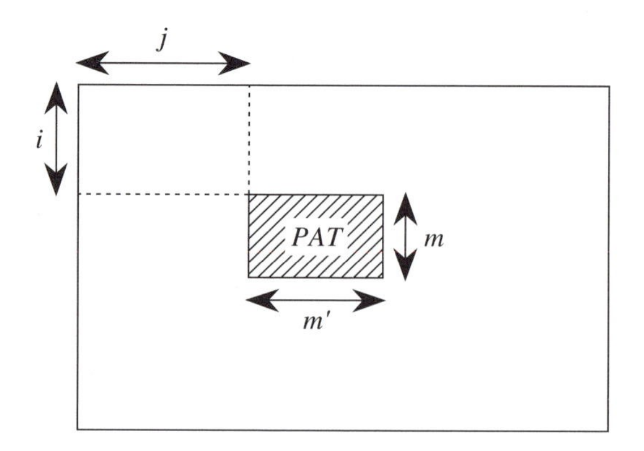

Figure 12.1: *PAT* occurs at position (i, j) in T.

12.1 Multi-pattern approach: Baker and Bird algorithms

The first solution to 2–D pattern matching is to translate the problem into a string-matching problem. The pattern is viewed as a set of strings, its columns. To locate columns of the pattern within columns of the text array requires searching for several patterns (see Figure 12.2). Moreover, the occurrences of patterns must be found in a particular configuration within rows: all columns of the pattern are to be found in the order specified by the pattern, and all ending on a same row of the text array. In this section, we use the Aho-Corasick approach (see Section 7.1) to solve the multi-pattern matching problem.

The strategy for searching for *PAT* in the text array T is as follows. Let Π be the set of all (distinct) columns of *PAT* (treated as words). We first build the string-matching automaton $G = SMA(\Pi)$ with terminal states (see Section 7.1). Each terminal state corresponds to a pattern of Π. Therefore, columns of the pattern are identified with states of the SMA automaton. There can be less than m terminal states because of possible equalities between columns. Then, the automaton is applied to each column of T. We generate an array T' of the same size as T, and in which the entries are states of G. The pattern *PAT* itself is replaced by a string *pat* over the set of states: the i-th symbol of *pat* is the state identified with the i-th column of *PAT*. The remainder of the procedure consists in locating *pat* inside the lines of T'. The strategy is illustrated by Figures 12.3, 12.4 and 12.5. This yields the subsequent result.

Theorem 12.1 Two-dimensional pattern matching can be done in $O(N\log|A|)$ time, where $N = n*n'$ is the size of the text array, and A is the alphabet.

Proof The time to build the automaton $SMA(\Pi)$ is $O(M\log|A|)$ where $M = m*m'$ is the size of the pattern *PAT*. The construction of the array of column numbers T' takes

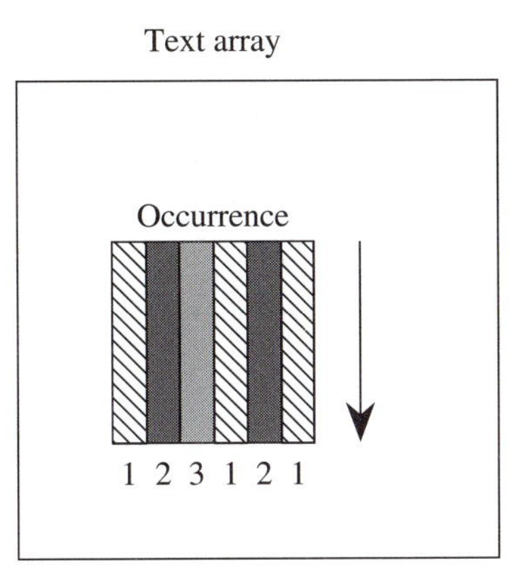

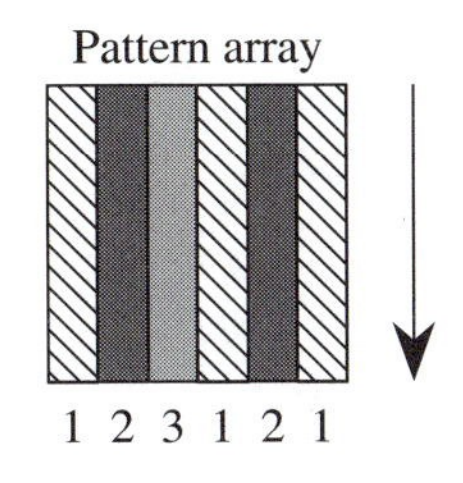

Figure 12.2: Two-dimensional pattern matching searching for columns of the pattern.

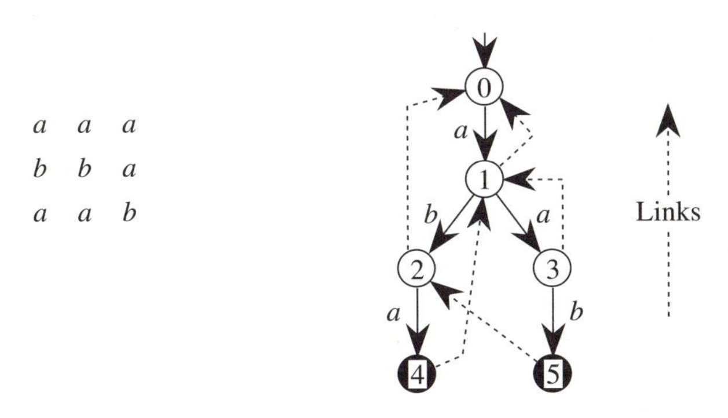

Figure 12.3: A pattern array, and the SMA automaton of its columns. Columns 1 and 2 are identified with state 4, column 3 with state 5.

a	*b*	*a*	*b*	*a*	*b*	*b*
a	*a*	*a*	*a*	*b*	*b*	*b*
b	*b*	*b*	*a*	*a*	*a*	*b*
a	*a*	*a*	*b*	*b*	*a*	*a*
b	*b*	*a*	*a*	*a*	*b*	*b*
a	*a*	*b*	*a*	*a*	*a*	*a*

1	0	1	0	1	0	0
3	1	3	1	2	0	0
5	2	5	3	4	1	0
4	4	4	5	2	3	1
2	2	3	4	4	5	2
4	4	5	3	3	4	4

Figure 12.4: A text array, and its associated array of states (according to the SMA automaton of Figure 12.3).

1	0	1	0	1	0	0
3	1	3	1	2	0	0
5	2	5	3	4	1	0
4	4	4	5	2	3	1
2	2	3	4	4	5	2
4	4	5	3	3	4	4

a	*b*	*a*	*b*	*a*	*b*	*b*
a	*a*	*a*	*a*	*b*	*b*	*b*
b	*b*	*b*	*a*	*a*	*a*	*b*
a	*a*	*a*	*b*	*b*	*a*	*a*
b	*b*	*a*	*a*	*a*	*b*	*b*
a	*a*	*b*	*a*	*a*	*a*	*a*

Figure 12.5: Occurrences of "445" in the array of states give occurrences of the pattern array in the original text array of Figure 12.4. Pattern "445" corresponds to the pattern array of Figure 12.3 (left).

$O(N\log|A|)$ time. The final search phase, string matching inside lines of T' takes $O(N)$ time. Thus, the result holds assuming that $M \leq N$. ♦

The above algorithm seems to be inherently dependent upon the alphabet. This is because of the automaton approach. The problem of the existence of a linear-time algorithm with a time complexity independent of the size of the alphabet (for two-dimensional pattern matching) is interesting, and important from a practical point of view. We show a partial solution in the next section (alphabet-independent searching phase, but alphabet-dependent preprocessing phase).

12.2 Periodicity approach: Amir-Benson-Farach algorithm

The algorithm of the present section is based on the idea of *duels*. The string-matching algorithm by duels presented in Chapter 3 for "one-dimensional" strings extends to the two-dimensional case. The advantage of this approach is to produce a two-dimensional pattern-matching algorithm in which the search phase takes linear time, independently of the alphabet. The two-dimensional settings for duels, witnesses, and consistency relation are necessary for adapting the string-matching algorithm by duels.

A period of the $m \times m'$ pattern array PAT is a non-null vector $p = (r, s)$ such that $-m < r < m$, $0 \leq s < m'$, and $PAT[i, j] = PAT[r+i, s+j]$ whenever both sides of the equation are defined. Note that the second component of a period is assumed to be a non-negative integer, which leaves us to consider that vector periods are always oriented from left to right. There are two categories of periods, see Figures 12.7 and 12.8, according to whether r is negative or not.

If there are close occurrences of the pattern in the text array, then there is an overlap of the pattern over itself, that is, a periodicity. If x and y are close positions of two occurrences PAT in the array T, assuming that y is to the right of x, the vector y-x is a period of the pattern.

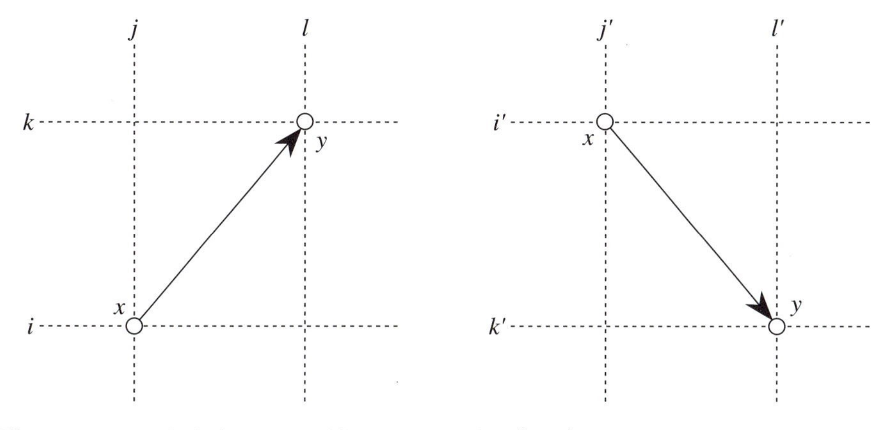

Figure 12.6: Orderings on positions $x <_b y$ (left), $x' <_t y'$ (right).

Positions in arrays are numbered top-down (rows) and left-to-right (columns). We define two partial orderings $<_b$ (for bottom) and $<_t$ (for top) on positions in the array T:

$(i, j) \leq_b (k, l)$ iff $i \geq k$ and $j \leq l$,
$(i, j) \leq_t (k, l)$ iff $i \leq k$ and $j \leq l$.

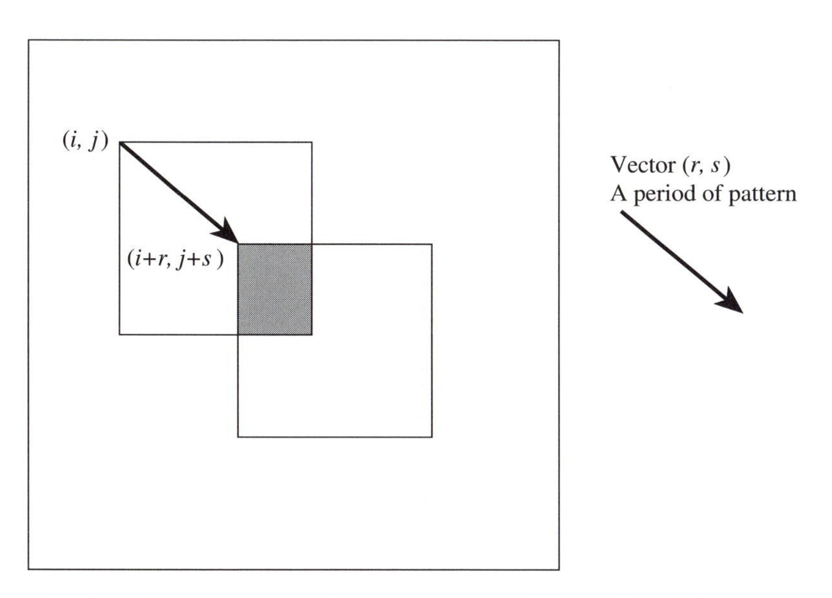

Figure 12.7: When $(i, j) <_t (k, l)$. If two occurrences of PAT overlap, PAT has a period $(r, s) = (k, l)-(i, j)$. Otherwise, a duel between (i, j) and (k, l) can be applied.

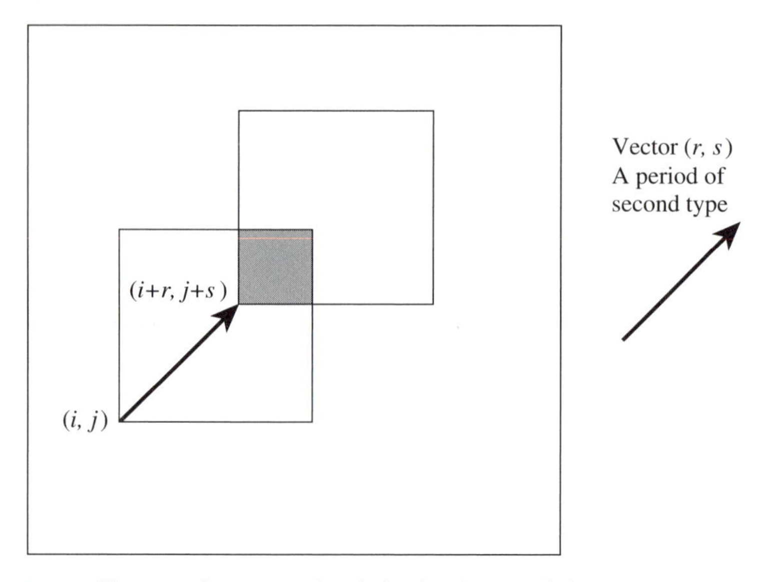

Figure 12.8: The second category of period, when $(i, j) <_b (k, l)$.

The relation $x <_b y$ means that position x is to the left and at the *bottom* of y. The relation $x <_t y$ means that position x is to the left and at the *top* of y. For example, we have $(i, j) \leq_b (k, l)$ and $(i', j') \leq_t (k', l')$ in Figure 12.6.

Making duels during the searching phase of the pattern-matching algorithm supposes that we have an analogue to the *witness table* considered for strings. For arrays, the two-dimensional witness table *WIT* is defined as follows:

$WIT[r, s]$ = *any position* (p, q) such that $PAT[p, q] \neq PAT[r+p, s+q]$,
$WIT[r, s] = 0$, if there is no such (p, q).

The definition is illustrated in Figures 12.9 and 12.10 for the two categories of vector (r, s) (depending on whether $r \geq 0$ holds or not).

A duel is only performed on close positions according to the following notion. A position (k, l) is said to be *in the range of* the position (i, j) (according to the size of *PAT*) iff $|k-i| < m$ and $|l-j| < m'$. In addition, two positions x and y such that y is to the right of x, are said to be *consistent* iff y is not in the range of x, or if $WIT[y-x] = 0$, which means that $y-x$ is a period of *PAT*.

Let us recall the notion of *duel*. If the positions x and y are not consistent, the pattern *PAT* cannot appear both at x and at y in the array T. In constant time, we can remove one of them as a candidate for a position of an occurrence of the pattern. Such an operation, called a *duel*, can be described as follows. Assume that positions x and y are not consistent, with y to the right of x. Let $z = WIT[y-x]$. Let $a = PAT[z]$, and $b = PAT[y-x+z]$. By definition of the witness table *WIT*, symbols a and b are distinct. Let c be the symbol $T[y+z]$. This symbol cannot be both equal to a and to b, so

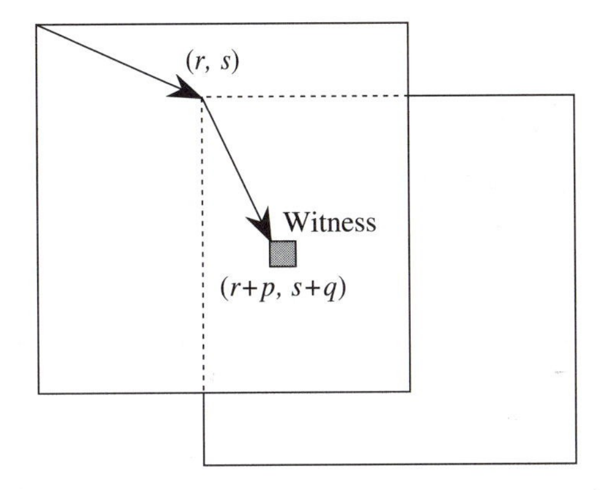

Figure 12.9: The first category of witnesses. The vector (r, s) is not a period, $(p, q) = WIT[r, s]$, and $PAT[p, q] \neq PAT[r+p, s+q]$.

at least one of the positions x, y is not a matching position for *PAT*. If $b \neq c$, the pattern cannot occur at position x. If $a \neq c$, the same holds for y. Therefore, comparing c with a and b permits us to eliminate (at least) one of the positions. Note that in some situations both positions could be eliminated, however, for simplicity of the algorithm, only one position is always removed at a time. This is a mere duplication of the strategy developed for "one-dimensional" string matching. Let *duel* be defined by

$$duel(x, y) = (\text{if } b = c \text{ then } x \text{ else } y).$$

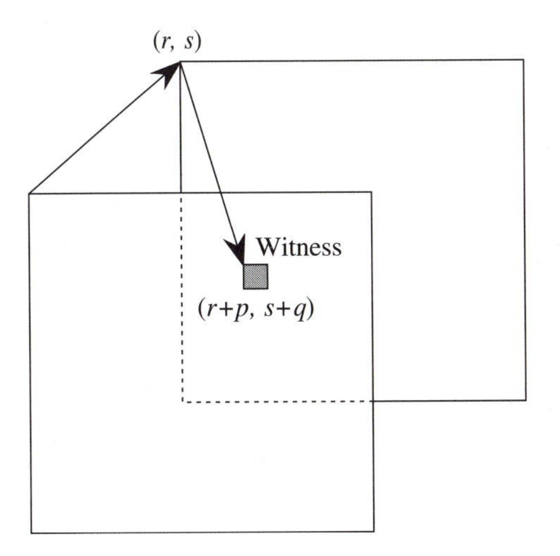

Figure 12.10: The second category of witnesses. The vector (r, s) is not a period, $(p, q) = WIT[r, s]$, and $PAT[p, q] \neq PAT[r+p, s+q]$.

The value *duel*(x, y) is the position that "survives" after the duel, the other position is eliminated.

We now describe the two-dimensional pattern-matching algorithm based on duels. We assume the witness table of the pattern *PAT* is computed. Its precomputation is sketched at the end of this section. The first step of the searching phase reduces the problem to a two-dimensional pattern matching for unary patterns, as if all entries of *PAT* were the unique symbol a.

We want to eliminate a set of candidate positions from the text array T in such a way that all remaining positions are pairwise consistent. Removed positions cannot be matching positions of the pattern. Then, with each position x on the text array we associate the value 1 iff, after duels, it corresponds to the symbol compatible with occurrences of the pattern placed at any position in the range of x. Otherwise, we associate 0 with position x. By doing so, we are left with a new text array consisting only of zeros and ones. Finally, we look for occurrences of an $m \times m'$ array containing only 1s. Therefore, the algorithm is essentially the same as in the one-dimensional case. But here, the relationship between positions is a bit more complicated. This is why relations $<_b$ and $<_t$ have been introduced.

The following property of consistent positions is crucial for the correctness of the algorithm.

Consistency property (transitivity): Let $x <_t y <_t z$, or $x <_b y <_b z$. If x, y are consistent, and y, z are consistent, then x, z are also consistent.

According to relations $<_b$ and $<_t$, consistency refines to *bottom consistency* and *top consistency*. A set of positions is *bottom consistent* iff for any two positions x, y of the set, such that $x <_b y$, the positions are consistent. Top consistent positions are defined similarly. It is clear that two elements are consistent iff they are top *and* bottom consistent. The same refers to sets of (pairwise) consistent positions.

Let R be a subrectangle of the text array T. The set S of positions in R is said to be *good* with respect to R if both positions in S are pairwise consistent, and there is no matching position within R–S.

Let k be a column of the text array T. In the searching algorithm, we maintain the following invariant. A good set of consistent positions in the columns k, k+1, ..., n' is known. First, we construct good sets of consistent positions separately in each column. This gives the invariant for $k = n'$. Then we satisfy the invariant for $k = n'-1$, $n'-2$, ..., 1. At completion we have a good set of consistent positions for the entire text array.

When processing the k-th column, we run through consistent positions of this column in a top-down fashion. We maintain the following invariant:

***inv*(x, z, k):** z is the leftmost consistent position in its own row; let R be the rectangle composed of rows above z, and columns k, ..., n'; the set S of all remaining positions in R is a good set in R.

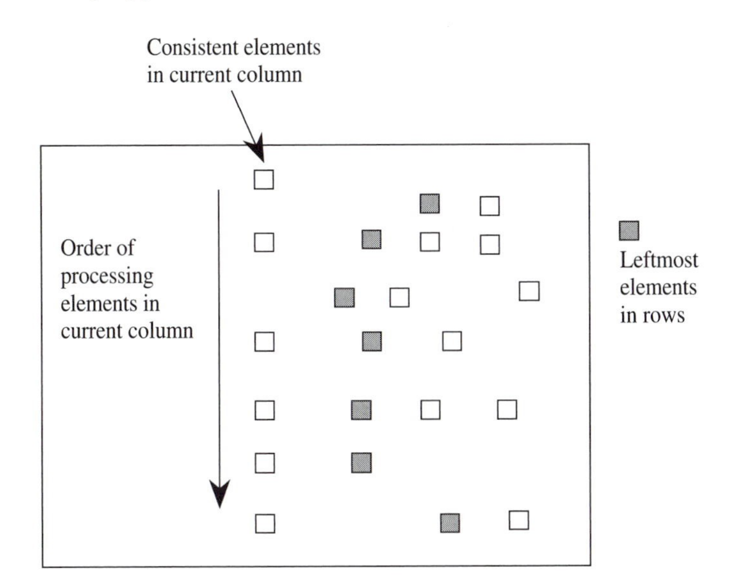

Figure 12.11: The situation when processing the next column. The current column contains positions mutually consistent within this column. Then, positions inconsistent with other columns are removed.

Let $x1$, $x2$ be two consecutive (in the top-down order), consistent positions in the k-th column. Figure 12.12 illustrates how the process goes from $inv(x1, z1, k)$ to $inv(x2, z2, k)$.

The set of consistent positions in columns $k+1, \ldots, n'$ is maintained as a set of stacks. These stacks correspond to rows. The positions in a given row, from left to right, are on their stack from top to bottom: the left most position in i-th row (in columns $k+1, \ldots, n'$) is at the top of the i-th stack. The duels of $x1$ against elements of the i-th row are executed using the same stack procedure as in the string-matching algorithm by duels (see Chapter 3).

Assume that $z1$ is in the $i1$-th row, and $x1$ is the $j1$-th row. Only positions in columns $k+1, \ldots, n'$ are considered (the duels are executed between them and $x1$). We process rows in the order $i1, i1+1, \ldots$; each row is processed from left to right, starting with the left most element. During the processing, we execute duels between $x1$ and considered positions.

Assume that we start a given phase with position $x1$ in the k-th column, and with position $z1$ in the $i1$ row (see Figure 12.12). The rows are processed top-down and left-to-right, starting with $z1$, and ending before or at the row containing $x1$. Initially $z = z1$. Then, assume that we consider a candidate z in a column to the right of $x1$. The basic operation is the duel between $x1$ and z. Three cases are possible:

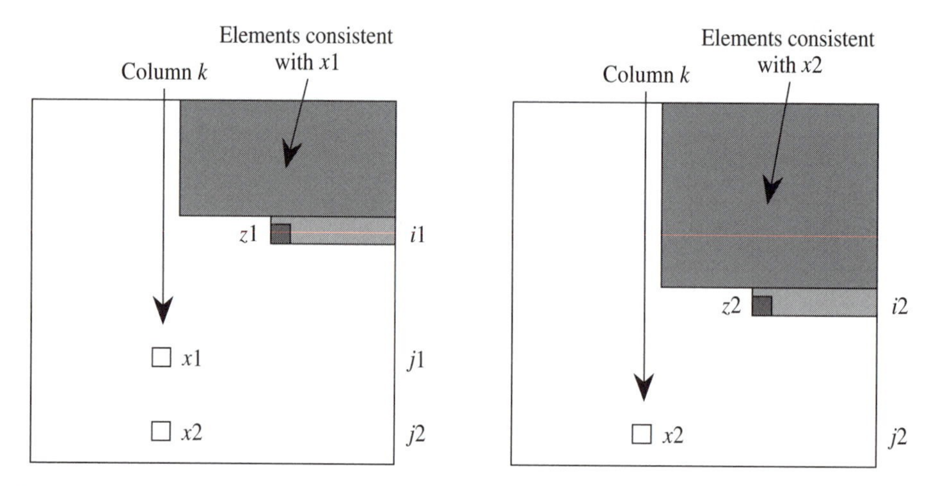

Figure 12.12: From $inv(x1, z1, k)$ to $inv(x2, z2, k)$.

(1) both $x1$ and z survive (they are consistent); the *crucial point* is that we know at this moment (due to transitivity of consistency) that all candidates to the right of z and in the same row as z are consistent with $x1$; we do not need to process them; we just go to the next row, starting with the left most candidate z (to the right of k-th column) in this row;

(2) z is "killed" by $x1$ in the duel; we process the next candidate z in the same row as last z; if there is no such candidate we simply go on to the next row;

(3) $x1$ is "killed" and z survives; then, the processing of $x1$ has been completed; $x1$ is removed as a candidate, $z1 = z$, and we start the next phase with the next candidate $x2$ below $x1$ in the same column as $x1$; if there is no such candidate, then the processing of the entire column has been completed; take $x1$ as the top most candidate in the $(k–1)$ column, and start processing column $k–1$.

Through this process, we obtain a set of consistent positions in the sense of the ordering $<_t$. If x, y are in this order, and are in the set, then they are consistent. After that, we again process the whole text array, but in a bottom-up manner, essentially performing the same algorithm as described above for the top-down ordering. The rows are again processed from left to right. The remaining set of positions is guaranteed to be bottom consistent. Thus, the final set S is a good consistent set.

The problem is reduced to "unary" pattern matching, in which the pattern consists only of one symbol, as follows. For each position x in the text array, find any position y in S such that x is in the range of y. Place the pattern at position y on the text array, and check if the symbol at x matches the corresponding symbol of the pattern. If "yes," associate "1" with position x. If "no," or if there is no such position y, associate "0" with x. In this way, we obtain a new array of zeros and ones. What remains is to search for a rectangular shape of size $m \times m'$ containing only 1s inside the new array. This is straightforward, and is left to the reader. The above discussion gives a proof of the following statement.

Theorem 12.2 If the witness table for the pattern array is computed, the search phase for the two-dimensional pattern matching can be done in linear time, independent of the size of the alphabet.

The computation of the witness table given as follows employs a suffix tree. It takes a time that depends on the size of the alphabet, though it is linear with respect to the length of the pattern. This is due to the construction of suffix trees. In the computation of the witness table, the basic operation consists of checking the equality of two subrows of the pattern. This is executed on the suffix tree for the set of rows of the pattern, and by preprocessing the tree for *LCA* queries.

Theorem 12.3 The witness table for an $m \times m'$ two-dimensional pattern can be computed in $O(mm' \log|A|)$ time, where A is the alphabet.

Proof Let us examine the situation in which the witness for the position (r, s) of *PAT* is to be computed. Let $k = m-s+1$. Assume that elements of the first and of the s-th columns are names of the rows of size k starting at position of these columns to the right. Denote the resulting columns by $C_1^{(k)}$ and $C_s^{(k)}$ (see Figure 12.13).

Let *ST* be the suffix tree of all rows of the pattern. It takes $O(mm' \log(|A|))$ time to build this tree. Preprocess the tree in order to answer *LCA* queries in constant time (see Section 5.7). Also consider the table *PREF* as defined in Chapter 3.

Claim The table *PREF* of the column $C_s^{(k)}$ with respect to $C_1^{(k)}$ can be computed in $O(m)$ time, once the tree *ST* is given.

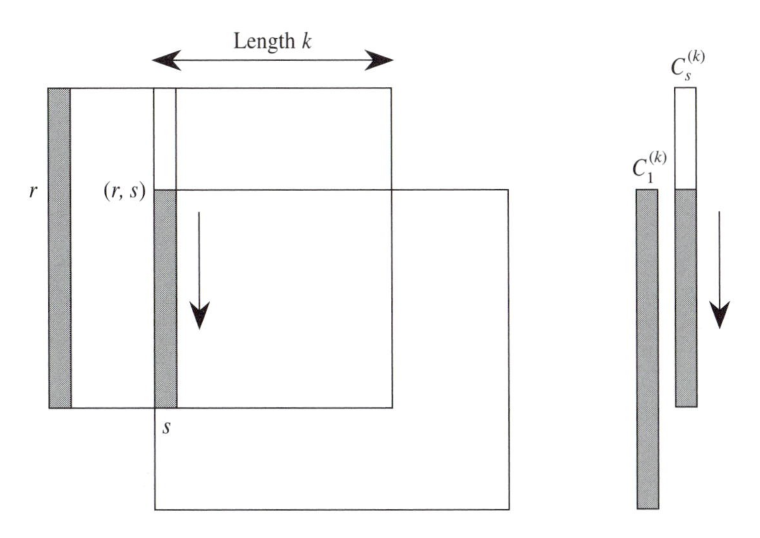

Figure 12.13

Proof (of the claim) The computation of this table essentially reduces to the computation of the table of border lengths (see Chapter 3). We don't need actually the names of entries of the columns $C_s^{(k)}$ and $C_1^{(k)}$. These names represent subrows of length k. It is sufficient to make comparisons in constant time, hence, it is also sufficient to be able to quickly check the equality between two subrows of the same size k. This can be executed using *LCA* queries about the rows of the pattern array. First assume that (r, s) is a period of the pattern. Then $PREF[r] = m\text{-}r$. Otherwise, $PREF[r]$ gives the index of the row where the witness position is. To find the witness position it is sufficient to find the longest common prefix of two subrows of length k. This can again be done using *ST* and *LCA* queries. This completes the proof. ♦

The end of the chapter presents a linear-time computation of witness table.

12.3 Matching with don't care symbols and non-rectangular patterns

Assume that the two-dimensional pattern contains a certain number of holes. Holes can be regarded as filled with a special symbol that matches any other symbol. It is the don't care symbol ∅ considered in Chapter 11 for approximate string matching. If the pattern is not rectangular, we can also complete it, adding enough don't care symbols so that it fits into an $m \times m'$ rectangle. By doing so, both questions become similar.

Theorem 12.4 Two-dimensional pattern matching with don't care symbols, and pattern matching of non-rectangular patterns can be done in $O(N\log^2 m)$ time (with an $m \times m'$ pattern, $m \geq m'$, and an $n \times n'$ text array, $N = nn'$).

Proof We linearize the problem. Let *PAT* be a non-rectangular pattern that fits into an $m \times m'$ rectangle, with $m \geq m'$. We consider windows of shape $2m \times 2m'$ on the text array (see Figure 12.14). We first solve the problem as if $n = 2m$. We define *Lin*(*PAT*) as a one-dimensional version of *PAT*. It is a string with don't care symbols of $O(m')$ size constructed as follows:

- place *PAT* inside a $2m \times 2m'$ shape S. All positions not occupied by *PAT* are filled with the don't care symbol ∅; then, concatenate the rows of S, starting from the top most row; within the string obtained in this way, remove the largest prefix, and the largest suffix containing only don't care symbols. The resulting string is *Lin*(*PAT*).

The basic property of the transformation *Lin* is:

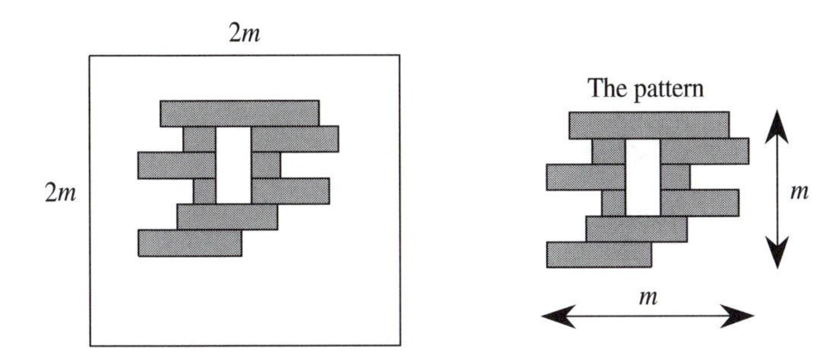

Figure 12.14: Searching a non-rectangular $m \times m'$ pattern within pieces of shape $2m \times 2m'$.

> *Lin*(*PAT*) is independent of the position where *PAT* is placed inside the shape *S*.

Let *T* be an $2m \times 2m'$ text array. Let *Lin*′(*T*) be the string obtained by concatenating all rows of *T*, starting from the top most row.

Then, searching *PAT* in *T* is equivalent to searching for *Lin*(*PAT*) inside of *Lin*′(*T*). This can be executed using methods for string matching with don't care symbols (see Section 11.4). There it is proven how to do it in $O(n\log^2 n)$ time, which here becomes $O(m^2\log^2 m)$.

A text array of size greater than $2m \times 2m'$ can be decomposed into such (overlapping) subarrays on which the above procedure is applied. The total time becomes then $O(N\log^2 m)$. This completes the proof. ♦

12.4 Matching with mismatches

The definition of a distance between two arrays is more complicated than for (one-dimensional) strings. Insertions or deletions of a symbol can result in an increase or decrease of the length of one row (column). Therefore, for simplicity, we concentrate here on the approximate pattern matching with only one edit operation: replacement of one symbol by another. This corresponds to unit-cost mismatches.

For two strings *x*, *y* denote by $MISM_k(x, y, i)$ the set of first (left-to-right) mismatch positions, up to *k*, between the substring $y[i...i+|x|-1]$ and *x*. We are not interested in more than *k* mismatches.

Lemma 12.5 Assume we are given two strings *x* and *y*, and their suffix trees with the *LCA* preprocessing. Then, the computation of $MISM_k(x, y, i)$ can be executed in $O(k)$ time for each position *i* in the text *y*.

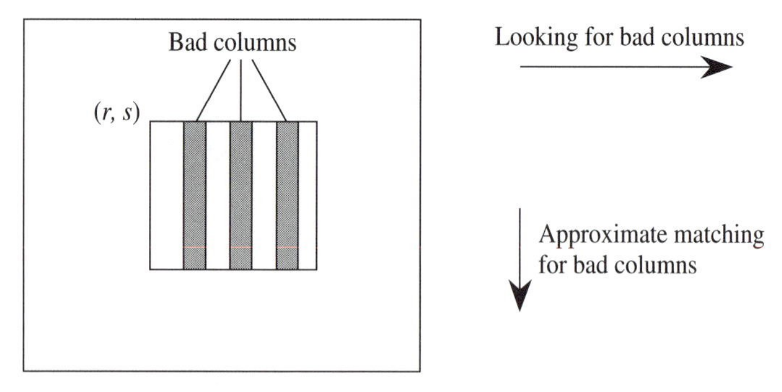

Figure 12.15: Approximate matching. There are at most k bad columns. If there is match with at most k mismatches, then the total number of mismatches in all bad columns cannot exceed k.

Proof First we find the longest common prefix of $y[i...n]$ and x. This is done using an *LCA* query for the leaves corresponding to y and x in the joint suffix tree for these both texts. In this way, we obtain the first mismatch position i_1. Then, we look for the longest common prefix of $x[i_1-i+1...m]$ and $y[i_1...n]$. This is again done by asking a suitable *LCA* query about leaves related to $x[i_1-i+1...m]$ and $y[i_1...n]$. We obtain the next mismatch position (if it exists). We continue in this way until k mismatch positions are found, or (in the case where there are less than k mismatch positions) all mismatch positions are found. The time is proportional to the number of *LCA* queries, that is $O(k)$. This completes the proof.♦

Theorem 12.6 Assume the alphabet is of a constant size. The problem of pattern matching with a fixed number k of mismatches within an $n \times n$ text array T can be solved in $O(kn^2)$ time.

Proof Let *PAT* be the $m \times m$ pattern, where $m \leq n$. The algorithm starts as in the exact two-dimensional pattern matching, by a multi-pattern string matching. The Aho-Corasick automaton for all columns of the pattern is built. Then, the automaton is applied to all columns of T to obtain a state array T'. The pattern array is replaced by a string of states *PAT'*.

Figure 12.15 illustrates how we check the approximate match at position (r, s) with at most k mismatches. Let y be the r-th row of T'. We compute $MISM_k(PAT', y, s)$. This produces all columns that contain at least one mismatch with the pattern *PAT* placed at (r, s). Let us call these columns the bad columns. We compute the total number of mismatch positions in bad columns with respect to the corresponding columns of the pattern (assuming it is placed at position (r, s)). We are only interested in at most a total of k mismatches. All mismatches are found by using the function *MISM*. The total complexity is proportional to the number of all *LCA* queries executed in the algorithm. We make at most k such queries. Hence, for a fixed position

(r, s) after the preprocessing, the complexity is $O(k)$. Since there is a quadratic number of positions, the total time complexity is as required. This completes the proof. ♦

12.5 Multi-pattern matching

In this section we consider a set of k square pattern arrays $X_1, X_2, \ldots, X_k$. For a given $n \times n$ text array T we want to check if any of these patterns occurs in T. This is the multi-pattern matching in two dimensions. Assume, for simplicity, that the size of the alphabet is constant. The strategy developed for the Karp-Miller-Rosenberg algorithm (see Chapter 8) yields a solution to the general multi-pattern problem that works in $O(n^2\log n)$ time. We omit the obvious proof.

Fact Two-dimensional multi-pattern matching can be solved in $O(n^2\log n)$ time using the KMR algorithm.

Indeed, the above result can be improved to $O(n^2\log k)$ time, where k is the number of patterns. Of course, k can be of the same order as n, and this does not provide a substantial improvement. But we are also interested in alternative algorithms and some interesting new ideas behind them that enrich the algorithmics of two-dimensional matching.

The natural alternative algorithm considered here is based on an extension of the Aho-Corasick string-matching automaton to the two-dimensional case. By the way, this also shows the important extension of the notion of *border*, *suffix*, and *prefix* to two-dimensional arrays. It also provides another pattern-matching algorithm for one pattern: it shows that searching the pattern along a fixed diagonal of the text array is reducible to one-dimensional string matching. Again, *LCA* preprocessing is crucial for the two-dimensional pattern-matching algorithm of the section.

First consider the simple case in which all the patterns are of the same shape. Assume that they are $m \times m$ arrays. The method of Section 12.1 generally facilitates this situation. This gives a linear-time algorithm when all patterns are of the same size.

Theorem 12.7 Two-dimensional multi-pattern matching can be solved in $O(N)$ time when the alphabet is fixed and all patterns are of the same size (where N is the total size of the problem).

Proof The algorithm works as follows. The Aho-Corasick machine is constructed for all columns of all patterns. Each pattern array is then transformed into a string of states. We obtain a set of strings $x_1, x_2, \ldots, x_k$. The text array T is replaced by the state array T' in the same way as in Section 12.1. Any multi-pattern string-matching algorithm then gives a solution. This gives a linear-time algorithm for this special case (fixed alphabet). ♦

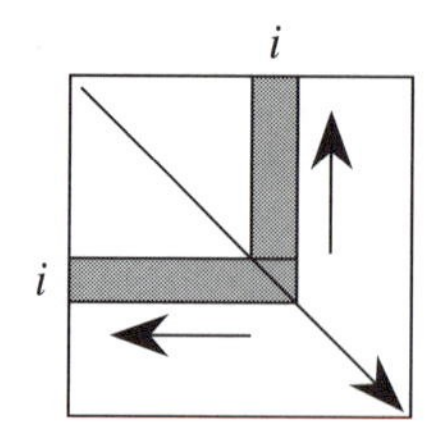

Figure 12.16: The i-th segment of a pattern array.

Next, we consider the general case, in which the patterns are square arrays of possibly different sizes. Again, the algorithm is an extension of the Aho-Corasick multi-pattern matching.

A prefix (resp. suffix) of a square array is a square subarray containing the left top corner (resp. right bottom corner) of the array. We construct the two-dimensional version of the Aho-Corasick multi-pattern automaton M as follows. Each pattern is considered as a string: its i-th letter is the i-th segment of the array. The i-th segment is composed of the upper part of the i-th column, and the left part of the i-th row, beginning both at the i-th position on the diagonal (see Figure 12.16). The states of M are prefixes of all the pattern arrays. The set of states is organized in a tree in which the nodes correspond to the two-dimensional prefixes of the patterns.

The edges outgoing a node at the depth i–1 are labeled by the names of the i-th segments of the patterns. We can give consistent names to i-th segments of all patterns in time $O(k\log k)$ for a given i, since there are at most k such segments, one for each pattern. The equality of two segments can be checked in constant time using an *LCPref* query (a longest common prefix query) after a suitable preprocessing of the tree in which the edge labels are names of segments.

After that, the failure table *Bord* on the tree is built as in Section 7.1. The notion corresponds to borders of square arrays as illustrated in Figure 12.17. They are the largest proper subarrays that are both prefix and suffix of the given array.

We say that a segment $\pi1$ is a part of a segment $\pi2$ iff the column part of $\pi1$ is a prefix of the column part of $\pi2$ and a similar relation holds between rows of the segments. Define the following relation == between two segments. Let $\pi1$, $\pi2$ be i-th and j-th segments, respectively, with $i \leq j$. Then, we write $\pi1 == \pi2$ iff either, both $i = j$ holds and the names of the segments are the same, or, $\pi1$ is a part of $\pi2$. The table *Bord* for the two-dimensional case is defined as for strings, except that relation == is considered instead of equality.

Theorem 12.8 Assume that the alphabet is fixed. Then the two-dimensional multi-pattern matching can be solved in time $O(N\log k)$ time, where k is the number of patterns.

Proof The two-dimensional pattern matching is essentially reduced to one-dimensional multi-pattern matching (see Figure 12.18). The equality = of symbols is replaced by the relation ==, and the corresponding table *Bord* works similarly. ♦

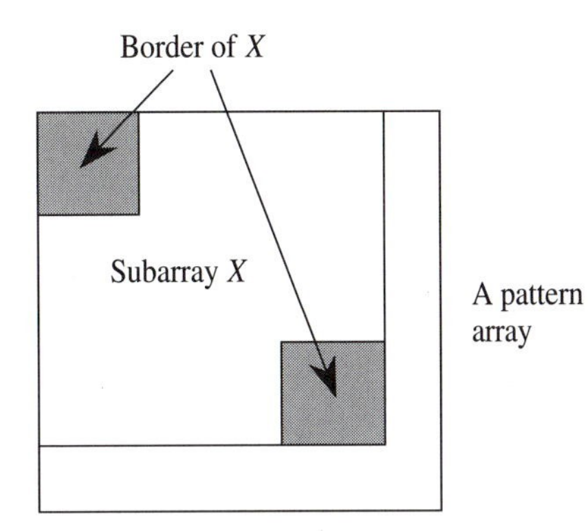

Figure 12.17: A border of a subarray X of a pattern.

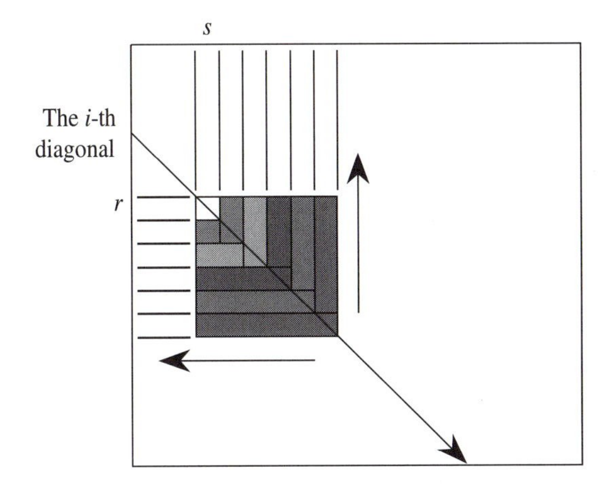

Figure 12.18: Checking the occurrence of a pattern at position (r, s) in the text array.

12.6 Matching by sampling

The concept of *deterministic sample* introduced in Chapter 3 for one-dimensional patterns is very powerful. Its wide applicability appears, for example, in the domain of parallel computations, leading to a constant-time parallel string matching. The aim of this section is to extend and use the concept of a deterministic sample to the two-dimensional case. Almost every non-periodic pattern has a deterministic sample for which the properties are analogous to those of one-dimensional patterns. The use of 2–D sampling (for short) gives solutions both to sequential computation requiring only small extra space, and to constant-time parallel computation for the two-dimensional pattern-matching problem.

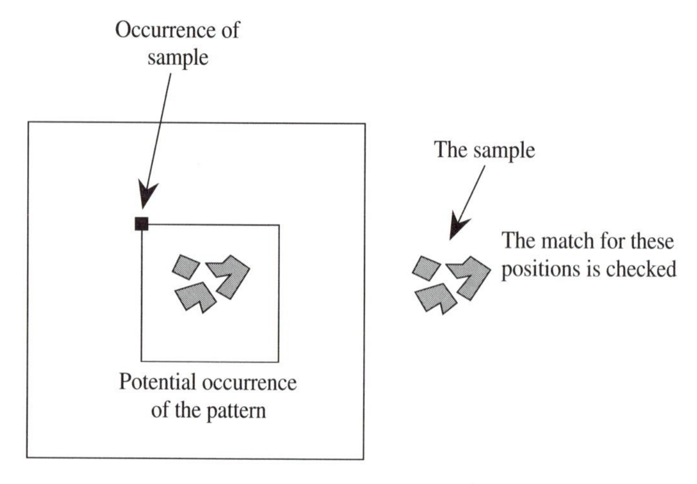

Figure 12.19: A deterministic sample S.

A *deterministic sample* S for PAT is a set of positions in the pattern PAT satisfying certain conditions. The sample S occurs at position $x = (i, j)$ in the text array iff $PAT[y] = T[x+y]$ for each y in S (see Figure 12.19).

The central idea related to samples is the *field of fire* of the sample S. Finding an occurrence of the sample in the text assures us that there is an $m/2 \times m/2$ square in the text, called the field of fire of the occurrence of S, where there is only one possible matching position of the pattern. This possible matching position is $z = (k, l)$ relative to the origin of the field of fire (see Figure 12.20). Let x be a position of S in the text array. Denote by $fofire(x, S)$ the corresponding field of fire: it is the $m/2 \times m/2$ subsquare of the text array at position x-z. The field of fire should satisfy the following condition.

Field of fire condition Whenever the sample S occurs at position x in the text array, then there is no occurrence of the pattern within the area $fofire(x, S)$ of the text array, except maybe at position x.

The size of a square S, denoted by $|S|$, is defined here as the length of its side. The deterministic sample S must be small but effective in "killing" other positions. It must satisfy the following conditions:

(*) $|S| \leq O(\log m)$,
(**) $|fofire(x, S)| \geq m/4$.

We consider only very particular samples. They are *special segments*: horizontal factor of length $4\log m$ at position $(m/2, m/2)$ in the pattern. In other words, it is the factor of length $4\log m$ starting at position $m/2+1$ in the $m/2+1$ row of the pattern

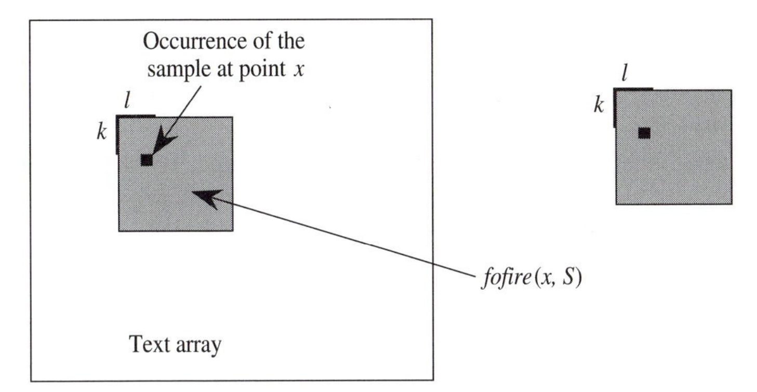

Figure 12.20: The field of fire of the sample S.

(see Figure 12.21). Moreover, we say that the pattern *PAT* is *good* if its special segment occurs only once in *PAT*. Note that any segment lying "far enough" away from the boundaries of the array would work as well.

Theorem 12.9 Assume that the alphabet contains at least two symbols. Then

- almost all patterns are good,
- for almost all patterns there is a logarithmic-size sample for which the field of fire is an $m/2 \times m/2$ square,
- both, the sample can be found, and the goodness of the pattern can be checked either in constant extra space and linear serial time, or in constant parallel time with $O(m^2)$ processors.

Proof The first point follows by simple calculations. If the pattern is good the special segment is the sample. It is of logarithmic size. Its field of fire is the left upper

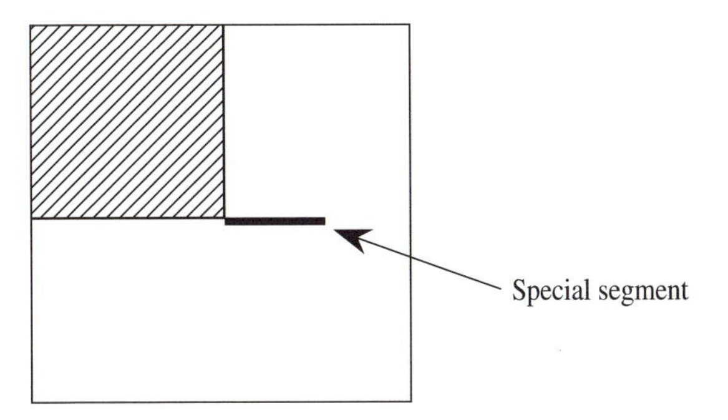

Figure 12.21: The special segment of *PAT*, and its field of fire.

$m/2 \times m/2$ quadrant of the pattern. If the sample occurs at position x in the text array, then no occurrence of the pattern in the text has position $x+(k, l)$ for $0 \leq k \leq m/2$, $0 \leq l \leq m/2$, and $(k, l) \neq (0, 0)$. In that case there would be two occurrences of the special segment in the pattern.

The last point follows from the fact that all occurrences of the special segment can be found within claimed complexities using algorithms for string matching of Chapters 13 and 14. This completes the sketch of the proof. ♦

Theorem 12.10 Assume that the alphabet contains at least two symbols. Then

- the two-dimensional pattern matching can be solved in constant extra space and linear serial time, for almost all patterns,
- the two-dimensional pattern matching can be solved in constant parallel time with $O(n^2)$ processors, for almost all patterns.

Sketch of the proof Only good patterns are considered, so results hold only for them. The whole text array is partitioned into $m/2 \times m/2$ windows. Within each window, we search for occurrences of the special segment with a serial algorithm working with the claimed complexities. For each occurrence of the segment (sample) we naively check if it corresponds to an occurrence of the whole pattern. This is done for all windows independently, window after window.

The proof of the second point is analogous to the above argument. But now, all windows are processed simultaneously with a constant-time parallel string-matching algorithm. ♦

12.7 An algorithm fast on the average

A natural problem related to pattern matching is designing algorithms that are fast in practice. Since the notion of "practice" is not formal, it is often considered for algorithms that are fast on the average. In this section, we construct an algorithm making $O(N\log(M)/M)$ comparisons on the average for the two-dimensional pattern matching (the pattern is an $m \times m$ array, the text is an $n \times n$, $N = n^2$, and $M = m^2$). All symbols appear with the same probability independently of each other in arrays. If M is of the same order as N, the algorithm makes only $O(\log N)$ comparisons on the average. The method described here is similar to the use of special segments in Section 12.6. For simplicity we assume that the alphabet has only two elements, and that each of the two symbols of the text is chosen independently with the same probability. Let r be equal to $4\log m$.

The algorithm is similar to the algorithm *fast_on_average* presented at the end of Chapter 4 as a variation of the Boyer-Moore algorithm for string matching.

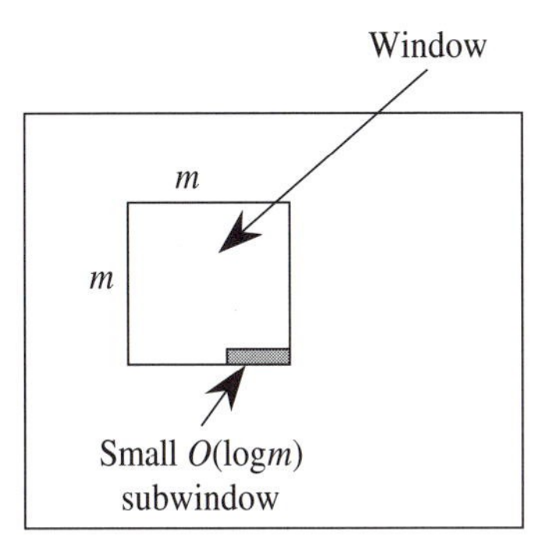

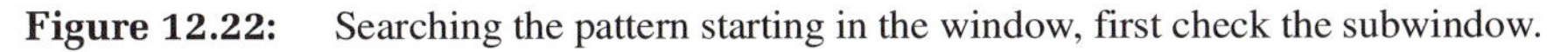
Figure 12.22: Searching the pattern starting in the window, first check the subwindow.

Informal description of the algorithm

- Partition the text array into windows of shape $m \times m$;
- the subwindow of a window consists of the last r positions of the lowest row of the window;
- first check if the text contained in the subwindow is a factor of any row of the pattern;
- if so, the search for an occurrence of the pattern having its left upper corner position in the window is done by any linear-time algorithm; the same procedure is applied to each window.

The suffix of length r of the last row of the pattern behaves like a fingerprint. It has only logarithmic size, by definition. But it is unlikely to appear in subwindows. The test at line 3 above can be done with the help of a suffix tree or a suffix dawg. On fixed alphabets, this takes $O(r)$ time. There are N/M windows, and a simple calculation shows the following (see end of Chapter 4).

Theorem 12.11 The two-dimensional pattern matching can be done with $O(n^2\log(m)/m^2)$ comparisons on the average, for fixed alphabets, after preprocessing the pattern.

12.8 Finding regularities in parallel

In this section we consider three problems on arrays. These problems consist of finding a largest subarray that:

1. repeats (occurs at least twice), or
2. is a common subarray of two given arrays, or
3. is symmetric.

The algorithmic solutions for these problems resemble the solutions for the corresponding one-dimensional problems.

Denote by *Cand*1(s) (resp. *Cand*2(s), *Cand*3(s)) a function in which the value is any subarray of size s that satisfies the condition (1) (resp. condition (2), condition (3)). If there is no such subarray then the value of the function is *nil*. The values of the function are *candidates* of size s. The problem is to find a non-*nil* candidate with maximum size s. More generally, we assume that we have a function *Cand*(s) satisfying the following monotonicity property:

$$Cand(s+1) \neq nil \Rightarrow Cand(s) \neq nil.$$

Denote by *Maxcand* a candidate of maximal size,

$$Maxcand = \{Cand(s): s \text{ is maximum such that } Cand(s) \neq nil\}.$$

Assume also that *Cand*(0) is some special value.

Lemma 12.12 If the value *Cand*(s) can be computed in $T(n)$ parallel time with $P(n)$ processors, then, the value of *Maxcand* can be computed in $O(T(n)\log n)$ time with the same number of processors.

Proof We can assume w. l. o. g. that n is a power of two, and that $Cand(n) = nil$. A variant of binary search can be applied. We look at $Cand(n/2)$. If it is *nil* then we try $Cand(n/4)$, otherwise we look at $Cand(n/2+n/4)$, and so on. In this way, after a logarithmic number of steps we obtain *Maxcand*. This completes the proof. ♦

The algorithm of Chapter 9 related to the computation of the dictionary of basic factors for strings has a natural counterpart for arrays. The naming technique applies as well. This is because an array of size $2k$ is composed of four (fixed finite number) subarrays of size k. Recall that a basic subarray has shape $k \times k$ with k a power of two. Each such subarray is given a name (or a number) through NUM_k.

Assume that the dictionary of basic subarrays is computed for the input array. Testing the equality of two subarrays in which the size is a power of two reduces to a mere comparison of their associated names. If the size of subarrays is not a power of two we can also test their equality in constant time. For that purpose, we define identifiers for subarrays in which the size s is not a power of two. Let k be the highest power of 2 less than s. The identifier of the $s \times s$ subarray at position $v1$ is $Ident(v1, s) = (NUM_k(v1), NUM_k(v2), NUM_k(v3), NUM_k(v4))$, where position $v2$, $v3$, and $v4$ are as in Figure 12.23. It is clear that, if the dictionary of basic subarrays is computed,

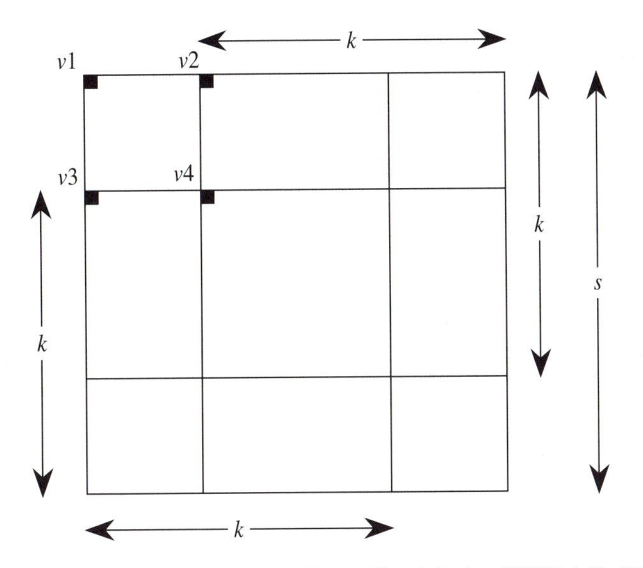

Figure 12.23: Identifier of a subarray of size s: $Ident(v1, s) = (NUM_k(v1), NUM_k(v2), NUM_k(v3), NUM_k(v4))$ where k is the highest power of 2 less than s.

the equality of two subarrays is equivalent to the equality of their identifiers, and can be checked in constant time. The key point is that identifiers are of constant size.

Identifiers of subarrays can be used to search for subarrays. This gives a straightforward parallel solution to the two-dimensional pattern-matching problem. The algorithm is simple but not optimal. A parallel optimal solution is presented in Chapter 14.

Theorem 12.13 The two-dimensional pattern-matching problem can be solved with n^2 processors in $O(\log^2 n)$ parallel time on the exclusive-write PRAM model, and in $O(\log n)$ parallel time on the concurrent-write PRAM model.

Proof Let *PAT* be the $m \times m$ pattern array, let T be the $n \times n$ text array. We create the dictionary of basic factors common to *PAT* and T. The identifier *ID* of the pattern is computed. Then we search in parallel for a subarray P' of T (of same size as *PAT*) in which the identifier equals *ID*. The search phase takes constant time, and the computation of the dictionary relies on algorithms of Chapter 9. This completes the proof. ♦

Observe that none of the linear-time sequential algorithms for two-dimensional pattern matching (see Section 12.1) is easily parallelizable. We now come to the problems of the present section. We apply Lemma 12.12 successively to functions *Cand*1, *Cand*2, and *Cand*3. The complexities of algorithms for these three problems are analogous to the complexity of the pattern-matching algorithm of Theorem 12.13.

Theorem 12.14 The largest repeated subarray of an $n \times n$ array can be computed with n^2 processors in $O(\log^2 n)$ parallel time on the exclusive-write PRAM model, and in $O(\log n)$ parallel time on the concurrent-write PRAM model.

Proof In view of Lemma 12.12, it is sufficient to show how to compute $Cand1(s)$ with n^2 processors in $O(\log n)$ time on the exclusive-write PRAM model, and in $O(1)$ time on the concurrent-write PRAM model.

Given identifiers of all subarrays, for each position x on the text array, it is a simple matter to compute in $O(\log n)$ time two positions with the same identifier. We can sort pairs $(Ident(x, s), x)$ lexicographically. Any two such consecutive pairs with the same identifier part in the sorted sequence give required positions. The model is the exclusive-write PRAM.

With concurrent writes, we can use an auxiliary *bulletin board* table, and proceed similarly as in the proof of Lemma 9.4. No initialization of the bulletin board is required, and this gives $O(1)$ time. This completes the proof. ♦

Theorem 12.15 The largest common subarray of two $n \times n$ arrays can be computed with n^2 processors in $O(\log^2 n)$ parallel time on the exclusive-write PRAM model, and in $O(\log n)$ parallel time on the concurrent-write PRAM model.

Proof In this case we compute, at the beginning, the common dictionary for both text arrays S, T. The remainder of the proof is essentially the same as the proof of Theorem 12.14. There are small technical differences. We sort pairs $(Ident(x, s), x)$ lexicographically. Now xs are positions in S and T. We can partition the sorted sequence into segments consisting of pairs having the same first component (identifier). In these segments it is now simple to look for two positions x, y such that x is a position in S and y is a position in T. This completes the proof. ♦

Theorem 12.16 The largest symmetric subarray of an $n \times n$ array can be computed with n^2 processors in $O(\log^2 n)$ parallel time on the exclusive-write PRAM model, and in $O(\log n)$ parallel time on the concurrent-write PRAM model.

Proof Let X be an $n \times n$ array. In view of Lemma 12.12, it is sufficient to compute $Cand3(s)$ with n^2 processors in $O(\log n)$ time on the exclusive-write PRAM model, and in $O(1)$ time on the concurrent-write PRAM model.

Let us compute the array T that results by reflecting each entry of X with respect to the center of the array. Denote by $Reflect(x, s)$ the position in T of the $s \times s$ subarray B, reflection of the subarray A occurring at position x in X (see Figure 12.24). We can now compute the common dictionary of basic subarrays of arrays X and T. The subarray A of X is symmetric iff $A = B$, which can be checked in constant time using s-identifiers at position (i, j) in X, and at position $Reflect((i, j), s)$ in T. The function $Reflect$ is easily computable in constant time. Once we know which $s \times s$ subarrays are symmetric we can easily choose one of them (if there are any) as the

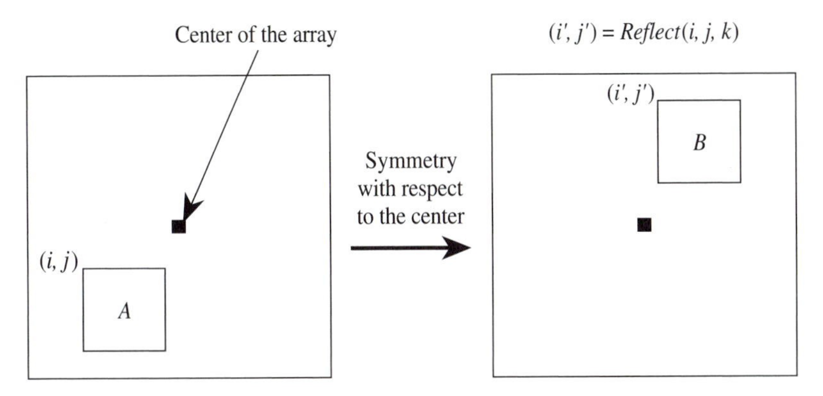

Figure 12.24: The $s \times s$ subarray A is symmetric iff $A = B$ iff $Ident((i, j), s) = Ident((i', j'), s)$.

value of $Cand3(s)$. Hence, $Cand3(s)$ can be computed within the required complexity bounds. This completes the proof. ♦

12.9 Simple geometry of two-dimensional periodicities

This section presents some theoretical tools for the Galil-Park 2d-pattern-matching algorithm of Section 12.11. The proofs of several simple facts are omitted and are left as exercises.

Let *PAT* be a two-dimensional pattern of shape $m \times m$, with its rows and columns numbered by 1, 2, ..., m. The vectors of *PAT* are denoted by π and β. We consider only two-dimensional vectors with integer components. Recall that a vector π is a *period* of *PAT* iff $PAT[x] = PAT[x+\pi]$, whenever both sides are defined. If both sides are defined for at least one point x, then π is a nontrivial period. We also write that π is a 1d-period to emphasize its one-dimensional status. The main difference between 1-dimensional and 2-dimensional pattern matching lies in the different structures of periods of patterns. In two dimensions, some periods are inherently two-dimensional and are called 2d-periods.

A pair $\mu = (\pi, \beta)$ of non-colinear vectors is a *2d-period* of *PAT* iff π and β are nontrivial periods, and each linear combination of π and β is a 1d-period of *PAT*. An equivalent formulation is: *PAT* can be extended to an infinite plane in which π and β are periods. By a linear combination we always mean a combination with integer coefficients, i.e., a vector $i.\pi+j.\beta$, in which i, j are integers. Two (or more) vectors are said to be colinear iff they are in the same direction, which, in this case, does not necessarily mean that one is an integer combination of the other (others). Let us denote by $Lattice(\mu)$ the set of all linear combinations of π, β. The elements of $Lattice(\mu)$ are called the lattice points. Therefore, the pair μ is a 2d-period iff all elements of $Lattice(\mu)$ as vectors, are periods.

A vector $\pi=(r, c)$ is said to be *small* iff its components r, c satisfy $|r|$, $|c| < d.m$, in which $d = 1/16$. A 2d-period is *small* iff both its components are small vectors. The pattern *PAT* is called *periodic* (*lattice-periodic or 2d-periodic*) iff it has any *small* 1d-periods (2d-periods).

Remark In one-dimensional string matching a linear combination of *small* 1d-periods is always a period. But this is not generally valid for two dimensions, even for non-negative combinations of colinear vectors (as well as for non-colinear vectors, of course). If all elements of the array *PAT* are the same letter except for a small number of elements closed to one fixed corner, then there is a great number of 1d-periods, but there is no nontrivial 2d-period. The parts around the corners are responsible for the irregularities.

The 2d-period $\mu = (\pi,\beta)$ is said to be *normal* iff π is a quad-I period as in Figure 12.7 and β is a quad-II period as in Figure 12.8.

Lemma 12.17 (normalizing lemma) If the pattern has a small 2d-period then it has a small *normal* 2d-period.

A notion of divisibility for 2d-periods, $\mu 1$, $\mu 2$, is introduced as follows:

$\mu 1 \mid \mu 2$ iff *Lattice*($\mu 1$) includes *Lattice*($\mu 2$).

We also introduce the notion of a *smallest* 2d-period μ=*period*(*PAT*). It is a fixed small 2d-period of *PAT* that "divides" each other's small 2d-period: in other words, $\mu \mid \mu'$ for each small 2d-period μ'. There are several ways to decide which μ is to be chosen, but any of them is good. Assume *PAT* is 2d-periodic. We define *period*(*PAT*) as a small *normal* 2d-period (π,β), in which π is a quad-I small period of a minimal length (in case of ties the most horizontal vector is chosen), and β is a quad-II small period of a minimal length (in case of ties the most vertical vector is chosen). Lemma 12.17 guarantees that this definition makes sense. It can be proven that any small 1d-period corresponds to a point in *Lattice*(*PAT*). This fact, together with Lemma 12.17, implies the following lemma.

Lemma 12.18 (2d-periodicity lemma)

(a) Assume $\mu 1$, $\mu 2$ are small 2d-periods of *PAT*. Then, there is a 2d-period μ such that $\mu \mid \mu 1$ and $\mu \mid \mu 2$.

(b) Assume *PAT* is lattice-periodic and π is a small vector. Then π is a 1d-period of *PAT* iff π is in *Lattice*(*period*(*PAT*)). Moreover, *period*(*PAT*) $\mid \mu$ for all small 2d-periods μ.

Observation Assume we know which points of *PAT* are small sources (see below). Then, if it exists, *period*(*PAT*) can be computed in $O(M)$ time independently of the alphabet.

Lemma 12.19 (overlap lemma) Assume the patterns PAT_1 and PAT_2 are 2d-periodic subsquares of the same rectangle, $period(PAT_1) = \mu 1$, and $period(PAT_2) = \mu 2$, in which $|\mu 1|, |\mu 2| < m'/2 \leq m$. If PAT_1 and PAT_2 overlap on an $m' \times m'$ square, then $\mu 1 = \mu 2$.

According to their periodicities, 2d-patterns are classified into four main categories:

- *non-periodic*: no small period at all,
- *lattice-periodic* (or *2d-periodic*): at least one small 2d-period,
- *radiant-periodic*: at least two non-colinear small 1d-periods, but not lattice-periodic,
- *line-periodic*: all periods in the same direction.

We have already defined quad-I periods and quad-II periods. We recall these definitions and introduce similar categories for so-called *sources*. The pattern is divided into four $m/2 \times m/2$ disjoint squares, called *quads*, and named quad I, quad II, quad III, and quad IV, according to the counterclockwise ordering, and starting at the upper left corner. Therefore, quad I corresponds to the upper left square corner, and quad II corresponds to the lower left square corner.

For technical reasons it is convenient for these categories to be disjoint. So we assume that horizontal vectors are not quad-II vectors, and vertical vectors are not quad-I vectors. As per this assumption, each 1d-period oriented from left to right is exactly of one type: a quad-I period or a quad-II period.

Let $Center_s(PAT) = PAT'$ be the central subarray of PAT that results after "peeling off" the s boundary columns and rows (from top, down, right, and left). The shape of such a subarray is $(m-2s) \times (m-2s)$. Next we state the key lemma used to avoid the radiant-periodic case in the main algorithm. Its proof is omitted.

Lemma 12.20 (radiant-periodicity lemma) Assume that PAT is radiant-periodic. Then $Center_{2d.m}(PAT)$ is not radiant-periodic.

A vector π can be identified as the point $\underline{\pi}$ of PAT, extremity of π, when its origin is at the quad-I top left corner (point $(0, 0)$) or at the quad-II corner (point $(m, 0)$) of the pattern. The point $\underline{\pi}$ is called the quad-I beginning point, or the quad-II beginning point, respectively, corresponding to π. If π is a period and the quad-I beginning point $\underline{\pi}$ is in PAT, then π is called a quad-I period, and $\underline{\pi}$ is called a quad-I *source*. Quad-II periods and sources are defined analogously. Using the terminology of sources, the periodicity type of pattern PAT can be characterized (equivalently) as follows:

- *non-periodic*: no small source;
- *lattice-periodic* or (*2d-periodic*): at least one quad-I source and one quad-II source;

- *radiant-periodic*: not lattice-periodic and at least two non-colinear small sources;
- *line-periodic*: all sources on the same line.

Let $\mu=(\pi, \beta)$ be a 2d-vector. We say that two points x, y are μ-equivalent iff $x-y$ is in $Lattice(\mu)$. A μ-path is a path consisting of edges that are vectors π, $-\pi$, β or $-\beta$. The processing of certain difficult patterns is executed by exploring some simple geometry of paths on a lattice generated by two vectors π, β belonging to the same quadrant. If we have two points x, y containing distinct symbols and we have a (π, β)-path from x to y within the pattern, then one of the edges of the path gives a witness of non-periodicity to one of the vectors π or β. This is due to the fact that the initial and terminating positions do not match, so there should be a mismatch "on the way" from x to y. The length of the path is the number of edges it contains.

Observation The basic difficulty with such an approach is the length of the path. It may be that μ is a small 2d-period, but the length of a shortest μ-path between two μ-connected points of an $m \times m$ square is quadratic. Consider, for example, $\mu=((m/2, 1), (1, 0))$, $x=(m/2, 0)$, and $y=(m/2, m-1)$.

Despite the previous observation, we can find useful short paths in some situations, as shown in the next lemma. Let $Cut\_Corners_s(A)$ be the part of array A without top-right and bottom-left corner squares of shape $s \times s$.

Lemma 12.21 (linear-path lemma) Assume π, β are quad-I vectors of size at most k. Let S be a subsquare of size $k \times k$ of a larger square A, and let x be the point that is the bottom-left corner or top-right corner of S. Assume x is inside $Cut\_Corners_{2k}(A)$. Then, there is a *linear-length* μ-path inside A from x to a point $y \neq x$ in S. Such a path can be computed in $O(k)$ time.

Proof We can assume, without loss of generality, that x is the quad-II corner of S, and that π and β are quad-I vectors of size at most k. Moreover, we can assume that the array A is of shape $3k \times 3k$, and S is the square of size $k \times k$ at the top-left corner of A. Then, x is the point of A at position $(k, 0)$. Assume that β is more a horizontal vector than π. We find a path and a point y by the algorithm GREEDY following.

We prove that the algorithm GREEDY terminates successfully after a linear number of iterations and generates the required path. Consider the lines $L_0, L_1, L_2, \ldots$, in which L_h is the line parallel to π and contains the points $x+h\beta$. Therefore, points $x+i\beta+j\pi$ (j integer) belong to the line L_i. If some line L_i cuts the two horizontal borders of S, or its two vertical borders, then the segment of the line that is inside S is longer than π. Thus, $x+i\beta+j\pi$ belongs to S for some (negative) integer j. If each line L_i cuts both a horizontal and vertical border of S then let i be such that lines L_i and L_{i+1} surround the diagonal segment of S; it can then be proven that, either there is a point $x+i\beta+j\pi$ in S or a point $x+(i+1)\beta+j'\pi$ in S. Values of variable y in the algorithm are points of a μ-path inside A because if "$y-\pi$ not in A," $y+\beta$ is in A. We have yet to

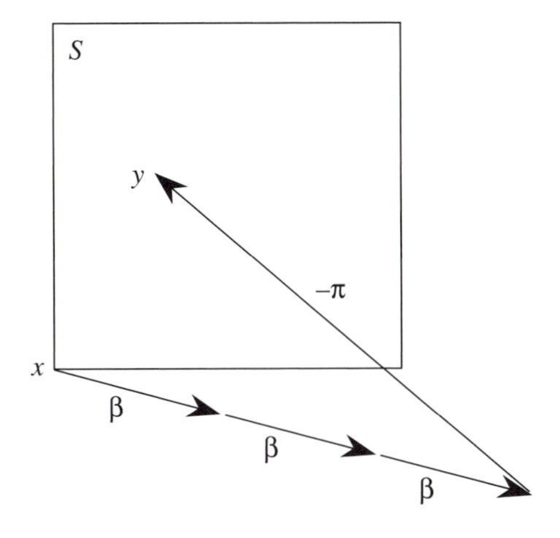

Figure 12.25: A short path from a given point x to some other point y of the square S.

explain why the path has linear length, which at the same time proves that the algorithm works in $O(k)$ time. Let $\pi=(r, c)$ and $\beta=(r', c')$. The point $x+r\beta-r'\pi$ is on the same column as x, and can be the point y if it is in S. It is clear that the μ-path followed by the algorithm is entirely (except for maybe the last edge) inside the triange $(x, x+r\beta, x+r\beta-r'\pi)$. Thus, the length of the μ-path followed by the algorithm is no longer than $r+r'$ which is $O(k)$. ♦

```
algorithm GREEDY
begin
    y := x;
    repeat
        if y − π is outside the area A then y := y + β
        else y := y − π;
    until y is in S;
end.
```

We introduce a special type of duels called here *long-duels*. Assume we have small quad-I vectors π, β, and a point x at distance at least $2k$ from quad-II and quad-IV corners. Assume also that if y is any point such that $y-x$ is a small quad-II vector, then $PAT[x] \neq PAT[y]$. The procedure *long_duel*(π, β, x) "kills" one of the vectors π, β, and finds its witness. It works as follows: a (π, β)-path from a given point x to some point y in the pattern is found by the algorithm of the linear-path lemma. The path consists of a linear number of edges. End points x and y contain distinct symbols. Therefore, one of the edges on the path gives a witness for π or β. In doing so, one of the potential small periods π or β is "eliminated" in linear time.

Theorem 12.22 (long-duel theorem) Assume we have a set X of small quad-I vectors, in which $|X| = O(m)$, and we are given a position x in *Cut_Cor-*

$ners_{2d.m}(PAT)$. Assume also that if y is any point such that $y-x$ is a small quad-II vector, then $PAT[x] \neq PAT[y]$. Then, in linear time, by using long duels, we can find witnesses for all small quad-I vectors (except maybe for a set of vectors on a same line L).

Proof We run the following statement:

```
while X is non-empty do begin
    take any element β from X; add β to Y, and delete β from X;
    while there are two non-colinear vectors π, β in Y do
        execute long-duel (π, β, x) and delete the "loser" from Y;
end;
```

We keep the elements of X that have not been eliminated so far in the set Y. Initially Y is empty. The invariant of the loop is: non-null witnesses for all elements not in the current sets X or Y are computed, and all elements of Y are on the same line L. Altogether, the execution time is $O(m^2)$ and alphabet independent. At the end, all remaining vectors (with non-null witnesses up to now) are on a same line L. This completes the proof. ♦

Let us call the algorithm of the long-duel theorem the *long-duel algorithm*. There is a natural analogue of the theorem for quad-II small vectors, and for quad-I and quad-III corners.

The crucial point in the processing of a difficult type of pattern (*corner* pattern) is the role played by the following suffix-testing problem: given m strings $x_1, \ldots, x_m$ of total size $O(m^2)$, compute the $m \times m$ table *Suf-Test* defined as follows:

Suf-Test$[i, j]$ = *nil* if the i-th string is a suffix of the j-th string,
Suf-Test$[i, j]$ = position of the right most mismatch otherwise.

The algorithm is given below as Algorithm *Suffix-Testing*. We sketch its rough structure to show that it runs in linear time independently of the alphabet. It is sufficient to compute for each pair i, j, the length $SUF[i, j]$ of the longest common suffix of x_i and x_j.

The algorithm can be easily implemented to work in $O(m^2)$ time. The main point in the evaluation of the time complexity is that if a position participates in a positive comparison (when two symbols match), then this position is never inspected again. When we process a given k and compute $SUF[k, j]$ for $j > k$, then we first look for $SUF[i, j]$, in which $i = MAX[j, k-1]$, and then for $SUF[i, k]$. These data are available at this moment, due to invariant. The word x_j is scanned backward starting from position $SUF[i, j]$. The pointers only go backward. This proves the next theorem.

```
Algorithm Suffix-Testing;
begin
    assume that strings x1, ..., xm are in increasing order of their lengths;
```

```
    { invariant(k) : for all i, j, 1 ≤ i ≤ k, 1 ≤ j ≤ m,
         SUF[i, j] is computed, and, for each j, 1 ≤ j ≤ m,
         we know MAX[j, k] = i, where i is the index i ≤ k which
         maximizes SUF[i, j]}
    make invariant(1);
    for k := 2 to m do
         make invariant(k) using invariant(k–1);
end.
```

Theorem 12.23 (suffix-testing theorem) The suffix-testing problem for m strings of total size $O(m^2)$ can be solved in $O(m^2)$ time, independently of the alphabet.

12.10* Patterns with large monochromatic centers

The alphabet-independent linear-time computation of 2d-witness tables is quite technical, hence the present and following sections may be considered optional. In this section, we present an alphabet-independent linear-time computation of witnesses for special patterns, the "large" central part of which is "monochromatic." The pattern *PAT* is called *mono-central* iff all symbols lying in $Center_k$ are equal to some letter a of the alphabet, and for $k<3/8m$. Then, the central subarray of *PAT* of size at least $m/4 \times m/4$ is *monochromatic*. A position containing a letter different from letter a is called a *defect*. Assume the existence of at least one defect (otherwise the preprocessing is trivial). Opposite corners of *PAT* are the corners lying on the same forward or backward diagonal of *PAT* (quad-I and quad-III, or quad-II and quad-IV corners). A *mono-central* pattern *PAT* is called a *corner* if there is a pair of opposite corners x, y of *PAT* such that each defect can be reached by at most two small vectors from x or y. The *corner* patterns are the most difficult with respect to their witness computation, because they can be radiant-periodic. The non-corner patterns are simpler to deal with, due to the following observation.

Observation If *PAT* is a periodic *non-corner mono-central* pattern, then *PAT* is non-periodic or line-periodic (therefore, *PAT* is not radiant-periodic).

The applicability of the *long-duel* algorithm follows from the next lemma.

Lemma 12.24 (subsquare lemma) Let us assume that *PAT* is mono-central, and that there is a defect inside the area $A = Cut\_corners_{2k}(PAT)$. Then there is a defect position x within A satisfying one of the following conditions:

1. x is a quad-I or quad-III corner of a $k \times k$ subsquare S containing no defect position strictly within S;

2. x is a quad-II or quad-IV corner of a $k \times k$ subsquare S containing no defect position at all, except x.

Proof Take a defect point in A closest to the center of *PAT*. ♦

Theorem 12.25 (non-corner theorem) The witness table for all small vectors of a mono-central non-corner pattern *PAT* can be computed in $O(m^2)$ time.

Proof Consider the defect z closest to the center of *PAT*. Assume without loss of generality that z is in quadrant II. The position z is the witness (of non-periodicity) for all small quad-II vectors, except perhaps for vertical vectors. The case of vertical and horizontal periodicities is very easy to process, therefore, we assume that all witnesses for vertical and horizontal vectors are computed and *PAT* is not vertically or horizontally periodic. The set of potential small quad-I periods is sparsified using vertical duels in columns. Afterward, in quadrant I we have only a linear number of candidates for small periods. Denote by X the set of these candidates. To compute witnesses for quad-I small vectors it is sufficient to find a point x satisfying the assumptions of the long-duel theorem. Take the defect x implied by the subsquare lemma. If condition (1) of this lemma holds, then x itself is the witness for all quad-I vectors that are not vertical or horizontal vectors. Otherwise, x is "good" to apply the *long-duel* theorem. The case of a horizontally- or vertically-periodic pattern can be easily processed. This completes the proof. ♦

Theorem 12.26 (corner theorem) Consider an $m \times m$ array *PAT* that is a mono-central corner pattern. Then, witnesses for all vectors of size at most $m/8$ can be computed in $O(m^2)$ time.

Proof Assume that opposite corners from the definition of corner patterns are quad-II and quad-IV corners. Then, all defects are closed to quad-II and quad-IV corners. These corners are separated by a large area of non-defects. Therefore, we can compute periods and witnesses separately with respect to each corner. Hence, without loss of generality, we can assume that all defects are close to the quad-II corner, and, in particular, that there is no defect that contains the same symbol a in quadrants I, III, and IV. Assume *PAT* contains at least one defect in quadrant II. *PAT* obviously has no small quad-II periods, since the right most defect gives witnesses against all quad-II vectors. We show how to compute witnesses for small quad-I vectors.

Let *PAT*1 be the following transformation of the pattern. In each row replace all symbols by a, except the right most non-a symbol of each row. Replace these right most non-a symbols by a special symbol \$. Let X be the set of positions containing the symbol \$; call them *special* positions.

For x in X denote by *string*(x) the word in *PAT* consisting of the part of the row containing x from the left side up to x (including x). Let π be a vector of size at most $m/4$. Then, it is easy to see the following:

π is not a period iff

1. π is not a period in *PAT*1, or

2. for two positions x, y in X we have $y-x=\pi$ and *string*(x) is not a suffix of *string*(y) (then, a witness for π is given by a mismatch between *string*(x) and *string*(y)).

The computation of all witnesses for vectors of size at most k in $PAT1$ is rather simple. Only quad-I vectors are to be processed. Assume there is no small vertical period. Then, the set of potential small quad-I periods is sparsified using duels in columns. Afterward a linear number of candidates remains. Each of them is checked against all (linear number) symbols at special positions in a naive way. The witnesses arising from condition (2) are computed directly using the *Suffix-Testing* algorithm. This completes the proof. ♦

We extend the definition of *defects*. Assume a mono-central pattern *PAT* has a lattice-periodic central subarray C of size at least $m/4$. We say that x is a *lattice-defect* iff x does not agree with (contains a symbol different from) any point y in C that is lattice-equivalent to x. Let *Mono*(*PAT*) be the pattern in which all positions that are not *lattice-defects* are replaced by the same special symbol. We omit the proof of the following simple lemma.

Lemma 12.27 (mono lemma) If a small vector π is in the lattice generated by the smallest period of C, then π is a period of *PAT* iff π is a period of *Mono*(*PAT*).

12.11* A version of the Galil-Park algorithm

Recall that the periodicity type of a subarray depends on its size. When we say that the witnesses for a given array or subarray are computed we mean the witnesses, if there are any, for all vectors that are small according to the size of the presently considered array.

Lemma 12.28 (line lemma) Assume we have a set S of points in a fixed quadrant of *PAT*, such that they are all on the same line L. Then, we can check which of them correspond to periods, and compute witnesses, wherever they are, in $O(m^2)$ time independently of the alphabet.

Proof The proof reduces to the computation of witness tables for m one-dimensional strings of size $2m$ each. Let L_i be all lines parallel to L; take m pairs of lines (L_i, K_i), in which K_i is parallel to L_i and the distance between K_i and L_i equals the distance between the point (0, 0) and L. Each line is taken as a string of symbols. For each i, lines L_i and K_i are concatenated, and witnesses for these strings are computed using a one-dimensional classical algorithm. ♦

Assume C is a central subarray of shape $s \times s$, in which $s \leq m$. Denote *Large_Extend*(C) = D, in which D is a central subarray of shape $2s \times 2s$. If ever $2s > m$, we define *Large_Extend*(C) = *PAT*. Observe that a small period with respect to *Large_Extend*(C), in case *Large_Extend*(C) ≠ *PAT*, means a vector of size at most $2d.s$ ($d = 1/16$, see Section 12.9). *Large_Extend*(C) is twice as large as C except maybe at the last iteration of the algorithm. The reason for such irregularity is that while making duels for small vectors in D we use mismatches in C, and we should guarantee that D is enough large with respect to C, and that vectors (taking part in a duel) starting in C do not go outside the pattern *PAT*. Also define *Small_Extend*(C) as a central subarray of shape $3/2s\times 3/2s$.

Due to Lemma 12.20, if *Large_Extend*(C) ≠ *PAT* and if it is radiant-periodic, then *Small_Extend*(C) is not radiant-periodic. This saves one case (radiant-periodic) in the algorithm: C is never radiant-periodic.

Before each iteration in GP algorithm the witnesses and periods are already known for a central subarray C the shape of which is $s\times s$. The witnesses for a larger central subarray D are computed, in which D = *Large_Extend*(C), or D = *Small_Extend*(C) in the case *Large_Extend*(C) is radiant-periodic and *Large_Extend*(C) ≠ *PAT*. In the latter case, D is of shape $3/2s\times 3/2s$, and Lemma 12.20 guarantees that D is not radiant-periodic. Then C is set to D and the next iteration starts.

```
Algorithm GP; { modified Galil-Park algorithm; computes witnesses for all
small vectors }
begin
  C := an initial constant-sized non radiant-periodic central subarray of PAT;
  compute the witness table of C in O(1) time;
  D := Large_Extend(C);
  while C ≠ PAT do begin {main iteration, C has s×s shape}
     if C is non-periodic then begin
          the witness table of C is used to make duels between
          candidates for small periods in D;
          after duelling, only a constant number of candidates remains;
          the witnesses for them are computed in a naive way;
     end else if C is line-periodic then begin
          consider the areas Q1, Q2 of candidates of small
          (with respect to D) periods in respectively quad I and
          quad II of D; divide each area into d.s×d.s subsquares;
          in each smaller subsquare do begin
               make duels between candidates using witnesses from C;
               only candidates on the same line survive,
               apply the algorithm from the Line Lemma;
          end;
     end else {C is lattice-periodic} begin
          let μ = period(C) ;
          for each small candidate period π ∈ Lattice(μ) do
               {use Lemma 1}
               find a μ-equivalent point y in C in quad I or quad II,
               then the witness corresponding to y gives easily a
               witness for x;
```

```
            for each small candidate period π ∈ Lattice(μ) do
                 compute witness of π in Mono(PAT); {Mono Lemma}
                 {use algorithms from Non-corner or Corner Theorems}
        end;
        if D ≠ PAT and D is radiant-periodic then
             D := Small_Extend(C)
        else begin C := D; D := Large_Extend(C) end;
    end; {main iteration}
end.
```

Three disjoint cases are considered in the algorithm depending on whether C is non-periodic, lattice-periodic, or line-periodic. The first case (non-periodic) is very simple. At each iteration we spend $O(r^2)$, in which r is the size of the actual array D; this size grows at least by a factor 3/2 at each iteration. Altogether, the time is linear with respect to the total size of the pattern, as the sum of a geometric progression.

When the witness table is eventually computed, the Amir-Benson-Farach searching phase of Section 12.2 can be applied. Altogether we have proven the following result.

Theorem 12.29 There is a 2d-pattern-matching algorithm, the time complexity of which is linear and independent of the alphabet (including the preprocessing).

Bibliographic notes

The simple linear-time (on fixed alphabets) two-dimensional pattern-matching algorithms of Section 12.1 have been found independently by Bird [Bi 77] and Baker [Ba 78].

The linear-time searching algorithm of Section 12.2 is from Amir, Benson, and Farach [ABF 92a]. It is quite surprising that this is the first alphabet-independent linear-time algorithm, because, in the case of strings, the first algorithm satisfying the same requirements is the algorithm of Morris and Pratt [MP 70]. The gap between these is more than twenty years! Furthermore, the preprocessing phase of the algorithm in [ABF 92a] is not alphabet-independent. Galil and Park have recently designed a global alphabet-independent linear-time algorithm for two-dimensional pattern matching [GP 92b]. The question of periodicity of two-dimensional patterns is discussed in several papers, particularly in [AB 92], [GP 92], and [RR 93].

The algorithm to search for non-rectangular patterns is from Amir and Farach [AF 91].

The powerful concept of a deterministic sample for strings was introduced by Vishkin in [Vi 91]. The wide applicability of this concept was recently shown by Galil [Ga 92], who designed a constant parallel-time string matching (with a linear number of processors). The sampling method of Section 12.6 is from Crochemore, Gasieniec, and Rytter [CGR 92].

The algorithm "fast on the average" is a simple application of the similar algorithm described in Chapter 4. The idea comes from [KMP 77]; see also [BR 90].

Almost optimal parallel algorithms of Section 12.8 are from Crochemore and Rytter [CR 91c].

The notion of a suffix tree on arrays is discussed by Giancarlo in [Gi 93]. Arrays are transformed into linear structures as in Section 12.5. The tree used in this section is essentially the suffix tree of Chapter 5.

The basic source for the classification of two-dimensional periodicities (Section 12.9) is the article by Amir and Benson [AB 92]. The rest of the chapter is an adaptation of the results of Galil and Park in [GP 92b].

Selected references

[ABF 92a] Amir, A., Benson, G., & Farach, M., "Alphabet-independent two-dimensional matching," in: (*Proc. 24th ACM Symp. on Theory of Computing*: 1992): 59–68.

[Ba 78] Baker, T.P., "A technique for extending rapid exact-match string matching to arrays of more than one dimension," *SIAM J.Comput. 7* (1978): 533–541.

[Bi 77] Bird, R.S., "Two-dimensional pattern matching," *Inf. Process. Lett. 6* (1977): 168–170.

[C-R 93b] Cole, R., Crochemore, M., Galil, Z., Gasieniec, L., Hariharan, R., Muthukrishnan, S., Park, K., & Rytter, W., "Optimally fast parallel algorithms for preprocessing and pattern matching in one and two dimensions," in: (*FOCS'93*, 1993): 248–258.

[GP 92b] Galil, Z., & Park, K., "Truly alphabet-independent two-dimensional matching," in: (*Proc. 33rd Annual IEEE Symposium on the Foundations of Computer Science*, 1992): 247–256.

13

Time-space optimal string-matching

In the chapter, we discuss the optimality of the string-matching problem according to both time and space complexities.

The string-matching algorithms of Chapter 3 make heavy use of the failure function *Bord*. This function gives the lengths of borders of pattern prefixes. This is equivalent to memorizing all the periods of pattern prefixes. More useful information to deal with shifts in the KMP algorithm would be the periods of words *ua*, for all prefixes *u* of *pat* and all possible letters of the alphabet *A* of the searched text. Indeed, this is realized by the minimal deterministic automaton recognizing the set of words ending with the pattern *pat*: the SMA automaton of Chapter 7. But the memory space needed to implement the automaton, in a straightforward way, is $O(|A|.|pat|)$, quantity, which depends on the size of the alphabet. This chapter presents three different solutions to string matching that all lead to algorithms running in linear time and using simultaneously only bounded space in addition to the text and the pattern. These algorithms are thus time-space optimal.

The first two presented solutions to time-space optimal string matching are GS and CP algorithms. These algorithms, called *factorized string matching*, may be considered as intermediate between KMP and BM algorithms. Like them, they run in linear time (including the preprocessing phase), but require only a bounded memory space compared to space linear in the size of the pattern for KMP and BM. The general scheme, common to GS and CP algorithms, is given in the next section.

The last section presents another more recent solution to optimal string matching. The algorithm uses a left-to-right scan of the pattern, as KMP does. It is based on a memoryless, lazy computation of periods. The great theoretical interest in scan-

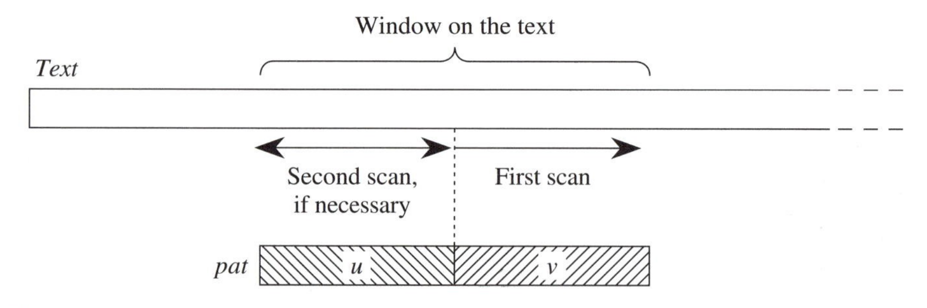

Figure 13.1: Scheme for factorized string matching.

ning the pattern from left to right is that the algorithm naturally finds all "overhanging" occurrences of the pattern in the text. And this yields a time-space optimal algorithm to compute the periods of a word.

13.1 Prelude to factorized string-matching

CP and GS algorithms have a common feature. Scans depend on a factorization uv of the pattern *pat*. The scan for a position of the window is conceptually divided into two successive phases. The first phase (right scan) consists in matching v only against *text*. The letters of v are scanned from left to right. When a mismatch is found during the first phase (right mismatch), there is no second phase and the pattern is shifted to the right. Otherwise, if no mismatch occurs during the first phase, that is, when an occurrence of v is found in *text*, the second phase starts (left scan). The left part u of the pattern is matched against the text. The word u can be scanned from right to left as in the Boyer and Moore approach, but the direction of this scan is not important. If a mismatch occurs at the second phase (left mismatch), the pattern is shifted to the right according to a rule different from the rule used in the case of right mismatch. After a shift the same process repeats until an occurrence of *pat* is eventually found, or the end of *text* is reached.

An important property of GS and CP algorithms is that the shifts they perform are computable in constant time and constant space. The scheme for factorized string matching is shown below (see also Figure 13.1).

```
Algorithm time-space-optimal-string-matching(text, pat);
{ n = |text|; m = |pat|; }
begin
   (u, v) := factorize(pat);
   pos := 0; i := |u|;
   while pos ≤ n-m do begin
        { right scan }
        while i < m and pat [i+1] = text [pos+i+1] do i := i+1;
        if i=m then begin
```

```
            { left scan }
            if text[pos+1, ..., pos+|u|] = u then return true;
        end;
        (pos, i) := shift(pos, i);
    end;
    return false;
end.
```

The preprocessing of the pattern consists of splitting it into two parts u and v ($pat = uv$). GS and CP algorithms differ greatly in the way they decompose the pattern, and also in the way they shift the pattern. We will informally explain what the main features of these algorithms are.

The decomposition of the pattern used in GS algorithm is called a *perfect factorization*. It gives a position inside the pattern from which at most one highly periodic factor starts (to the right). For the shifts, the algorithm distinguishes between two cases according to whether a highly periodic prefix of the right part of the factorization is found in the text or not.

CP algorithm uses a decomposition of the pattern called a *critical factorization*. This factorization gives a position inside the pattern at which the local period coincides with the global period of the pattern. The rule for shifts takes into account the period of the entire pattern, which is a great advantage from the point of view of practical time complexity.

The precise description of these algorithms is given in following sections. However, the following lines summarize the rules applied by the algorithms to shift the pattern along the text.

Shift in GS:

$shift(pos, i) =$ **if** $v[1 \ldots i-|u|]$ not highly periodic **then** $(pos+\lfloor(i-|u|)/3\rfloor+1), |u|)$
else $(pos+p, i-p)$,

where p is the unique small period of prefixes of v.

Shift in CP:

$shift(pos, i) =$ **if** $i<m$ **then** $(pos+i-|u|, |u|)$ **else** $(pos+p, \max(|u|, i-p))$,
where p is $period(pat)$, the smallest period of the pattern.

The main conclusion of the present chapter is summarized in Theorem 13.1 as follows.

Theorem 13.1 The string-matching problem can be solved in $O(|text|+|pat|)$ time using constant space in addition to the pattern and the text. The constants involved in "O" notation are independent of the size of the alphabet.

13.2 GS algorithm: Search phase

This section is devoted only to the search phase of first time-space optimal string-matching algorithm of the chapter, GS algorithm. The preprocessing phase is treated in Section 13.3.

GS algorithm can be regarded as a space efficient implementation of KMP algorithm (see Chapter 3). Its space efficiency is based on properties of repeated factors inside the pattern *pat*. While CP algorithm deals with squares (repetitions) from their center, GS algorithm treats repetitions (of higher order) from their left end. In the context of KMP algorithm, the idea behind saving on space is to avoid storing all the periods of prefixes of the pattern. Only small periods are memorized, where "small" is relative to the length of the prefix. Approximations of other periods are computed as necessary during the search phase.

In this section we consider an integer $k > 1$. It is a parameter of the method we are discussing, and typically we shall consider $k = 3$. Let x be a non-empty word. A primitive word w is called a *k-highly repeating prefix* of x (a *k–hrp* of x, in short) if w^k is a prefix of x. Recall that a word w is said to be *primitive* if it is not a (proper) power of any word. Primitive words are non-empty, so are k–hrps.

Example Consider the word *abaababaabab*. It has two 2–hrps, namely *aba* and *abaab*. The periods of its prefixes are shown in the next array.

		a	*b*	*a*	*a*	*b*	*a*	*b*	*a*	*a*	*b*	*a*	*b*
Prefix lengths:	0	1	2	3	4	5	6	7	8	9	10	11	12
Periods:	—	1	2	2	3	3	3	5	5	5	5	5	5
Exponents:	—	1	1	1.5	1.3	1.7	2	1.4	1.6	1.8	2	2.2	2.4

Only four prefixes have an exponent greater than or equal to 2. The shortest of them, *abaaba*, is the square of the first 2–hrp. The three other prefixes are "under the influence" of the second 2–hrp *abaab*. ♦

When w is a k–hrp of x, $period(w^2) = |w|$. Thus, we can consider the longest prefix z of x that has period $|w|$. We define *the scope of* w as the interval of integers $[|w^2|, |z|]$. Note that, by definition, any prefix of x which length falls within the scope of w has period $|w|$. Some shorter prefixes may also have the same period, but we do not address them. Figure 13.2 shows the structure of scopes of k–hrps of a word x. It so happens that scopes are disjoint (see proof of Lemma 13.3).

Example (continued) Inside the word *abaababaabab*, the scope of *aba* is [6, 6], and the scope of *abaab* is [10, 12]. ♦

We first present a simple string-matching algorithm based on the notion of scopes of k–hrps. The algorithm is a version of KMP algorithm. During a run of the

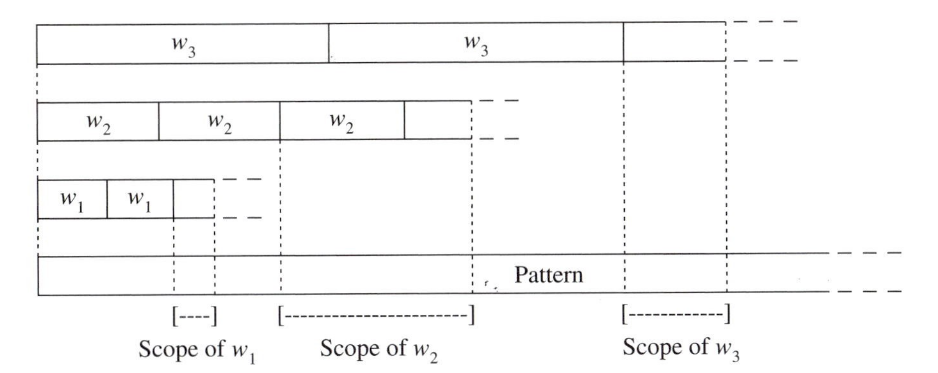

Figure 13.2: Scopes of three highly repeating prefixes (w_1, w_2, w_3).

algorithm, lengths of shifts are computed with the help of scopes of k–hrps. We assume that these intervals have been computed earlier.

```
Algorithm Simple-Search;
{ An O(r)-space version of KMP, where r is the number }
{ of k-highly repeating prefixes of pat and }
{ ([L_j, R_j] : j=1,...,r) is their sequence of scopes }
begin
    pos := 0; i := 0;
    while pos ≤ n-m do begin
        while i < m and pat[i+1] = text[pos+i+1] do i := i+1;
        if i = m then report match at position pos;
        if i belongs to some [L_j, R_j] then begin
            pos := pos+L_j/2; i := i-L_j/2;
        end else begin
            pos := pos+⌊i/k⌋+1; i := 0;
        end;
    end;
end.
```

Inside algorithm *Simple-Search*, the test "i belongs to some $[L_j, R_j]$" can be implemented in a straightforward way so that it requires $O(1)$ space, and so that its time does not affect the asymptotic time complexity of the whole algorithm.

Lemma 13.2 If the pattern *pat* has r k–hrps, algorithm *Simple-Search* runs in $O(|text|)$ time using $O(r)$ space. It makes less than $k.|text|$ symbol comparisons.

Proof To evaluate the time performance of the algorithm it can be shown that the value of expression $k.pos+i$ is strictly increased after each symbol comparison. ♦

The correctness of algorithm *Simple-Search* is a direct consequence of the following lemma. It essentially gives a lower bound on periods of prefixes which length does not belong to any scope of a highly repeating prefix.

Lemma 13.3 Let $([L_j, R_j] \,/\, j=1,\ldots,r)$ be the sequence of scopes of all k–hrps of a word x. Then any non-empty prefix u of x satisfies:

- $period(u) = L_j/2$ if $|u| \in [L_j,R_j]$ for some j,
- $period(u) > |u|/k$ otherwise.

Proof First we prove that two different scopes $[L_1, R_1]$ and $[L_2, R_2]$ are disjoint. Let w_1 and w_2 be their corresponding k–hrps, and assume that, for example, $|w_1| < |w_2|$. We show that $R_1 < L_2$. Let z be the prefix of length R_1 of x. If $L_2 \le R_1$, the square w_2^2 is a prefix of z. Therefore, the periodicity lemma applies to periods $|w_1|$ and $|w_2|$ of w_2^2. It implies that $\gcd(|w_1|, |w_2|)$ is a period of this prefix. Since $|w_1| < |w_2|$, $\gcd(|w_1|, |w_2|) < |w_2|$, we get a contradiction with the primitivity of w_2.

Let u be a non-empty prefix of x. If $|u|$ does not belong to any scope of k-highly repeating prefix, then obviously $period(u) > |u|/k$.

Assume that $|u|$ belongs to some $[L_j,R_j]$. It is then a prefix of some power w_j^e ($e \ge k > 1$) of the k-highly repeating prefix w_j. The quantity $|w_j| = L_j/2$ is a period of u. Moreover, it is the smallest period of u by the periodicity lemma again. Then, $period(u) = L_j/2$. ♦

The preprocessing phase required by algorithm *Simple-Search* is presented as follows. Its correctness is left to the reader. It is an adaptation of algorithm *Simple-Search*, and it works as if the pattern is being searched for inside itself. It also runs in linear time.

Algorithms *Simple-Search* and *Scopes* require $O(r)$ extra space to work. This space is used to store the r scopes of k–hrps of the pattern *pat*. A simple application of the periodicity lemma shows that, for $k \ge 3$, any word x has no more than $\log_{k-1}|x|$ k–hrps. This relies on the following fact: if u and v are two k–hrps of x and $|u| < |v|$, then the even stronger inequality holds: $(k-1)|u| < |v|$. For $k = 2$, the same kind of logarithmic bound also holds (see next examples). So, algorithms *Simple-Search* and *Scopes* together solve the string-matching problem in linear time with logarithmic extra space.

```
Algorithm Scopes
{ Compute the list of scopes of all k-hrp's of pat }
begin
    SCOPE := empty list;
    pos := 1; i := 0;
    while pos+i < m do begin
        while pos+i < m and pat[i+1] = pat[pos+i+1] do i := i+1;
        if pos ≤ (pos+i)/k then add [2*pos, pos+i] to end of SCOPE;
```

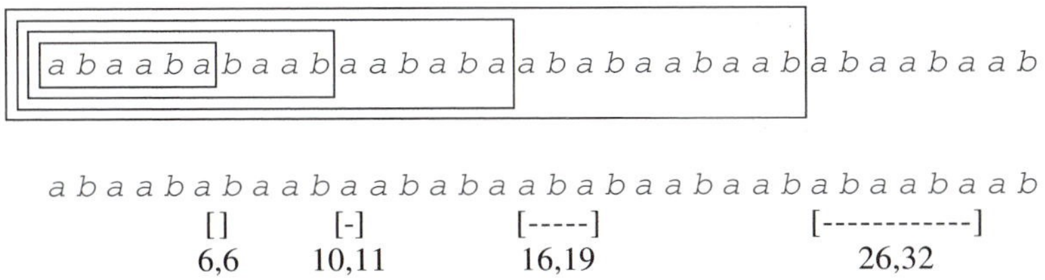

```
a b a a b a|b a a b|a a b a b a|a b a b a a b a a b|a b a a b a a b

a b a a b a b a a b a a b a b a a b a b a a b a a b a b a a b a a b
        []       [-]          [-----]              [------------]
       6,6      10,11          16,19                    26,32
```

Figure 13.3: Four scopes in a Fibonacci word. Fibonacci words have a logarithmic number of 2–hrps.

```
        if i belongs to some [L_j, R_j] in SCOPE then begin
            pos := pos+L_j /2; i = i-L_j /2;
        end else begin
            pos := pos+⌊i/k⌋+1; i := 0;
        end;
    end;
    return the list SCOPE;
end.
```

Example Let $k = 2$, and consider the Fibonacci word Fib_9 displayed in Figure 13.3 (recall that the sequence of Fibonacci words is defined by $Fib_1 = b$, $Fib_2 = a$, and $Fib_i = Fib_{i-1}Fib_{i-2}$ for $i > 2$). Figure 13.3 shows the scopes of its 2–hrps. It can be proven that the number of 2–hrps of a Fibonacci word F is $\Theta(\log|F|)$ (see following example). ♦

Example We consider words having the maximum number of square prefixes (indeed, maximum number of 2–hrps). They are the squares of words h_i, $i \geq 1$, defined inductively by

$h_1 = a$, $h_2 = aab$, $h_3 = aabaaba$, and
$h_i = h_{i-1}\, h_{i-2}$, for $i > 3$.

For example, $h_5^2 = aabaabaaabaabaaba.aabaabaaabaabaaba$. Its square prefixes are h_1^2, h_2^2, h_3^2, h_4^2, and h_5^2 itself. In other words, its 2–hrps are h_1, h_2, h_3, h_4, and h_5. Note that h_5^2 has the same length as the Fibonacci word Fib_9 (see previous example). But, this latter word has only four 2–hrps.

It can be proven that words shorter than h_i^2 start with less than i squares (of primitive words), or, equivalently, with less than i 2–hrps. Moreover, for all $i \geq 4$, word h_i is the unique word starting with i 2–hrps. ♦

We are now ready to present GS algorithm. Let $k = 3$. The idea is to scan the pattern from a position where at most one k–hrp starts. Fortunately, such a position always exists because words satisfy a remarkable combinatorial property stated in the next theorem and proven in Section 13.3.

Theorem (*perfect factorization theorem*) Any non-empty word x can be factorized into uv such that

- $|u| < 2.period(v)$, and
- v has at most one 3–hrp.

The example of pattern $aaa...a$ shows that we cannot require that the right part v of the factorization has no 3–hrp. A factorization uv of x that satisfies the conclusion of the theorem is called a *perfect factorization*. After a preprocessing phase aimed at computing a perfect factorization, GS algorithm presented below follows the scheme introduced in Section 13.1, and uses the algorithm *Simple-Search*.

```
Algorithm GS; { informal description }
{ search for pat in text }
begin
   (u, v) := perfect factorization of pat;
   find all occurrences of v in text with Simple-Search;
   for each position q of an occurrence of v in text do
        if u ends at q then report the match at position q-|u|;
end.
```

The following theorem shows that GS algorithm is time-space optimal.

Theorem 13.4 GS algorithm computes all positions of occurrences of *pat* within *text*. The algorithm runs in $O(|pat|+|text|)$ time, and uses a bounded memory space. The number of letter comparisons made during the search phase is less than $5|text|$.

Proof The correctness mainly comes from that of algorithm *Simple-Search*. We assume that the first statement of GS algorithm runs in $O(|pat|)$ time. This assumption is satisfied by the preprocessing algorithm of Section 13.3.

According to Lemma 13.2 the search for v in *text* uses at most $3.|text|$ symbol comparisons. The distance between two consecutive occurrences of v in *text* is not less than $period(v)$. Thus, the condition $|u| < 2.period(v)$ implies that a given letter of *text* is matched against a letter of u (during the test "u ends at q ?") at most twice, which globally gives at most $2.|text|$ more symbol comparisons. ♦

13.3 Preprocessing the pattern perfect factorization

This section is devoted to the preprocessing phase of GS algorithm presented in the previous section. The preprocessing consists of computing a perfect factorization of the pattern. The method relies on a constructive proof of the perfect factorization theorem. We first prove that such a factorization exits, and then, we analyze the com-

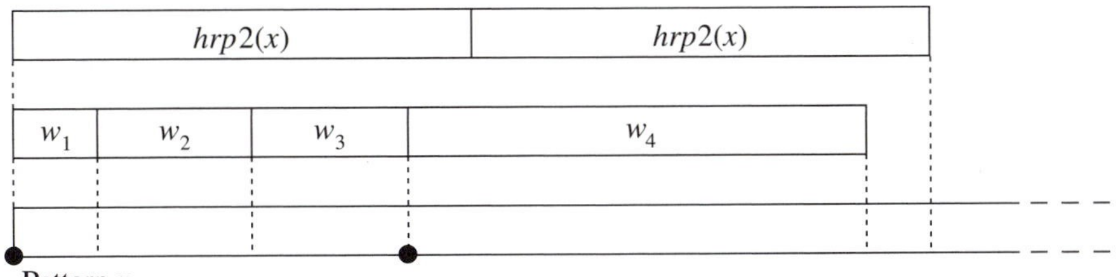

Figure 13.4: The first two testing points • . The working factor *w4* has the same length as hrp2(x).

plexity of the factorization algorithm. The parameter k of Section 13.2 is set to 3 throughout the present section. Therefore, the hrps (highly repeating prefixes) that are considered should be understood as 3–hrps, but all the results are also valid for k–hrps with $k > 3$. Thus we are interested in cubes occurring in the pattern. If the pattern is cube-free, no preprocessing for GS algorithm is even needed.

In the following, we denote by hrp1(x) and hrp2(x), respectively, the shortest and the second shortest hrp of a given non-empty word x, when they are defined.

The structure of the factorization algorithm is based on a sequence $W(x)$ of hrp1s. The elements of the sequence $W(x) = (w_1, w_2, \ldots, w_r)$ are called *the working factors of* x, and are defined as follows. The first element w_1 is hrp1(x) if x starts with a cube, and is x itself otherwise. Let $x = w_1x'$, and assume that x' is a non-empty word. Then w_2 is defined in the same manner on x', and so on, until there is no hrp. The last element of the sequence is a suffix of x that does not start with a cube. In particular, the sequence is reduced to x itself if it does not start with a cube.

We shall see that there is a perfect factorization uv of x of the form $u = w_1w_2 \ldots w_j$, $v = w_{j+1} \ldots w_r$. Therefore, a natural procedure for factorizing x would be to compute the positions of working factors, and test, at each position, whether there are two hrps starting there. The procedure would stop as soon as at most one hrp is found starting at the current position. Unfortunately, this could cost a quadratic number of comparisons. In order to keep the time complexity linear, the test is done only at some specific positions, called *testing points* (see Figure 13.4). The first testing point on x is the position 0. When x has no second hrp, there is no more testing point, and the factor u of the perfect factorization of x is empty. Assume now that x has a second hrp, z = hrp2(x). Then, the second testing point is the last position of a working factor within z. In other words, the second testing point is $|w_1w_2 \ldots w_j|$, where j is the largest integer such that $|w_1w_2 \ldots w_j| < |z|$. The remainder of the sequence of testing point (see Figure 13.5) is defined in the same way on the rest of the word x (i.e., on $w_{j+1} \ldots w_r$).

The preprocessing phase of GS algorithm, which reduces to the computation of a perfect factorization of the pattern, is shown as follows. It computes the sequence of positions i of working factors from left to right, and checks at each testing point whether a second hrp (or cube) starts.

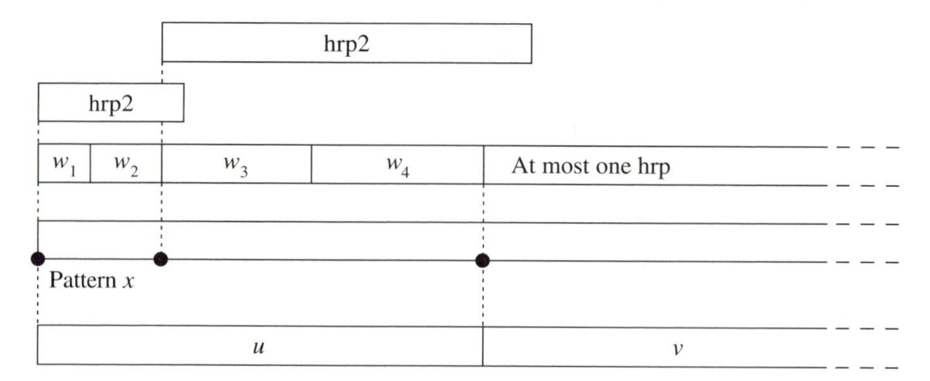

Figure 13.5: The sequence of working factors $w_1, w_2, \ldots$ leads to testing points• , and eventually to a perfect factorization uv of x.

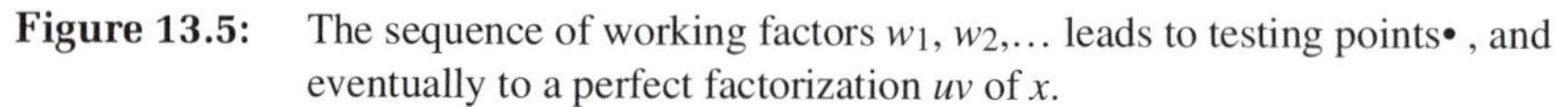

```
Algorithm Perfect_fact;
{ preprocessing pat for GS algorithm; m = |pat|}
{ computes a perfect factorization of pattern pat}
begin
   i := 0; h1 := |hrp1(pat)|; h2 := |hrp2(pat)|;
   while h1 and h2 exist do begin
      i := i+h1; h1 := |hrp1(pat [i+1...m])|;
      if h1 ≥ h2 then h2 := |hrp2(pat [i+1...m])|;
   end;
   return factorization (pat [1...i], pat [i+1...m]) of pat;
end.
```

Theorem 13.5 Algorithm *Perfect_fact* computes a perfect factorization of the pattern *pat*. It runs in $O(|pat|)$ time and requires constant extra space.

Proof The proof strongly relies on Lemma 13.6 following.

Correctness A straightforward verification shows that variable i runs through the positions of working factors of *pat*, $w_1, w_2, \ldots, w_r$. At the end of the execution of the algorithm, the value of i is the last testing point on *pat*, $|w_1w_2\ldots w_{r-1}|$. The output is the factorization uv defined by $u = w_1w_2\ldots w_{r-1}$ and $v = w_r$. By construction, v starts with at most one hrp. Thus, what remains to be proven is that the condition on lengths is also satisfied, $|pat[1\ldots i]| < 2.period(pat[i+1\ldots|pat|])$, or, equivalently, to prove the inequality $|w_1w_2\ldots w_{r-1}| < 2.period(w_r)$.

The property is obviously true if the sequence $(w_1, w_2, \ldots, w_r)$ is reduced to *pat* itself, because in this situation u is the empty word. Thus, we assume $r > 1$, which implies that the sequence of hrp2s computed by the algorithm is not empty. Let it be $(z_1, z_2, \ldots, z_t)$ (note that $0 < t < r$). We certainly have $|u| < |z_1|+|z_2|+\ldots+|z_t|$. We also have $|z_t| \leq period(v)$ because the inverse would contradict part (c) of Lemma 13.6. Furthermore, by part (d) of Lemma 13.6, we get $|z_j| < |z_t|/2^{t-j}$. The conclusion follows:

$$|u| < |z_1|+|z_2|+\ldots+|z_t| < 2.|z_t| \leq 2.period(v),$$

This ends the proof of correctness of the algorithm.

Complexity First note that algorithm *Scopes* of Section 13.2 can be used to compute hrp1s (resp. hrp2s). The trick is simply to stop the execution of the algorithm the first time it discovers the first hrp (resp. the second hrp). By doing so, the computation of hrp1(y) for a given word y takes $O(|\text{hrp1}(y)|)$ time if hrp1(y) exists, and $O(|y|)$ otherwise. The same is true for hrp2s. The extra space needed to compute these values is constant because the list *SCOPE* used in algorithm *Scopes* contains at most one element. Thus, the global algorithm also needs only constant extra space.

The time complexity of the algorithm may be analyzed as follows. The total cost of computation of hrp1s, as per the above argument, is proportional to the total length of working factors (including the last one that may have no hrp), which is exactly $|pat|$. The same argument also shows that the total cost of computation of hrp2s is proportional to their total length, plus the length of v (last test): $|z_1|+|z_2|+\ldots+|z_t|+|v|$. We have already seen above that $|z_1|+|z_2|+\ldots+|z_t| < 2.|z_t|$, and, since $2.|z_t| < |pat|$ (indeed $3.|z_t| \leq |pat|$), the previous sum is less than $2.|pat|$. The time complexity of the algorithm is thus linear as expected. ♦

The following lemma may be regarded as a constructive version of the perfect factorization theorem. Part (d), which generally asserts that each hrp2 is at least twice as long as the preceding one, is the key point used both in the correctness and in the time analysis of algorithm *Perfect_fact* (proof of Theorem 13.5).

Lemma 13.6 Let $W(x) = (w_1, w_2, \ldots, w_r)$ be the sequence of working factors for x.

(a) If $j < k$, then w_j is a prefix of w_k; lengths of w_is are in non-decreasing order.

Assume that both hrp2(x) exists, and x has a second testing point $i = |w_1 w_2 \ldots w_j|$.

(b) hrp2(x) is at least twice as long as w_1, $|\text{hrp2}(x)| > 2.|w_1|$.

(c) If $j < r$, the next working factor is no shorter than hrp2(x), $|w_{j+1}| \geq |\text{hrp2}(x)|$.

(d) If hrp2($x[i+1\ldots|x|]$) exists, it is at least twice as long as hrp2(x).

Proof Parts (a) and (b) are mere applications of the periodicity lemma (see Chapter 2). Part (d) is a consequence of (b) and (c) together. Therefore, what remains is to prove part (c). Let h be hrp2(x). We want to show that w_{j+1} cannot be shorter than h.

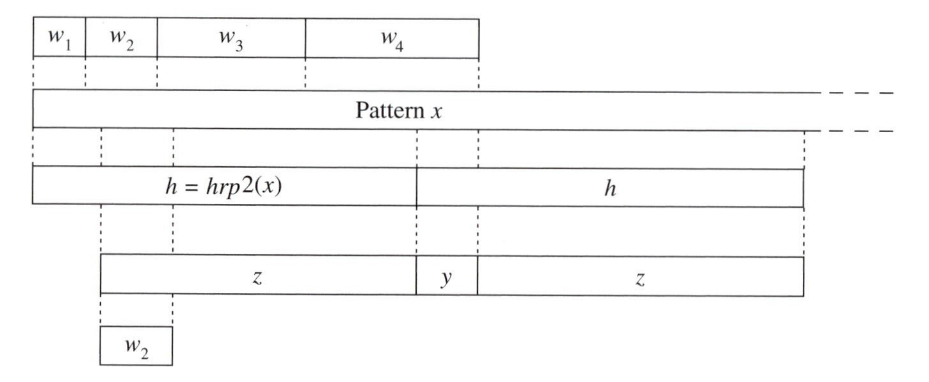

Figure 13.6: Proving (c), i.e., on the figure $|w4| \geq |h|$. Case 1: impossible because $w2$ is primitive.

The assumptions imply that w_{j+1} overlaps the boundary between the first two occurrences of h, as shown in Figures 13.6 and 13.7. Assume that $|w_{j+1}| < |h|$ holds. Let y and y' be the non-empty words defined by the equations $h = w_1w_2...w_jy'$, and $w_{j+1} = y'y$. Because we assume that w_{j+1} is shorter than h, y is a prefix of h, so that we can consider the word z defined by $h = yz$. We now focus our attention on the occurrence of the word $g = zy$, rotation of h, occurring at position $|y|$ in x.

Case 1 (Figure 13.6). We first consider the situation in which the beginning of g inside h falls properly within some w_k. Integer k is less that j+1 because $|w_{j+1}| < |g|$. Since $h = \text{hrp2}(x)$, h^3 is a prefix of x. Therefore, another occurrence of g immediately follows the occurrence we consider. Because w_{j+1} is an hrp1, w_{j+1} is a prefix of g. Part (a) implies that w_k is a prefix of both w_{k+1} and w_{j+1}, and therefore is also a prefix of g. This eventually implies that w_k is an internal factor of w_kw_k, a contradiction to the primitivity of w_k (consequence of the weak periodicity lemma).

Case 2 (Figure 13.7). The second situation is one in which g (= zy) = $w_k...w_jw_{j+1}$. Again, hypothesis $|w_{j+1}| < |h|$ leads to $k < j+1$. This implies $2|w_k| \leq |h|$, and then, the word w_k is an hrp of zyz. Thus, just after the occurrence of g that is considered, w_k is still an hrp1. This proves, through part (a), that $w_{j+1} = w_k$. But then, g is a nontrivial power of w_k, and its rotation $h = \text{hrp2}(x)$ is not primitive, a contradiction.

This completes the proof of Lemma 3.14. ♦

Note We can also consider perfect factorizations with the parameter k = 2. But then, not all words have a perfect factorization. A counterexample is given by the word $x = (aabaabab)^4$. The longest suffix of x having only one 2–hrp (or one square prefix) is v = $baababaabaabab$ of period 8. The corresponding prefix u = $(aabaabab)^2aa$ has length 18. With exponent e instead of 4, we can make the ratio

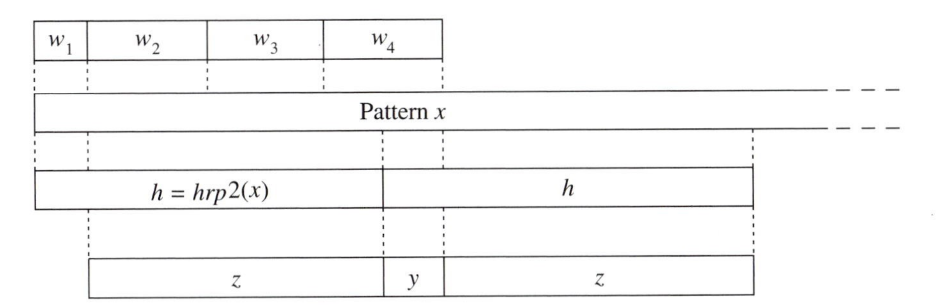

Figure 13.7: Proving (c), i.e., on the figure $|w_4| \geq |h|$. Case 2: $|w_2| = |w_3| = |w_4|$, impossible because $h = \text{hrp2}(x)$ is primitive.

$|u|/period(v)$ as large as necessary, provided the e chosen is large enough. Therefore, there is no statement equivalent to the perfect factorization theorem for squares.

13.4 CP algorithm: Search phase

This section deals with the search phase of the second time-space optimal string-matching algorithm of the chapter, CP algorithm. The preprocessing phase of the whole algorithm, and the proof of the combinatorial theorem on which CP algorithm is based, are given in the next section. First we will introduce the notions on words related to the design of CP algorithm, and afterward, we will present its search phase.

Let x be a non-empty word, and let uv be a factorization of x. Denote by l the length of u ($0 \leq l \leq |x|$). A non-empty word w is called a *repetition for the factorization uv*, or a *repetition at position l* in x, if the following two conditions are satisfied:

- w is a suffix of u, or u is a suffix of w,
- w is a prefix of v, or v is a prefix of w.

The generic situation for a repetition is one in which the cut between factors u and v of x is the center of a square occurring in x (see Figure 13.8). The other cases correspond to a cut close to the ends of x, or equivalently to an overflow of w. Note that the word $w = vu$ is a repetition for uv, so that any factorization of x has a repetition.

The length r of a repetition for the factorization uv is called a *local period for uv*, or a *local period of x* at position l. The smallest possible value of r is called *the local period for uv*, and denoted by $r(u, v)$. It is convenient to reformulate the definition of a local period as follows. A positive integer r is a local period of x at position l if $x[i] = x[i+r]$ for all indices i such that $l-r+1 \leq i \leq l$, and such that both sides of the equation are defined (see Figure 13.8). The word w of the previous definition, repetition at position l, is then defined by

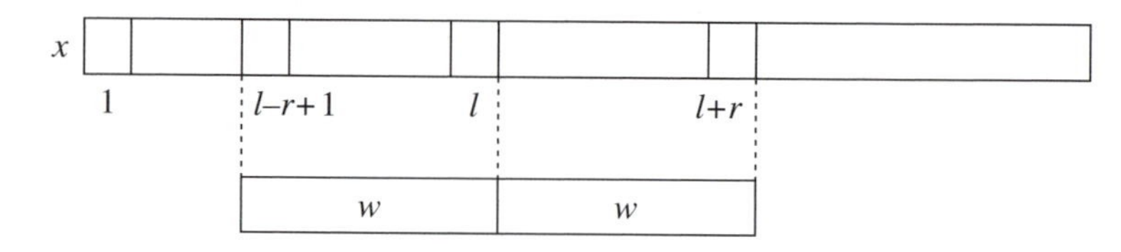

Figure 13.8: A local period r, and repetition w at position l in x.

$w[i] = x[l+i]$ or $w[i] = x[l-r+i]$

(for $1 \le i \le r$) according to whatever expression is defined on the right-hand side.

Example Consider the word $x = abaabaa$. Its local period at position 3 is $r(aba, abaa) = 1$, which corresponds to the repetition $w = a$. At position 2, $r(ab, aabaa) = 3$, and the shortest repetition is $w = aab$.

The local period of the word $x = aababab$ at position 6 is $r(aababa, b) = 2$ with repetition $w = ba$. Finally, the shortest repetition at position 2 is $r(aa, babab) = 7$, corresponding to $w = bababaa$. ♦

It is a natural consequence of the definitions that, when $x = uv$,

$1 \le r(u, v) \le period(x)$.

A factorization uv of x that satisfies $r(u, v) = period(x)$ is called a *critical factorization*, and the position $l = |u|$ is called a *critical position* of x. At a critical position, the local period coincides with the smallest period of the whole word. The string-matching algorithm of this section relies on the combinatorial property of words stated in the following theorem. Its proof is given in Section 13.5.

Theorem (*critical factorization theorem*) Any word x has at least one critical factorization uv (i.e., $r(u, v) = period(x)$). Moreover, u can be chosen such that $|u| < period(x)$.

Example The (smallest) period of the word *aabababaab* is 7. Its local periods are given in the next array.

	a	a	b	a	b	a	b	a	a	b	
Positions:	0	1	2	3	4	5	6	7	8	9	10
Local periods:	1	1	7	2	2	2	2	7	1	3	1

This shows that factorizations (*aa*, *bababaab*) and (*aababab*, *aab*) are critical. ♦

CP algorithm is designed according to the scheme of Section 13.1. In the following, it is assumed that uv is a critical factorization of *pat* that satisfies $|u| <$

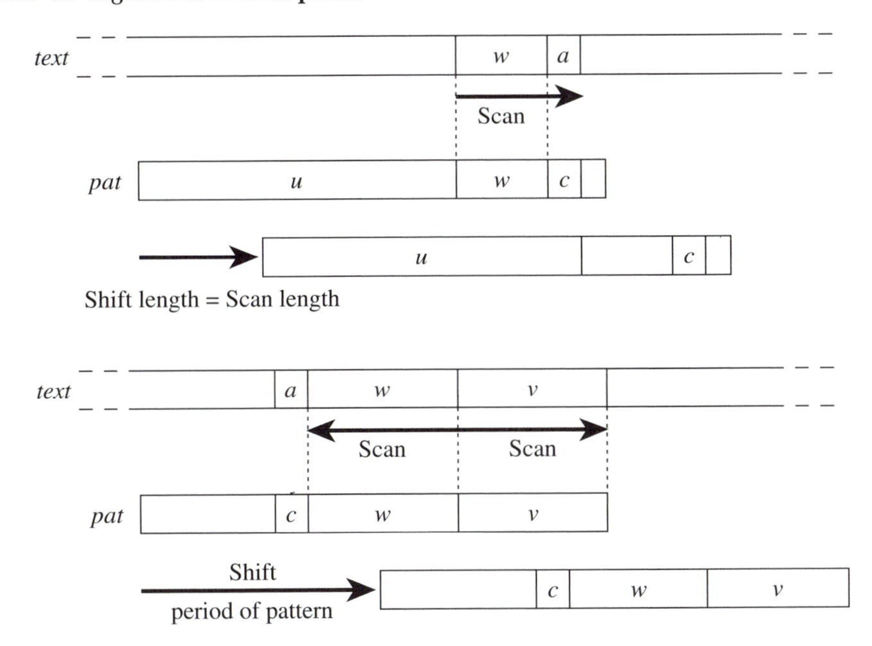

Figure 13.9: CP string matching: shift after a mismatch on the right (top), and after a mismatch on the left (bottom).

period(*pat*). The rules that guide shifts in CP algorithm are relatively simple. In case of a mismatch during the right scan, the shift pushes the cut of the critical factorization to the right of the letter of the text that caused the mismatch. Otherwise, the length of the shift is the period of *pat* (see Figure 13.9).

The searching algorithm has an additional feature: prefix memorization. This trick has already been used in Chapter 4 to improve on the worst-case time complexity of BM algorithm (see Section 4.3). It is used here with the same purpose.

The first version of CP algorithm is presented below as CP1 algorithm. It assumes that both the period *period*(*pat*) and a critical factorization uv with $|u| <$ *period*(*pat*) have been computed previously. Precomputations are discussed in the next section.

```
Algorithm CP1; { informal description, see Figure 13.9 }
{ uv is a critical factorization of pat }
begin
   align left ends of pat and text;
   while not end of text do begin
      { right scan } scan v from left to right against text;
      if mismatch then shift pat of the length of the scan
      else begin
         { left scan } scan u against text,
         and report possible match;
```

```
            shift pat of length period(pat);
         end;
      end;
   end.

Algorithm CP1;
{ search for pat in text; m = |pat|; n = |text| }
{ p = period(pat) }
{ uv is a critical factorization of pat such that |u| < p }
begin
   pos := 0; s := 0;
   while (pos+m ≤ n) do begin
      i := max(|u|, s)+1;
      { right scan }
      while (i ≤ m and pat [i] = text [pos+i]) do i := i + 1;
      if (i ≤ |pat|) then begin
         pos := pos+i-|u|; s := 0;
      end else begin
         j := |u|;
         { left scan }
         while (j > s and pat [j] = text [pos+j]) do j := j - 1;
         if (j ≤ s) then report match at position pos;
         pos := pos + p; s := m - p;
      end;
   end;
end.
```

The algorithm uses four local variables: i, j, s, and *pos*. The variables i and j are used as cursors on the pattern to perform the scan on each side of the critical position respectively (see Figure 13.10). The variable s is used to memorize a prefix of the pattern that matches the text at the current position, given by the variable *pos* (see Figure 13.11). Variable s is updated at every mismatch. It can be set to a nonzero value only in the case of a mismatch occurring during the scan of the left part of the pattern.

Before proving the correctness of the algorithm, we first present a property of a critical factorization of a word.

Lemma 13.7 Let uv be a critical factorization of a non-empty word x. If w is both a suffix of u and a prefix of v, then $|w|$ is a multiple of $period(x)$.

Proof The property trivially holds if w is the empty word. Otherwise, since $period(x)$ is also a period of w, this word can be written $(yz)^e y$ with $|yz| = period(x)$, z non-empty, and $e > 0$. If y is non-empty, it is a repetition for uv. But $|y| < period(x)$ contradicts the fact that uv is a critical factorization (see Figure 13.12). Hence, $|w| = e.period(x)$ as expected. ♦

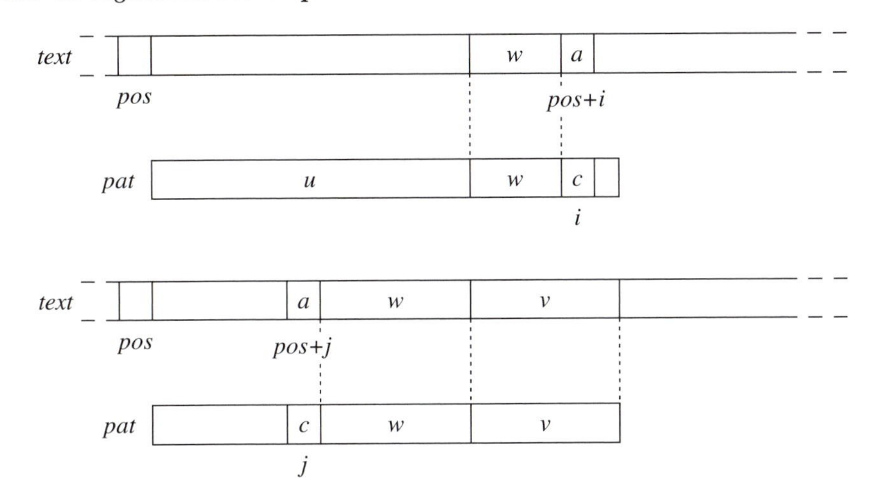

Figure 13.10: The role of variables *pos*, *i*, *j*.

Lemma 13.8 (correctness of CP1) CP1 algorithm computes the positions of all occurrences of *pat* within *text*.

Proof Let $q_1, q_2, \ldots, q_K$ be the successive values of the variable *pos* during a run of CP1 on inputs *pat* and *text*. It is clear that the algorithm reports position q_k if and only if it is a position of an occurrence of *pat* in *text*. Therefore, it still must be shown that no matching position is missed:

(*) any position of an occurrence of *pat* in *text* is some q_k.

Before coming to the proof of (*), the reader may note that the following property is an invariant of the main "while" loop:

$$pat[s'] = text[pos+s'],\ 1 \le s' \le s,$$

i.e., the prefix of *pat* of length s occurs in *text* at position *pos*.

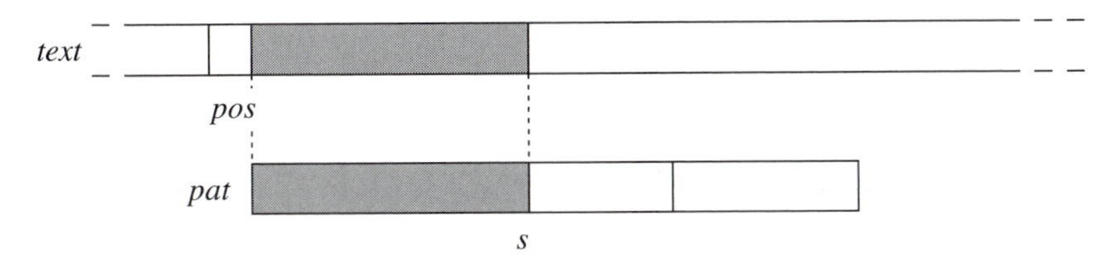

Figure 13.11: The role of variable s: prefix memorization.

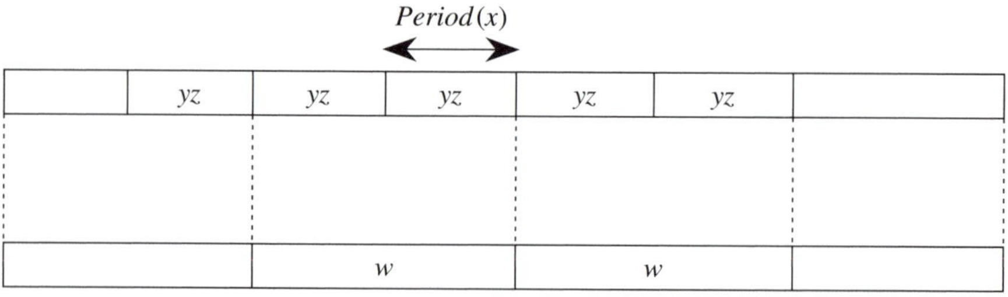

Figure 13.12: A repetition w for a critical factorization.

Proof of (*) We prove that no position q, strictly between two consecutive values of *pos*, can be a matching position. Let q be a matching position such that $q_k < q$. Consider the step of the main "while" loop in which the initial value of *pos* is q_k. Let i' be the value assigned to the variable i by the right scan, and let s' be the value of s at the beginning of the step. We now consider two cases according to a mismatch occurring in the right part or in the left part of the pattern. In both cases, we use the following argument: if, after a mismatch, the shift is too small, then its length is a multiple of the period of the pattern, which implies that the same mismatch recurs.

Case 1 (Figure 13.13). Let $l = |u|$. If $i' \le |pat|$ a mismatch has occurred during the right scan, and we have:

$$pat[l+1\ldots i'-1] = text[q_k+l+1\ldots q_k+i'-1], \text{ and } pat[i'] \neq text[q_k+i'].$$

Let w be the word $text[q_k+l+1\ldots q+l]$. If $q < q_k + i'-l$, the above equation implies

$$pat[l+1\ldots l+q-q_k] = w.$$

Since q is a position of *pat* in *text* the suffixes of length $\max(1, l+1-q+q_k)$ of w and of *pat* $[l+1-q+q_k\ldots l]$ coincide. The quantity $q-q_k$ is then a local period at the critical position $|u|$ and, according to Lemma 13.7, $q-q_k$ is a multiple of *period*(*pat*). Therefore, $pat[i'] = pat[i'-q+q_k]$. But, since q is a matching position, we have $pat[i'-q+q_k]$

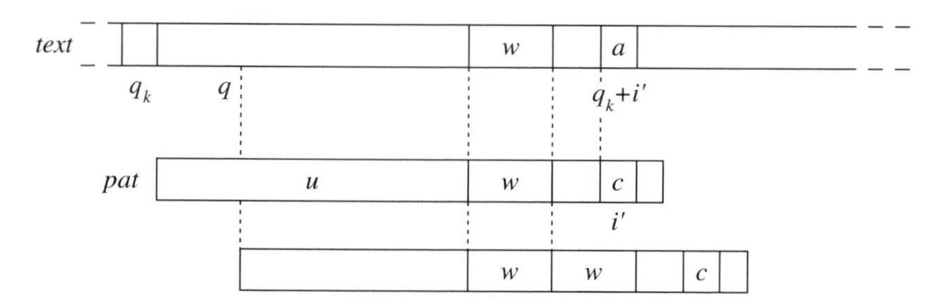

Figure 13.13: Case of a right mismatch.

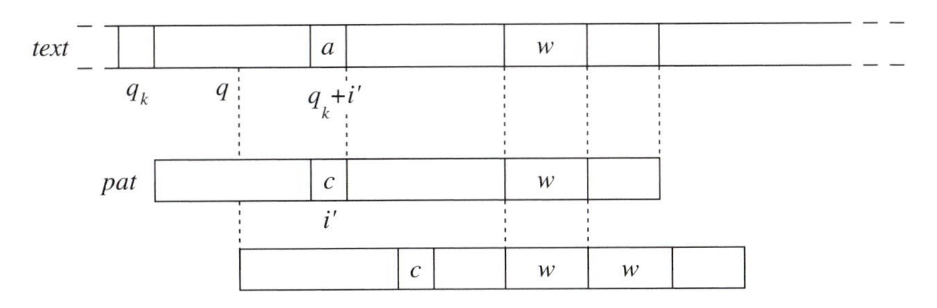

Figure 13.14: Case of a left mismatch.

= $text[q+i'-q+q_k]$ which presents a contradiction with the mismatch $pat[i'] \neq text[q_k+i']$. This proves $q \geq q_k + i' - |u|$.

So far, we have proven that, if a mismatch occurs during the right scan, q is greater than or equal to $q_k+i-|u|$, quantity which is exactly q_{k+1}.

Case 2 (Figure 13.14). If no mismatch is met during the right scan, the right part v of *pat* occurs at position $q_k+|u|$ in *text*. The word $w = text[q_k+|u|+1 \ldots q+|u|]$ then occurs in *pat* at position $|u|$. Since q is a matching position, w also occurs at the left of position $|u|$. Thus, $|w|$ is a local period at the critical position $|u|$, and $|w| \geq period(pat)$. We get $q-q_k \geq period(pat)$. Since $q_{k+1} = q_k+period(pat)$ again, in this second case, the inequality $q \geq q_{k+1}$ holds. This proves assertion (*), and ends the proof of the theorem. ♦

The time complexity of CP1 algorithm is proportional to the number of comparisons between letters of *pat* and *text*. This number is bounded by $2.|text|$ as shown by the following lemma.

Lemma 13.9 The execution of CP1 algorithm uses less than $2.|text|$ symbol comparisons.

Proof Each comparison done during the right scan strictly increases the value of *pos+i*. Hence, their number is at most $|text| - |u|$, because expression *pos+i* has initial value $|u|+1$ and terminal value $|text|$ in the worst case.

During the left scan, comparisons are done on letters of the left part of the pattern. After the left scan, the length of the shift is *period(pat)*. Since, by assumption, the length of u is less than *period(pat)*, two different comparisons performed during left scans are executed on letters of *text* occurring at different positions within *text*. Then, at most $|text|$ letter comparisons are performed during all left scans.

This gives the upper bound $2.|text|$ to the number of letter comparisons. ♦

CP1 algorithm uses the period of the pattern. A previous computation of this period is possible with KMP algorithm. But since we want an algorithm that is globally linear in time with constant extra space, it is desirable to improve on the precomputation of the period of the pattern. There are two ways to achieve this goal. One way is a direct computation of the period by an algorithm working in linear time and constant space. Such an algorithm is described in Section 13.6. The approach described here is different. It avoids the use of the period of the pattern in some situations. A variant CP2 is designed for that purpose. The key point is that the period of the pattern is actually needed only when it is small, which roughly means that the period is less than half the length of the pattern. Therefore, CP2 algorithm is intended to be used when the period of the pattern is large. This approach has the additional advantage of keeping the maximal number of comparisons spent by the algorithm relatively small.

CP2 algorithm is described below. It differs from CP1 in two aspects. First, the "prefix memorization" is no longer used. Second, it treats mismatches occurring during left scans (left mismatches) differently. In this situation, instead of shifting the pattern *period*(*pat*) places to the right, it is only shifted q places to the right with $q \leq$ *period*(*pat*). Since q can be less than *period*(*pat*), prefix memorization is indeed impossible.

The correctness of CP2 algorithm is straightforward from that of CP1. Note that, if we choose $q = 1$, we get a kind of naive algorithm that is absolutely inefficient. In fact, CP2 is to be applied with an integer q satisfying the additional condition $q > \max(|u|, |v|)$ (recall that uv is a critical factorization of *pat*). With this assumption, the maximal number of symbol comparisons used by CP2 is less than $2.|text|$, as it is for CP1.

```
Algorithm CP2;
{ search for pat in text; m = |pat|; n = |text| }
{ search phase without p = period(pat) }
{ q satisfies 0 < q ≤ p }
{ uv is a critical factorization of pat such that |u| < p }
begin
   pos := 0;
   while (pos+m ≤ n) do begin
      i := |u|+1;
      { right scan }
      while (i ≤ m and pat [i] = text [pos+i]) do i := i + 1;
      if (i ≤ m) then
         pos := pos + i-|u|;
      else begin
         j := |u|;
         { left scan }
         while (j > 0 and pat [j] = text [pos+j]) do j := j - 1;
         if (j = 0) then report match at position pos;
         pos := pos + q;
      end;
   end;
end.
```

Lemma 13.10 CP2 algorithm computes the positions of all occurrences of *pat* inside *text*. Furthermore, if the parameter q of the algorithm satisfies $q > \max(|u|, |v|)$, the number of letter comparisons used by CP2 is less than $2.|text|$.

Proof One can prove the first assertion by reproducing a simplified version of the proof of correctness of CP1 (Lemma 13.8).

First we prove that the total number of comparisons performed during right scans is bounded by $|text|$. Consider two consecutive values k and k' of the sum $pos+i$. If these values are obtained during the same scan, then $k' = k+1$ because i is increased by one unit. Otherwise, *pos* is increased either by instruction "$pos := pos + i - |u|$" or by instruction "$pos := pos + q$". In the first case, $k' = k - |u| + |u| + 1 = k+1$ again. In the second case, $k' \geq k+q-|u|$, and the assumption $q > \max(|u|, |v|)$ implies $k' \geq k+1$. Since comparisons performed during right scans strictly increase the value of $pos+i$ (which has initial value $|u|+1$ and final value at most $|text|+1$), the claim has been proven.

We show that the number of comparisons performed during left scans is also bounded by $|text|$. Consider two values k and k' of the sum $pos+j$ obtained during two consecutive left scans, respectively. Let p be the value *pos* has during the first of these two scans. Then $k \leq p+|u|$, and $k' \geq p' = p+q$. The assumption $q > \max(|u|, |v|)$ implies $k' \geq k+1$. Thus, no two letter comparisons made during left scans are performed on a same letter of *text*, which proves the claim. The total number of comparisons is thus bounded by $2.|text|$. ♦

The complete CP string-matching algorithm is shown below. It calls for a procedure to compute a critical factorization *uv* of the pattern, suitable for CP1 and CP2 algorithms. Moreover, as we shall see in Section 13.5, the procedure computes the period of the right part *v* without any additional cost. After this preprocessing phase, a simple test allows us to decide which version of the search phase is to be run, CP1 or CP2.

```
Algorithm CP { search for pat in text }
begin
   (u, v) := critical factorization of pat such that |u| < period(pat);
   p := period(v);
   if (u is a suffix of v[1...p]) then
        { p = period(pat) }
        run CP1 algorithm on text using u, v, and period p
   else begin
        q := max(|u|, |v|) + 1;
        run CP2 algorithm on text using u, v, and parameter q;
   end;
end.
```

Theorem 13.11 CP algorithm computes the positions of all occurrences of *pat* within *text*. The algorithm runs in $O(|pat|+|text|)$ time, and uses a bounded memory space. The number of letter comparisons made during the search phase is less than $2.|text|$.

Proof We assume that the first instruction of CP algorithm runs in $O(|pat|)$ time with bounded extra space. An algorithm satisfying this condition is presented in Section 13.5.

We have to prove that, after the first statement and the succeeding test, conditions are met to realize the search phase either with CP1 or with CP2 algorithms.

Assume first that u is a suffix of $v[1...p]$. Then, obviously, the integer p is also a period of *pat* itself. Thus, CP works correctly because CP1 does (see Lemma 13.8).

Assume now that u is a not a suffix of $v[1...p]$. In this situation the correctness of CP depends on that of CP2. The conclusion comes from Lemma 13.10 if we prove that $q = \max(|u|, |v|)+1$ satisfies $q \le period(pat)$. And, since $|u| < period(|pat|)$ by assumption, it remains to be shown that $|v| < period(|pat|)$. Equivalently, because uv is a critical factorization, we show that v is shorter than the local period for uv.

We prove that there is no non-empty word w such that wu is a prefix of *pat*. Assume the contrary, that is, wu is prefix of *pat*. If w is non-empty its length is a local period for uv, and then, $|w| \ge period(pat) \ge period(v)$. We cannot have $|w| = period(v)$ because u is not a suffix of $v[1...p]$. Neither can we have $|w| > period(v)$ because this would lead to a local period for uv strictly less than $period(v)$, a contradiction. This proves the assertion, and ends the correctness of CP algorithm.

The conclusions on the running time of CP algorithm, and on the maximum number of comparisons executed by CP algorithm come readily from Lemmas 13.9 and 13.10, and from the assumption made at the beginning of the proof on the first instruction. ♦

Note We shall see that the first instruction of CP algorithm uses less than $4.|pat|$ symbol comparisons of the kind "<, =, >". Since the test "u is a suffix of $v[1...p]$?" takes at most $|pat|/2$ comparisons, the preprocessing phase globally uses less than $4.5.|pat|$ comparisons. ♦

We end this section by presenting examples of the behavior of CP algorithm. The examples are presented in Figure 13.15 in which the letters of the pattern scanned during the search phase of the algorithm are underlined.

On pattern $pat = a^n b$ the algorithm finds its unique critical factorization (a^n, b). The search for *pat* within the text $text = b^m$ uses $2|text|/|pat|$ comparisons, as does BM algorithm. Both algorithms attempt to match the last two letters of *pat* against the letters of *text*, and shift *pat* $n+1$ places to the right as shown in Figure 13.15 (i).

The pattern $pat = a^n ba^n$ has period $n+1$. Its critical factorization computed by CP algorithm is (a^n, ba^n). CP algorithm behaves like CP1, and uses $2|text|-2e-2$ comparisons to match *pat* in $text = (a^n ba)^e a^{n-1}$ (see Figure 13.15 (ii)). The same number

```
bbbbbbbbb . . .      aabaaabaaabaa . . .      aaabaaabaaabaa
 aab                 aabaa                    aaabaa
    aab                 aabaa                     aaabaa
                         aabaa                        aaabaa

  (i)                    (ii)                     (iii)
```

Figure 13.15: Behavior of the search phase of CP algorithm.

of comparisons is reached when searching for $pat = a^n b a^{n-1}$ inside $text = (a^n b)^e a^{n-1}$, but, in this latter case, the algorithm behaves like CP2 (see Figure 13.15 (iii)).

13.5* Preprocessing the pattern: Critical factorization

In this section we are interested in the computation of critical factorizations, which is the central part of the preprocessing of CP algorithm. Among the existing proofs of the critical factorization theorem for x, one relies on the property that if $x = ayb$ (a, b are letters), then a critical factorization of x comes either from a critical factorization of ay, or from a critical factorization of yb. This leads to a quadratic algorithm. Another proof relies on the notion of a Lyndon factorization. It leads, via the use of a linear time string-matching algorithm, to a linear-time algorithm for computing a critical factorization. This method is not suitable for our purpose, because our aim is to incorporate the algorithm into a string-matching algorithm.

The proof of the critical factorization theorem presented here gives a method both practically and algorithmically simple for computing a critical position, and that, in addition, uses only constant additional memory space. The method relies on a computation of maximal suffixes that is presented afterward.

Remark A weak version of the critical factorization theorem occurs if one makes the additional assumption that the inequality $3.period(x) \le |x|$ holds. Indeed, in this case, one may write $x = l.w.w.r$ where $|w| = period(x)$, and w is chosen alphabetically minimal among its cyclic shifts. This means, by definition, that w is a Lyndon word (see Chapter 15). One can prove that a Lyndon word is unbordered. Consequently the factorization (lw, wr) is critical. ♦

The present proof of the theorem (in the general case) requires two orderings on words that are recalled first. Each ordering $\le$ on the alphabet A extends to an *alphabetical ordering* on the set A^*. It is defined as usual by $x \le y$ if either

- x is a prefix of y, or
- x is *strongly less* than y, denoted by $x << y$ (i.e., $x = w.a.x'$, $y = w.b.y'$ with w, x', y' words of A^* and a, b two letters such that $a < b$).

The following theorem illustrates the existence of critical factorizations adapted to CP algorithm. The statement involves two alphabetical orderings on words. The first one is $\leq$ induced by a given ordering $\leq$ on the alphabet. The second ordering on A^*, called the *reverse ordering* and denoted by $\subseteq$, is obtained by reversing the order $\leq$ on A.

Remark The ordering $\subseteq$ is not the inverse of $\leq$. For example, on $A = \{a, b\}$ with $a < b$, we have both $abb \leq abbaa$ and $abb \subseteq abbaa$. In fact, it is easy to see that the intersection of the orderings $\leq$ and $\subseteq$ is exactly the prefix ordering. Which means that, for any words x and y, inequalities $x \leq y$ and $x \subseteq y$ are equivalent to x is a prefix of y. ♦

Theorem 13.12 Let x be a non-empty word on A. Let $x = uv = u'v'$, where v (resp.v') is the alphabetically maximal suffix of x according to the ordering $\leq$ (resp. $\subseteq$).

If $|v| \leq |v'|$, then uv is a critical factorization of x. Otherwise, $u'v'$ is a critical factorization of x. Moreover, $|u|, |u'| < period(x)$.

Proof We first rule out the case in which the word x has period 1, that is to say, when x is a power of a single letter. In this case any factorization of x is critical.

We now suppose that $|v| \leq |v'|$, and prove that uv is a critical factorization. The other case ($|v'| < |v|$) is symmetrical. Indeed, $|v| < |v'|$ because x contains at least two different letters. Let us first prove that $u \neq \varepsilon$. Indeed, let $x = ay$ with a in A. If $u = \varepsilon$, then $x = v = v'$, and both inequalities $y \leq x$ and $y \subseteq x$ are satisfied by definitions of v and v'. Thus, y is a prefix of x from which comes $period(x) = 1$, contrary to the hypothesis.

Let w be the shortest repetition for uv ($|w| = r(u, v)$). We distinguish four cases according to whether w is a suffix of u or vice-versa, and whether w is a prefix of v or vice-versa.

Case 1 w is both a suffix of u, and a prefix of v. The word v can be written wz ($z \in A^*$). Since wv and z are suffixes of x, by the definition of v we have $wv \leq v$ and $z \leq v$. The first inequality can be rewritten $wwz \leq wz$ and implies $wz \leq z$. The second inequality gives $z \leq wz$. We then obtain $z = wz$, from which comes $w = \varepsilon$, a contradiction.

Case 2 w is a suffix of u, and v is a prefix of w. The word w can be written vz ($z \in A^*$). But then, vzv is a suffix of x strictly greater than v, a contradiction with the definition of v.

Case 3 (Figure 13.16): u is a suffix of w, and v is a prefix of w.

The integer $|w|$ is a period of x because x is a factor of ww. The period of x cannot be shorter than $|w|$ because this quantity is the local period for uv. Hence $period(x) = |w|$, which proves that the factorization uv is critical.

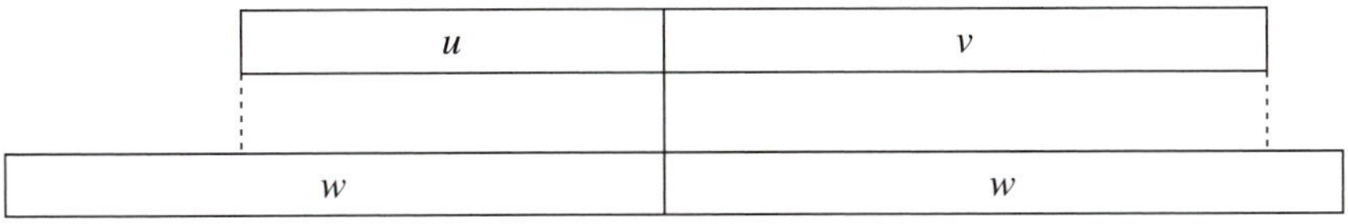

Figure 13.16: Case 3.

Case 4 (Figure 13.17): u is a suffix of w and w is a prefix of v.

Let $w = zu$ and $v = wy$ ($z, y \in A^*$). Since $|w|$ is the local period for uv, as in the previous case we only need to prove that $|w|$ is a period of x.

Let u'' be the non-empty word such that $u = u'u''$ (recall that hypothesis $|v| < |v'|$ implies that u' is a proper prefix of u). Since $u''y$ is a suffix of x, by the definition of v', we get $u''y \subseteq v' = u''v$, hence $y \subseteq v$. By the definition of v, we also have $y \leq v$. By the remark above, these two inequalities imply that y is a prefix of v. Hence, y is a border of v, and v is thus a prefix of w^e for some integer $e > 0$. Then, x is a factor of w^{e+1}, which shows that $|w|$ is a period of x as expected. This ends Case 4.

The proof shows that cases 1 and 2 are impossible. As a consequence, $|u|$ is less than the local period. We then get $|u| < period(x)$ because the factorization uv is critical. We also get $|u'| < period(x)$ when $|u'| < |u|$. The same argument holds symmetrically under the assumption $|v'| < |v|$. This completes the entire proof. ♦

According to Theorem 13.12, the computation of a critical factorization reduces to that of maximal suffixes. More accurately, it requires the computation of two maximal suffixes corresponding to reversed orderings of the alphabet.

The remainder of this section is devoted to the preprocessing part of CP algorithm, which is given below. The design of this phase of CP algorithm is a direct result of Theorem 13.12. The algorithm calls the procedure *MS* that essentially computes the maximal suffix of a word. For a word x, it returns both (u, v) such that $x = uv$ and v is the maximal suffix of x, and $p = period(v)$. The next algorithm runs in linear time and uses only a bounded memory space because, as we shall see afterward, algorithm MS has the same performance.

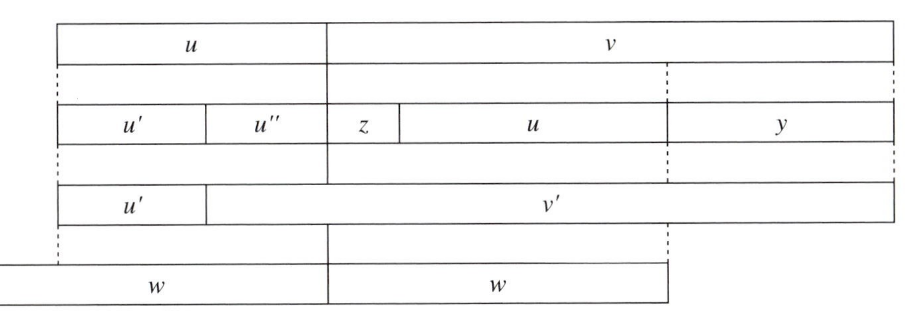

Figure 13.17: Case 4.

```
Algorithm { preprocessing of pat for CP algorithm }
{ compute a critical factorization of pat }
begin
   (u, v, p) := MS(pat) according to ≤ ;
   (u', v', p') := MS(pat) according to ⊆;
   if (|v| < |v'|) then
        return (u, v) and p the period of v;
   else
        return (u', v') and p' the period of v';
end.
```

Example Consider the word *abaabaa* of period 3. Its maximal suffix for the usual ordering on letters is *baabaa*. The factorization (*a*, *baabaa*) is not critical because its local period is 2 (repetition *ba*). According to the reverse ordering, the maximal suffix becomes *aabaa*. The factorization (*ab*, *aabaa*) is critical.

The word *ababaabbababa* has period 8. Its maximal suffixes for usual and reverse orderings are *bbababa* and *aabbababa*, respectively. Their associated factorizations (*ababaa*, *bbababa*) and (*abab*, *aabbababa*) are both critical. ♦

We end this section with a brief description of algorithm MS. The algorithm below is strongly related to the Lyndon factorization of a word presented in Chapter 15. The reader should refer to this chapter to develop a proof of correctness of the algorithm.

```
Algorithm MS(x) { computes Maxsuf(x) and its period }
{ operates in linear and constant space }
begin
   ms := 0; j := 1; k :=1; p := 1;
   while (j + k ≤ |x|) do begin
      a' := x [ms+k]; a := x [j+k];
      if (a < a') then begin
         j := j+k; k := 1; p := j-ms;
      end if (a = a') then
         if (k ≠ p) then
            k := k+1
         else begin
            j := j+p; k := 1;
         end
      else { a > a' } begin
         ms := j; j := ms+1; k := 1; p := 1;
      end;
   end;
   return (x [1...ms], x [ms+1...|x|], p);
end.
```

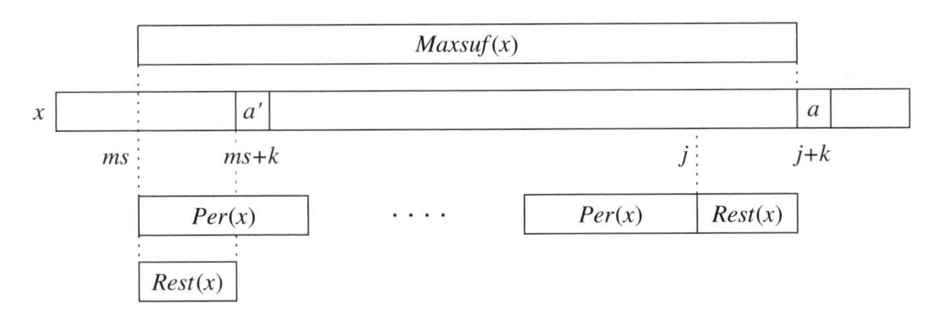

Figure 13.18: Variables in MS algorithm.

Let *Maxsuf*(x) be the suffix of x that is maximal for alphabetic ordering. We consider the words f,g, and the integer $e > 0$ such that *Maxsuf*(x) $= f^e g$ with $|f| =$ *period*(*Maxsuf*(x)), and g is a proper prefix of f. We define *Per* and *Rest* by *Per*(x) $= f$, *Rest*(x) $= g$. Note that *Rest*(x) is a border of *Maxsuf*(x), and that *Border*(*Maxsuf*(x)) $= f^{e-1}g$, even when $e = 1$.

The interpretation of the variables ms, j, k occurring in the algorithm MS is indicated on Figure 13.18. The integer p is the period of *Maxsuf*(x) that is also the length of *Per*(x). The integer ms is the position of *Maxsuf*(x) in x, and j is the position of the last occurrence of *Rest*(x) in *Maxsuf*(x).

Example Let $x = abcbcbacbcbacbc$. Then, for the usual ordering, *Maxsuf*(x) $=$ $cbcbacbcbacbc$, which is also $(cbcba)^2cbc$.

The maximal suffix of xa is $cbcbacbcbacbca$ = *Maxsuf*(x)a, which is a border-free word. The maximal suffix of xb is $cbcbacbcbacbcb$ = *Maxsuf*(x)b. This word has the same period as *Maxsuf*(x).

Finally, the maximal suffix of xc is cc. ♦

The correctness of MS algorithm essentially relies on Lemma 13.13 (see Chapter 15 for a proof of a similar result). Its time complexity is stated in Lemma 13.14.

Lemma 13.13 Let x be a word and a be a letter. Let a' be the letter such that *Rest*(x)a' is a prefix of *Maxsuf*(x). Then the triple (*Maxsuf*(xa), *Per*(xa), *Rest*(xa)) is equal to

(*Maxsuf*(x)a, *Maxsuf*(x)a, ε)	if $a < a'$,
(*Maxsuf*(x)a, *Per*(x), *Rest*(x)a) or (*Maxsuf*(x)a, *Per*(x), ε)	if $a = a'$,
(*Maxsuf*(*Rest*(x)a), *Per*(*Rest*(x)a), *Rest*(*Rest*(x)a))	if $a > a'$.

Lemma 13.14 MS algorithm runs in $O(|x|)$ time with constant additional memory space. It makes less than $2.|x|$ letter comparisons.

Proof The value of the expression $ms + j + k$ is increased by at least one unit after each symbol comparison. The result then follows from inequalities $2 \leq ms + j + k \leq 2.|x| + 1$. ♦

13.6 Optimal computation of periods

In this section we develop yet another time-space optimal string-matching algorithm, which naturally extends to the computation of periods of word with the same performance.

GS and CP algorithms factorize the pattern *pat* into *uv* according to some property of the periodicities existing within the pattern. The search phase is guided thereafter by the search for the right part *v* of the pattern. Both algorithms tend to avoid some periodicities of the pattern to make the search faster. This contrasts with both KMP algorithm and MPS algorithm in which the first phase amounts to computing periods of the pattern. The present algorithm adopts the same strategy as the latter. It computes periods to realize shifts. This is executed "on the fly" during the search phase using maximal suffix computations, and periods are not stored. Moreover, no preprocessing is necessary.

When a shift is to be performed, the strategy is to compute the period of the scanned segment of the text (including the mismatch letter). The algorithm does not always find the exact period of this segment, but in any case computes an approximation of it. The approximation is sufficient for producing an overall linear-time algorithm, and the computation requires only a bounded extra memory space.

```
Algorithm { preliminary version of algorithm P }
{ search for pat in text; no preprocessing needed }
begin
   pos := 0; i := 0;
   while (pos ≤ n-m) do begin
        while i < m and text[pos+i+1] = pat[i+1] do i := i+1;
        if i = m then report match at position pos;
        (u, v, p) := MS(pat[1...i]text[pos+i+1]);
        if (u suffix of v[1...p]) then begin
             pos := pos+p; i := i-p;
        end else begin
             pos := pos+max(|u|,⌊(i+1)/2⌋)+1; i := 0;
        end
   end;
end.
```

The algorithm above resembles KMP algorithm. It makes use of a left-to-right scan of the pattern against the text. When during the execution, a mismatch is encountered or an occurrence of the pattern is discovered, the algorithm shifts the pattern to the right. The shift is computed as follows. Let y be the longest prefix of the pattern found at the current position in the text. Let also b be the letter in the text that immediately follows the occurrence of y. Then, the algorithm tries to make a shift of length as close as possible to $period(yb)$ (in the above algorithm, yb is $pat[1...i]text[pos+i+1]$). This quantity corresponds to the best possible shift in such a situation. The computation of an approximation of $period(yb)$ is executed after the computation of the maximal suffix of yb. This is analogous to what is done at the preprocessing phase of CP algorithm in Section 13.5.

The correctness of the algorithm amounts to proving that the length of the shift is not larger than the period of $x = yb$. But, this is a direct consequence of Lemma 13.16 below, which relates the factorization of x to its maximal suffix, and to the period of this suffix. To state it we need to introduce the notion of MS-factorization of a non-empty word x.

Let v be the maximal suffix of x (according to the alphabetical ordering). Let u be such that $x = uv$. Considering its (smallest) period, the word v can be written $w^e w'$ where $e \geq 1$, $|w| = period(v)$, and w' is a proper prefix of w. Recall that the pattern x is non-empty so that words u, v, w and w' are well-defined. The sequence (u, w, e, w') is called the *MS-factorization of* x (MS stands for maximal suffix).

The MS-factorization of x into $uw^e w'$ yields a rather precise information on the period of x. Among several interesting properties, it is worth noting that w is border-free (it has no smaller period than its length). The inequality $period(x) > |u|$ is a rather intuitive property of the maximal suffix. Because of this, the suffix must start within the first period of the word (see Theorem 13.12).

Lemma 13.15 Let $uw^e w'$ be the MS-factorization of a non-empty word x, and $v = w^e w'$ be the maximal suffix of x. Then, these five properties hold:

(1) the word w is border-free.

(2) if u is a suffix of w, $period(x) = period(v)$,

(3) $period(x) > |u|$,

(4) if $|u| \geq |w|$, $period(x) > |v| = |x| - |u|$,

(5) if u is not suffix of w, and $|u| < |w|$, $period(x) > \min(|v|, |uw^e|)$.

Proof

(1) Assume that $w = zz' = z''z$ for three non-empty words z, z' and z''. The word $zw^{e-1}w'$ is a suffix of x distinct from v, therefore, it is greater than v according to the alphabetical ordering. The inequality $zw^{e-1}w' < v$ rewrites as $zw^{e-1}w' < zz'w^{e-1}w'$, and implies $w^{e-1}w' < z'w^{e-1}w'$. Moreover, $zw^{e-1}w'$ is not a prefix of v because if it were, the smallest period of v would be $|z''|$, less than $|w|$, contrary to the definition of w. Since $w^{e-1}w'$ is not a prefix of $z'w^{e-1}w'$, there is a word y, and letters a and b such that simultaneously ya is a prefix of $w^{e-1}w'$, yb is a prefix of $z'w^{e-1}w'$, and $a < b$. Since ya is also a prefix of v, we then get $v < z'w^{e-1}w'$, a contradiction with the definition of v. Thus,

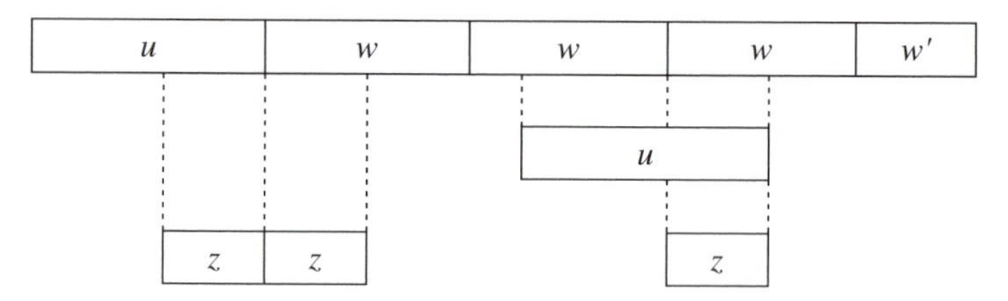

Figure 13.19: Impossible: no suffix of u can be a prefix of w.

w cannot be properly written simultaneously as zz' and $z''z$. This means that w is border-free, or, equivalently, $period(w) = |w|$.

(2) When u is a suffix of w, $|w|$ is obviously a period of the whole word x. The smallest period of x cannot be less than the smallest period of its suffix v. Since this period is precisely $|w|$, we get the conclusion $period(x) = period(v) = |w|$.

(3) We prove that $period(x) > |u|$. Otherwise, if $period(x) \leq |u|$, there is an occurrence of v in x distinct from its occurrence as suffix. In other words, x can be written $u'vv'$ with $|u'| < |u|$, and $|v'| > 0$. But the suffix vv' is then alphabetically greater than v, a contradiction with the definition of v.

(4) (Figure 13.19) Assume that $|u| \geq |w|$. We prove that $period(x) > |v|$ (= $|x|$-$|u|$). If the inverse holds, there is an occurrence of u in x distinct from the prefix occurrence. This occurrence overlaps v. Then, taking into account that $|u| \geq |w|$, there is a non-empty word z that is both a prefix of w and a suffix of u. The former property shows that v can be written zz' (for some word z'), and the latter property shows that zv is a suffix of x. The maximality of v implies $v > zv$, that is $zz' > zv$. But this yields $z' > v$, a contradiction with the definition of v.

(5) (Figure 13.20) Assume that u is not a suffix of w, and that $|u| < |w|$. Assume also, ab absurdo, that $period(x) \leq \min(|v|, |uw^e|)$. Let z be the prefix of x of length $period(x)$. The word x itself is then a prefix of zx. From $period(x) \leq |v|$ we deduce that the word zu is a common prefix to x and zx. And from $period(x) \leq |uw^e|$ we know that u overlaps w^e. If u overlaps the boundary between two ws, or the boundary between the last w and w', the same argument as used in case (4) applies and leads to a contradiction. The remaining situation is one in which u is a factor of w, as shown in Figure 13.20. The last occurrence of w in the prefix zuw of zx overlaps an occurrence of w in the prefix uw^e of x. Note that these occurrences of w cannot be equal or adjacent because u is not suffix of w. This presents a contradiction with the border-freeness of w stated in (1).

This completes the proof of the lemma. ♦

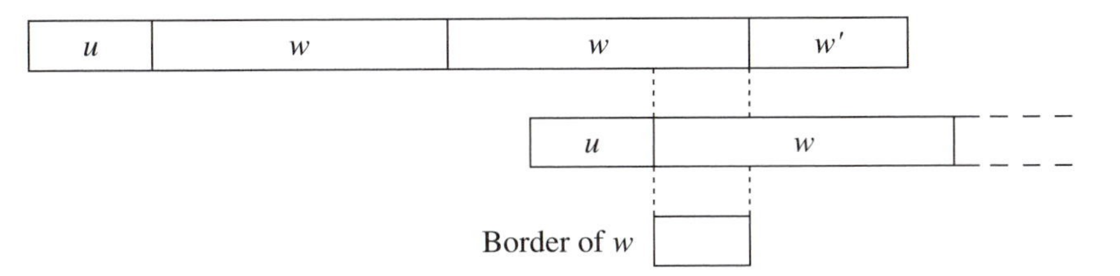

Figure 13.20: Impossible because w is border-free.

Note One may note that, in case (4), u cannot be a suffix of w, because this would imply $u = w$ which is found to be impossible by (3). ♦

An immediate consequence of Lemma 13.15 is that condition (2) is certainly true for words having a small period, that is, having a period not greater than half their length ($period(x) \leq |x|/2$). The computation of the smallest period of these words can then be deduced from a computation of their maximal suffixes (and MS-factorization), together with a test "is u a suffix of w ?". This fact is used in CP algorithm (see Section 13.4) to approximate smallest periods of patterns. The following lemma summarizes the correctness of algorithm P. It provides an accurate approximation of the smallest period of the word x.

Lemma 13.16 Let uw^ew' be the MS-factorization of a non-empty word x, and let v $(= w^ew')$ be the maximal suffix of x.

- If u is a suffix of w, $period(x) = period(v) = |w|$.
- Otherwise, $period(x) > \max(|u|, \min(|v|, |uw^e|)) \geq |x|/2$.

Proof This is in essence a corollary of Lemma 13.15.

When u is not a suffix of w, statements (3), (4), and (5) of Lemma 13.15 illustrate that the inequality $period(x) > \max(|u|, \min(|v|, |uw^e|))$ holds. What remains is to prove that the last quantity is greater than $|x|/2$.

The inequality is trivially satisfied if $|u| \geq |x|/2$. Otherwise, quantity $|v|$, which is equal to $|x|-|u|$ is greater than $|x|/2$. And $|uw^e|$, equal to $|x|-|w'|$, is also greater than $|x|/2$ because $|w'| < |w|$.

Then, $\min(|v|, |uw^e|) > |x|/2$, which ends the proof. ♦

Examples Bounds on the smallest period given in Lemma 13.15 are sharp. We list a few examples of patterns that give evidence of this fact. We consider words on the alphabet $\{a, b, c\}$ with the usual ordering ($a < b < c$).

Let x be $aaaaba$. The maximal suffix of x is v=ba in which the smallest period is 2. In the MS-factorization uw^ew' of x, $u = aaaa$, $w = ba$, and w' is empty. Then, *period*(x) = 5 = $|u|$+1. As stated in Lemma 13.15 case (3), *period*(x) is greater than $|u|$, but only by one unit.

The word $x = aababa$ meets case (4) of Lemma 13.15. Here $u = aa$, $w = ba$, and w' is empty. The smallest period of x, 5, is exactly one unit more than the length of the maximal suffix *baba*.

We exhibit two examples for case (5) of Lemma 13.15. The first example is x = *acabca*. We have $u = a$, $w = cab$, and $w' = ca$. Then, the quantity min($|v|$, $|uw^e|$) is $|uw^e| = 4$. The smallest period of x is *period*(x) = 5. The word satisfies *period*(x) = $|uw^e|+1 = |v|$. The second example is $x = ababbbab$. In this case, $u = aba$, $w = bbba$, and $w' = b$. This is a reverse situation in which min($|v|$, $|uw^e|$) = $|v|$ = 5. The smallest period of x is now 6, and we have *period*(x) = $|v|+1 < |uw^e|$. ♦

Algorithm MS may be implemented to run in linear time (see Section 13.5). But even under this assumption, algorithm P does not always run in linear time. Quadratic complexity is due to computations of maximal suffixes. For example, if the pattern is $a^{m-1}b$ and the text is a long repetition of the only letter a, at each position in the text the maximal suffix of a^m is re-computed from scratch, leading to an $O(|pat|.|text|)$ time complexity.

Indeed, quadratic behavior of the algorithm is only reached with highly periodic patterns (or, more precisely, with patterns having a highly periodic prefix). But, with such types of patterns, entire re-computations of maximal suffixes are not necessary. Therefore, the trick to reducing the running time is to save as much of the work as possible done in computing the maximal suffix. We introduce a new algorithm for that purpose. It is called *Next_MS*, and shown below. It is a mere transformation of algorithm MS of Section 13.5. The tuple of variables of MS, (ms, j, k, p), is made accessible to the string-matching algorithm, so that it can control its values. The tuple (ms, j, k, p) is called the *MS-tuple of* x. It is related to the MS-factorization (u, w, e, w') of x by the equations:

$$ms = |u|,\ j = |uw^e|,\ k = |w'|+1,\ p = |w| = period(v).$$

Figure 13.21 illustrates this situation. The value of ms is the position in x of its maximal suffix v, and j is the position of the rest w'. The period of v, which is also the length of w, is given by p.

The string-matching algorithm is shown below as algorithm P (for Periods). The difference with the preliminary version lies in the computation of the MS-tuple (ms, j, k, p). In this version, the MS-tuple is still initialized, as in the preliminary version, at the beginning of a run of the algorithm and after each shift, except in one situation where it meets with a highly periodic pattern. In this case the variable j of the MS-tuple is simply decreased by the period p, length of the shift. Algorithm P contains another modification it computes all *overhanging occurrences* of the pattern *pat* within the text *text* (when *text* is prefix of *z.pat*). This is realized by a change on the test of the main "while" loop, and by the technical instruction "**if** *pos*+*i* = *n* **then** *i* := *i*−1" in which the purpose is to avoid considering symbols beyond the end of the text.

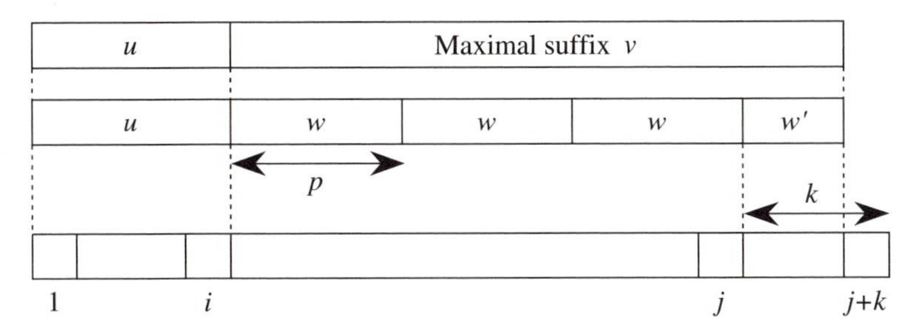

Figure 13.21: The MS-tuple (ms, j, k, p) of the pattern.

```
Algorithm P;
{ search for pat in text }
{ time-space optimal; no preprocessing needed }
begin
   pos := 0; i := 0; (ms,j,k,p) := (0,1,1,1);
   while (pos ≤ n) do begin
        while pos+i+1 ≤ n and i+1 ≤ m and text[pos+i+1] = pat[i+1]
        do i := i+1;
        if pos+i = n or i = m then report match at position pos;
        if pos+i = n then i := i-1;
        let scanned be pat[1...i]text[pos+i+1];
        (ms,j,k,p) := Next_MS(pat[1...i]text[pos+i+1],(ms,j,k,p));
        if pat[1...ms] suffix of
                   the prefix of length p of pat[ms+1...i]text[pos+i+1] then
             if j-ms > p then begin { shift = period in an hrp }
                   pos := pos+p; i := i-p; j := j-p;
             end else begin { shift = period outside an hrp }
                   pos := pos+p; i := i-p; (ms,j,k,p) := (0,1,1,1);
             end
        else begin { shift close to period }
             pos := pos+max(ms, min(i-ms,j))+1; i := 0;
             (ms,j,k,p) := (0,1,1,1);
        end;
   end;
end.
```

Provided algorithm *Next_MS* is proved to be correct, the correctness of algorithm P relies on Lemma 13.16 (as the correctness of the preliminary version does), and on Lemma 13.17 below. Before proving it, we first explain the main idea through an example.

Example Consider the pattern pat = *babbbabbbab*. Its maximal suffix is v = *bbbabbbab*. Both words have smallest period 4. With the notation of MS-factorizations, we have $u = ba$, $w = bbba$, and $w' = b$. The exponent of w in v is $e = 2$. If the

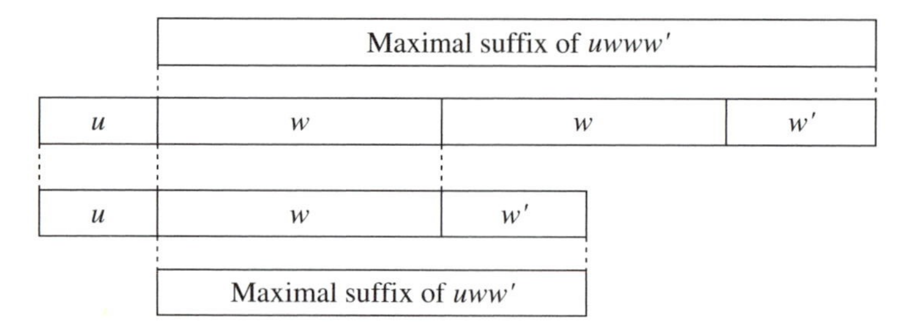

Figure 13.22: Highly periodic pattern.

last four letters of *pat* are deleted, corresponding, for example, to a shift of 4 positions to the right, we are left with the word $x_1 = babbbab$. Its MS-factorization is $u_1 = ba$, $v_1 = bbbab$, $w_1 = bbba$ and $w_1' = b$. Note that the second factorization is produced from the first one by pruning one occurrence of w. Indeed, this generalizes to any word for which the exponent e of the MS-factorization is greater than 1 (see Figure 13.22). The above result does not necessarily hold when $e = 1$. Let, for example, *pat* be *bbabbbabb*. It has period 4, like its maximal suffix $v = bbbabb$. Deletion of the last four letters yields $x_1 = bbabb$, which is its own maximal suffix and has period 3. This situation is quite different than the previous one. ♦

Lemma 13.17 Let (u, w, e, w') be the MS-factorization of a non-empty word x ($w^e w'$ is the maximal suffix of $x = uw^e w'$). If u is a suffix of w and $e > 1$, then $(u, w, e{-}1, w')$ is the MS-factorization of $x' = uw^{e-1}w'$. In particular, the word $w^{e-1}w'$ is the maximal suffix of x'.

Proof Let $v = w^e w'$ be the maximal suffix of x. Any proper suffix of v of the form $zw^{e-1}w'$ (with $z \neq \varepsilon$) is less than v itself by definition. However, since w is border-free (see Lemma 13.15), the longest common prefix of w and z is shorter than z. This leads to $w > z$ and proves that w is greater than all its proper suffixes. This further proves that $w^{e-1}w'$ is its own maximal suffix. And, since hypothesis $period(x) = |w|$ is equivalent to "u is a suffix of w," this also proves that $w^{e-1}w'$ is the maximal suffix of $uw^{e-1}w'$.

The equation $period(w^{e-1}w') = |w|$ is another consequence of the border-freeness of w stated in Lemma 13.15. Thus, $(u, w, e-1, w')$ is the MS-factorization of x' as announced. ♦

A pattern x that satisfies hypothesis of Lemma 13.17 has a rather small period, and may be considered as highly periodic. Its smallest period is not greater than half its length. Conversely, if the smallest period of x is not greater than $|x|/3$, the conclusion of Lemma 13.17 applies.

We now explain how Lemma 13.17 is used to improve on the complexity of the preliminary version of the string-matching algorithm. When a shift executed according to a period leaves a match between the text and the pattern of the form $uw^{e-1}w'$, we avoid computing the next maximal suffix from scratch. We are better off exploiting the known factorization of the match. Within algorithm P the test "$j–ms > p$" is equivalent to condition "$e > 1$" of Lemma 13.17. By doing so, we get a linear-time algorithm. Below is the modification of Algorithm MS used in algorithm P.

```
Algorithm Next_MS (x [1...m], (ms,j,k,p));
begin
   while (j + k ≤ n) do begin
      if (x [i+k] = x [j+k]) then begin
         if (k = p) then begin
            j := j+p; k := 1;
         end else k := k+1;
      end else if (x [i+k] > x [j+k]) then end
         j := j+k; k := 1; p := j-ms;
      end else begin
         ms := j; j := ms+1; k := 1; p := 1;
      end;
   end;
   return (ms, j, k, p);
end.
```

Lemma 13.18 The number of letter comparisons executed during a run of algorithm P on words *pat* and *text* is less than $6.|text|+5$. This includes comparisons made during *Next_MS* calls.

Proof First consider the test for prefix condition in algorithm P (third "if"). The comparisons executed on the prefix $pat[1...i]$ of the pattern can be charged to $text[pos+1...pos+i]$. Whichever shift follows the test, the value of *pos* increases by more then *ms*. Thus, never again are the comparisons charged to the factor $text[pos+1...pos+i]$ of the text. The total number of these comparisons is thus bounded by $|text|$.

We next prove that each other letter comparison executed during a run of algorithm P (including comparisons done during calls of *Next_MS*) leads to a strict increment of the value of expression $5pos+i+ms+j+k$. Since its initial value is 3, and its final value is $5|text|+8$, this proves the claim.

Whatever the result of the letter comparison within algorithm *Next_MS* is, the value of $ms+j+k$ increases by 1, and even by more than 1 when the last line of the algorithm is executed. It is worthwhile to note that the inequality $k \leq j–ms$ always holds.

Successful comparisons at the first instruction of algorithm P trivially increase i, and consequently the expression $5pos+i+ms+j+k$. On the same line, there is at most one unsuccessful comparison. This comparison is eventually followed by a shift. We examine successively the three possible shifts in the order they appear in algorithm P.

The effect of the first shift is to replace $5pos+i+ms+j+k$ by $5(pos+p)+(i–p+1)+ms+(j–p)+k$. The expression is thus increased by $3p+1$, which is greater than 1.

The second shift is executed when $j–ms > p$ does not hold. This means indeed that $j–ms = p$ ($j–ms$ is always a multiple of p). We also know that $ms < p$ (see Lemma 13.15, case (2)). One can observe on algorithm *next_mS* that $k \le p$. Now, immediately after the shift, i is decreased by $p–1$, ms and k are decreased by less than p, and $j = ms+p$ is decreased by less than $2.p$. Since pos is replaced by $pos+p$ the value of $5pos+i+ms+j+k$ is increased by more than 1.

Finally, consider the effect of the third shift on the expression. If s is the value of $\max(ms, \min(i–ms, j))+1$, the increment of expression $5pos+i+ms+j+k$ is $I = 5s–(i–1)–ms–(j–1)–(k–1) = 5s–i–ms–j–k+3$, that is, $5s–2i–ms+2$, because $j+k = i+1$. The value s is greater than or equal to both ms and $i/2$. Therefore, $I \ge 2$, which proves that the third shift increases the value of $5pos+i+ms+j+k$ by more than 1.

The effect of the second "if" in algorithm P, that can decrease i by one unit, is of no importance in the preceding analysis. ♦

Since algorithm P computes all overhanging occurrences of the pattern within the text, it can be used, in a natural way, to compute all periods of a word. Each (overhanging) position of x inside x itself (except position 0) is a period of word x. Following, we give a straightforward adaptation of algorithm P that computes the smallest period of a word. It can easily be extended to compute all periods of its input. As a consequence of Lemma 13.18, we get the following result.

Theorem 13.19 The periods of a word x can be computed in $O(|x|)$ time with a constant amount of space in addition to x.

```
function Per(x);
{ time-space optimal computation of period(x) }
begin
   per := 1; i := 0; (ms,j,k,p) := (0,1,1,1);
   while per+i+1 ≤ |x| do begin
       if x [per+i+1] = x [i+1] then i := i+1
       else begin
           (ms,j,k,p) := Next_MS(x [1...i]x [per+i+1],(ms,j,k,p));
           if (x [1...ms] suffix of
                    the prefix of length p of x [ms+1...i]x [per+i+1]) then
               if (j-ms > p) then begin
                    per := per+p; i := i-p; j := j-p;
               end else end
                    per := per+p; i := i-p; (ms,j,k,p) := (0,1,1,1);
               end
           else begin
               per := per+max(ms, min(i-ms,j))+1; i := 0;
               (ms,j,k,p) := (0,1,1,1);
           end;
       end;
   end;
   return(per);
end.
```

Bibliographic notes

The first time-space optimal string-matching algorithm is from Galil and Seiferas [GS 83]. The same authors have designed other string-matching algorithms requiring only a small memory space [GS 80], [GS 81]. In the original article [GS 83], the perfect factorization theorem is proved for the parameter $k \geq 4$. The present proof, also valid for $k = 3$, and the exposition of Sections 13.2 and 13.3, is from [CR 94].

Algorithm CP of Sections 13.4 and 13.5 is from Crochemore and Perrin [CP 91]. The present proof of the critical factorization theorem is also from [CP 91]. The theorem is originally from Cesari, Duval, and Vincent (see [Lo 83] Chapter 8). The algorithm used to compute the maximal suffix of a string is adapted from an algorithm of Duval [Du 83] (see also Chapter 15).

The time-space optimal computation of periods of a string given in Section 13.6 is from [Cr 92]. The result is announced in [GS 83]. Although the proof contains a small flaw, the idea leads to another time-space optimal computation of periods (see [CR 94]).

Selected references

[CP 91] Crochemore, M., & Perrin, D., "Two-way string matching," *J. ACM* 38, 3 (1991) 651–675.

[CR 94] Crochemore, M., & Rytter, W., "Cubes, squares and time-space efficient string searching," *Algorithmica* (1994). To appear.

[GS 83] Galil, Z., & Seiferas, J., "Time-space optimal string matching," *J. Comput. Syst. Sci.* 26 (1983) 280–294.

[Lo 83] Lothaire, M., *Combinatorics on Words*, Addison-Wesley, Reading, Mass., 1983.

14

Time-processors optimal string matching

There are two types of optimal parallel algorithms for the string-matching problem. The first type uses the combinatorics of periods in texts in an essential way. In the second type, no special combinatorics of texts is necessary. In this chapter, we first follow the approach of Vishkin's algorithm. Then, the Galil's sieve method is shown. It produces a constant-time optimal algorithm. These algorithms are of the first type. Next, we present the suffix-prefix algorithm of Kedem, Landau and Palem (KLP). This algorithm is of the second type. It is essential for the optimality of this algorithm that we use the model of parallel machine with concurrent writes (CRCW PRAM). KLP algorithm can be viewed as a further application of the dictionary of basic factors (see Chapter 9). The suffix-prefix string matching introduced there is extended to the case of two-dimensional patterns at the end of the section. The chapter ends with a sketch of the so-called splitting technique and its application to string matching.

14.1 Vishkin's parallel string matching by duels and by sampling

Historically, the first optimal parallel algorithm for string matching used the notion of expensive duels (see Chapter 3). Moreover, it was optimal only for fixed alphabets. The notion has been strengthened to the more powerful operation of *duel* that leads to an optimal parallel string matching. Unfortunately, preprocessing the table of witnesses required by the method is rather complicated. Therefore, we simply state the result, but omit the proof and refer the reader to the references at the end of this chapter.

Theorem 14.1 The smallest period and the witness table of a pattern of length m, can be computed in $O(\log m)$ time with $m/\log m$ processors of a CRCW PRAM, or in $O(\log^2 m)$ time with $m/\log^2 m$ processors of a CREW PRAM.

Suppose that v is the shortest prefix of the pattern that is a period of the pattern. If the pattern is periodic (vv is a prefix of the pattern) then $vv-$ *is called the non-periodic part of the pattern (*$v-$ *denotes the word* v with the last symbol removed). We omit the proof of the following lemma, which justifies the name "non-periodic part" of the pattern:

Lemma 14.2 If the pattern is periodic (it is twice as long as its period) then its non-periodic part is non-periodic.

The witness table is relevant only for the non-periodic pattern. So, it is easier to deal with non-periodic patterns. We prove that such assumption can be done without loss of generality, by ruling out the case of periodic patterns.

Lemma 14.3 Assume that the pattern is periodic, and that all occurrences (in the text) of its non-periodic part are known. Then, we can find all occurrences of the whole pattern in the text

(i) in $O(1)$ time with n processors in the CRCW PRAM model,

(ii) in $O(\log m)$ time with $n/\log m$ processors in the CREW PRAM model.

Proof We reduce the general problem to unary string matching. Let $w = vv-$ be the non-periodic part of the pattern. Assume that w starts at position i in the text. By a segment containing position i we mean the largest segment of the text containing position i and having a period of size $|v|$. We assign a processor to each position. All these processors simultaneously write 1 into their positions if the symbol at distance $|v|$ to the left contains the same symbol. The last position containing 1 to the right of i (all positions between them also contain ones) is the end of the segment containing i. Similarly, we can compute the first position of the segment containing i. It is easy to compute it optimally for all positions i in $O(\log m)$ time by a parallel prefix computation (see Chapter 9). The constant time computation on a CRCW PRAM is more advanced; we refer the reader to [BG 90]. Some tricks are used by applying the power of concurrent writes. This completes the proof. ♦

Now we can assume that the pattern is non-periodic. We consider the witness table used in Chapter 3 in two sequential string-matching algorithms: by duels and by sampling. The parallel counterparts of these algorithms are presented.

Recall that a position l on the text is said to be *in the range of* a position k iff $k < l < k + m$. We say that two positions $k < l$ on the text are *consistent* iff l is not in the range of k, or if $WIT[l-k] = 0$. If the positions are not consistent, then, in constant time we can remove one of them as a candidate for a starting position of the pattern using the operation *duel* (see Chapter 3).

Let us partition the input text into *windows* of size $m/2$. Then, the duel between two positions in the same window eliminates at least one of them. The position that "survives" is the value of the duel. Define the operation $\otimes$ by $i \otimes j = duel(i, j)$. The operation $\otimes$ is "practically" associative. This means that the value of $i_1 \otimes i_2 \otimes i_3 \otimes \ldots \otimes i_{m/2}$ depends on the order of multiplications, but all values (for all possible orders) are equivalent for our purpose. We need any of the possible values.

Once the witness table is computed, the string-matching problem reduces to instances of the parallel prefix computation problem. We have the following algorithm.

```
algorithm Vishkin_string_matching_by_duels;
begin
    consider windows of size m/2 on text;
    /* sieve phase */
    for each window do in parallel
        /* ⊗ can be treated as if it were associative */
        compute the surviving position i1 ⊗ i2 ⊗ i3 ⊗ .. ⊗ im/2,
        where i1, i2, i3, ..., im/2 are consecutive positions
        in the window;
  /* naive phase */
    for each surviving position i do in parallel
        check naively an occurrence of pat at position i
        using m processors;
end.
```

Theorem 14.4 Assume we know the witness table and the period of the pattern. Then, the string-matching problem can be solved optimally in $O(\log m)$ time with $O(n/\log m)$ processors of a CREW PRAM.

Proof Let $i_1, i_2, i_3, \ldots, i_{m/2}$ be the sequence of positions in a given window. We can compute $i_1 \otimes i_2 \otimes i_3 \otimes \ldots \otimes i_{m/2}$ using an optimal parallel algorithm for the parallel prefix computation (see Chapter 9). Then, in a given window, only one position survives; this position is the value of $i_1 \otimes i_2 \otimes i_3 \otimes \ldots \otimes i_{m/2}$. This operation can be executed simultaneously for all windows of size $m/2$. For all windows, this takes $O(\log m)$ time with $O(n/\log m)$ processors of a CREW PRAM.

Afterward, we have $O(n/m)$ surviving positions altogether. For each of them we can check the match using $m/\log m$ processors. Again, a parallel prefix computation is used to collect the result, that is, to compute conjunction of m Boolean values (match or mismatch, for a given position). This takes again $O(\log m)$ time with $O(n/\log m)$ processors.

Finally, we collect $O(n/m)$ Boolean values using a similar process. It gives an optimal parallel algorithm working with the announced complexities. This completes the proof. ♦

The consequence of Lemma 14.1 and Theorem 14.4 together is the basic result of this section stated as the following corollary.

Corollary 14.5 There is an $O(\log^2 n)$ time parallel algorithm that solves the string matching problem (including preprocessing) with $O(n/\log^2 n)$ processors of a CREW PRAM.

The idea of deterministic sampling was originally developed for parallel string matching. In Chapter 3, a sequential use of sampling is shown. Recall the definition of the good sample. A sample S is a set of positions in the pattern. A sample S occurs at position i in the text iff $pat[j] = text[i+j]$ for each j in S. A sample S is called a *deterministic sample* iff it is small ($|S| = O(\log m)$), and it has a large field of fire—a segment $[i-k...i+m/2-k]$. If the sample occurs at i in the text, then positions in the field of fire of i, except i, cannot be matching positions. The important property of samples is that if we have two positions of occurrences of the sample in a window of size $m/2$, then one "kills" the other: only one can possibly be a matching position.

```
algorithm Vishkin-string-matching-by-sampling;
begin
    consider windows of size m/2 on text;
    /* sieve phase */
    for each window do in parallel begin
        for each position i in the window do in parallel
            kill i if the sample does not occur at i;
        kill all surviving positions in the window,
            except the first and the last;
        eliminate one of them in the field of fire of the other;
    end;
    /* naive phase */
    for each surviving position i do in parallel
        check naively an occurrence of pat starting at i
        using m processors;
end.
```

In the sieve phase, we use $n\log m$ processors to check a sample match at each position. In the naive phase, we have $O(n/m)$ windows, and, in each window, m processors are used. This proves the following.

Theorem 14.6 Assume we know the deterministic sample and the period of the pattern. Then, the string-matching problem can be solved in $O(1)$ time with $O(n\log m)$ processors in the CRCW PRAM model.

The deterministic sample can be computed in $O(\log^2 m)$ time with n processors using a direct parallel implementation of the sequential construction of deterministic samples presented in Chapter 3. But faster sampling methods are possible (see Bibliographic notes).

14.2* Galil's sieve

The constant-time parallel sampling algorithm of the last section is not optimal except within a logarithmic factor ($\log m$). In order to produce an optimal searching phase, we apply a method to eliminate all but $O(n/\log m)$ candidate positions in constant time with n processors. We call it the Galil's sieve. After applying the sieve with the sampling algorithm of the last section, only $O(n/\log m)$ positions need to be checked as sample matches. Hence, we can do the search phase of the string matching in constant time with an optimal number of processors (linear number of processors). This is stated in the next theorem.

Theorem 14.7 There is a preprocessing of the pattern in $O(\log^2 m)$ time with m processors such that the search phase on any given text can be executed in constant time with a linear number of processors.

The proof of Theorem 14.7 relies on the next lemmas of the section. The present proof consists of the description of Galil's sieve. It is based on a simple combinatorial fact described below. Let T be an $r \times r$ zero-one array. We say that a j-th column hits a i-th row iff $T[i, j] = 1$. A set of columns hits a given row if at least one of its columns hits this row (see Figure 14.1).

Lemma 14.8 (Hitting-set lemma) Assume we have an $r \times r$ zero-one array T such that each row contains at least s ones. Then, there is a set H of $O(r/s \log r)$ columns that hits all the rows of T.

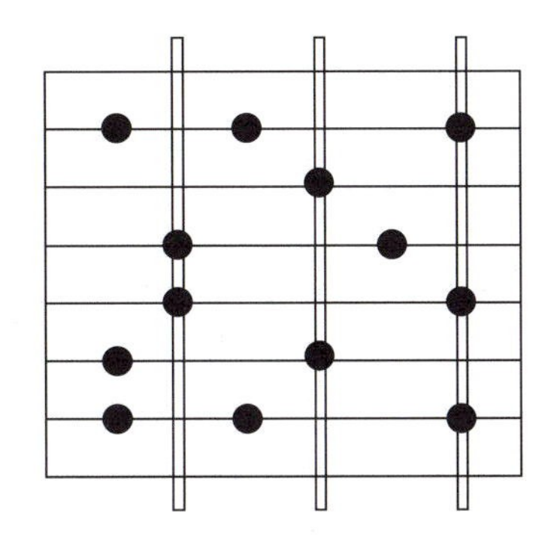

Figure 14.1: A hitting-set of columns the three columns hit all rows.

Proof The required set H is constructed by the following algorithm.

```
algorithm Hitting-set;
begin
    H := empty set; R := set of all rows of the array;
    while R non-empty do begin
        choose a column j hitting the largest number of rows of R;
        delete from R the rows hit by the j-th column;
        add the j-th column to H;
    end;
    return H;
end.
```

It is sufficient to prove the following:

the number of iterations of the algorithm "Hitting-set" is $O(r/s \log r)$.

Let *total* be the total number of ones in all rows of R. Since there are r columns (containing altogether all ones), the selected j-th column contains at least *total*/r ones. We remove at least *total*/r rows, each containing at least s ones. This implies that we remove at least $s.total/r$ ones. By doing so, the total number of ones decreases at each iteration by a factor at least (1–s/r). It is easy to calculate that there is a number $k = O(r/s \log r)$, such that $(1-s/r)^k < 1/r^2$. Initially, we have globally at most r^2 ones. Hence, after k iterations there is no one, and R is empty. This completes the proof. ♦

Let us consider occurrences of the pattern that start in a "window" of size $m/4$ on the text (see Figure 14.2). It is enough to remove all but $O(m/\log m)$ candidate positions in such a window. Let us fix this window. Let us partition the pattern into four quarters. The window corresponds to the first quarter of the pattern. Let us call the two middle quarters the *essential part of the pattern*. Let z be a factor of size $\log m/4$ of the essential part, that has the highest number of occurrences within this part. The segment of length $m/2$ of the text immediately following the window is called the *essential part of the text.*

We assume for simplicity that the alphabet is binary. If the alphabet is unbounded, it has no more than n letters, and the sieve method is considered for small subwords of lengths doubly logarithmic.

Lemma 14.9 If an occurrence of the pattern starts in the window, then we can find one occurrence of z within the essential part of the text in constant time with m processors.

Proof There are at least $s = m^{0.5}$ occurrences of z within the essential part of the pattern, since there are $m/2$ occurrences and at most $2^{\log m/4}$ possible subwords of length $\log m/4$. Let us consider all possible occurrences of the pattern starting in the window and the relative occurrences of z in the essential part. Consider $m/4$ shifts of

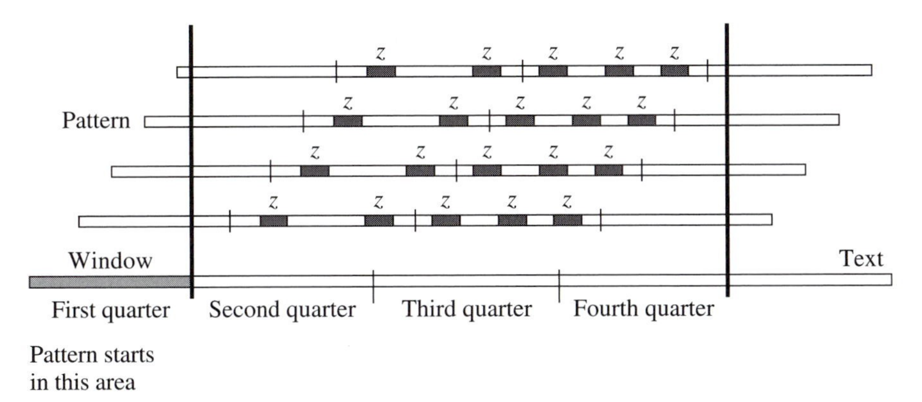

Figure 14.2: The factor z in the essential part of the pattern.

the pattern, and, in each of them, put 1 at starting positions of z. We have $m/4$ rows, each of them with $\sqrt{m}$ ones. According to the hitting-set lemma there is a set H of positions (corresponding to columns in the hitting-set lemma) in the essential part of the text such that, if there is any occurrence of the pattern starting in the window, then there is an occurrence of z in the essential part of text starting at a position in H. The set H is small, $|H| = O(m/\log m)$, in fact it is even of a lower order, due to the hitting-set lemma. Therefore, it is sufficient to check for an occurrence at positions in the hitting-set. This can be done on a CRCW PRAM in constant time with $|z|$ processors for each position in H. Altogether a linear number of processors is adequate. This completes the proof. ♦

We introduce the idea of *segments* related to the factor z (in text or in pattern). A *segment* of an occurrence of z in a given word w is the maximal factor of w containing z, and having the same period as z. We leave the proof of the following technical fact to the reader.

Lemma 14.10 Assume we are given the period of z and an occurrence of z in a given text. Then the segment of the occurrence of z can be found in constant time with n processors of a CRCW PRAM.

Lemma 14.11 (Galil's sieve lemma) Assume that the pattern is non-periodic, and both the subword z and its hitting-set H of size $m/\log m$ are constructed. We can then eliminate all but $O(m/\log m)$ candidate positions for matching positions in a given window in constant time with m processors.

Proof We must remove all but $O(m/\log m)$ candidates from the window. First, we find one occurrence of z in the essential part of the text. This can done optimally according to Lemma 14.9. If there is no occurrence then all positions in the window

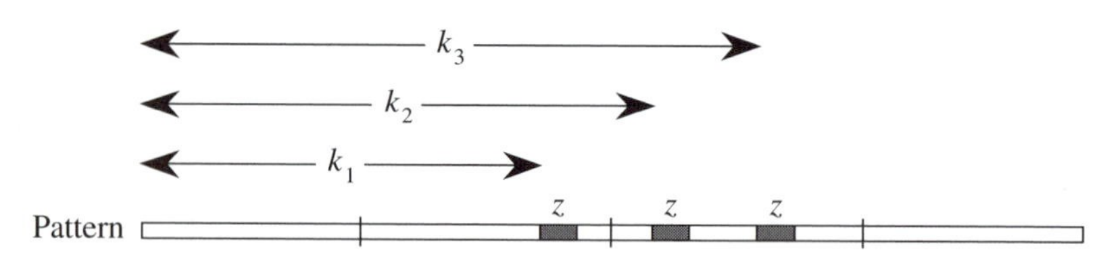

Figure 14.3: Positions of z in the pattern.

can be eliminated. Therefore, we assume that there is an occurrence, and we find one. Now there are two cases based on whether z is periodic or non-periodic.

Case 1 (z is non-periodic) This is the simpler case. Let $k_1, k_2, \ldots, k_p$ be the sequence of positions of occurrences of z in the pattern (see Figure 14.3). There are at most $O(m/|z|)$ occurrences due to the non-periodicity of z. This number is $O(m/\log m)$ because $|z| = \log m$.

Now, if we found an occurrence of z at position i in the essential part of the text, then the possible candidates for matching positions of the pattern in the window are only $i-k_1, i-k_2, \ldots, i-k_p$ (see Figure 14.4). All positions in the window that do not appear in this sequence are eliminated. This completes the proof in the non-periodic case.

Case 2 (z is periodic) We cannot directly apply the same argument now when z is periodic, because the number of occurrences of z can be larger than $O(m/\log m)$. However, we shall deal with extended occurrences of z: the segments of z. Let us fix z. It is easily apparent that two segments cannot overlap at more than $|z|/2$ positions. Hence, the number of segments in the pattern is $O(m/\log m)$. For a given z found in the essential part of the text we compute its segment. Then the end of this segment

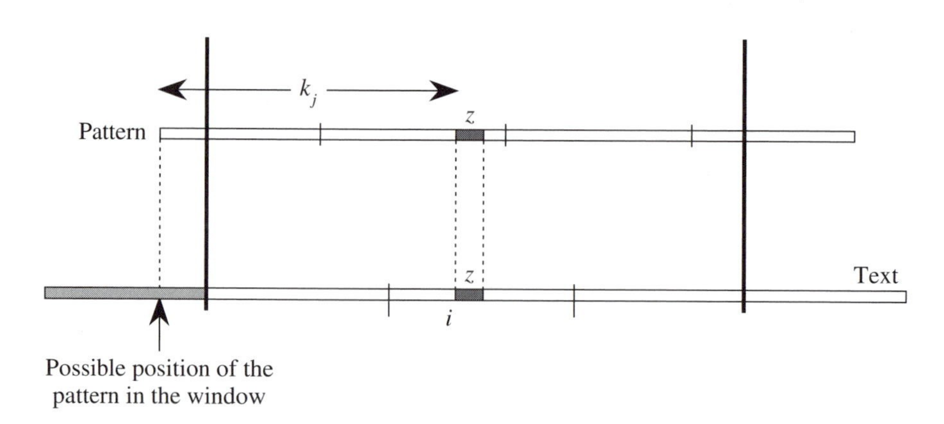

Figure 14.4: Possible matching position $i-k_j$.

corresponds to an end of some segment in the pattern, or the beginning of the segment corresponds to the beginning of some segment in the pattern. We use the same argument as in the first case. This completes the proof. ♦

The conclusion of this section, the corollary of the previous lemma, yields a constant-time string-matching algorithm.

Theorem 14.12 The search of a preprocessed pattern in a text of length n can be done in constant time with $O(n)$ processors in the CRCW PRAM model.

14.3 The suffix-prefix matching of Kedem, Landau, and Palem

The method discussed in the present section is based on the naming (or numbering) technique, that is intensively used by KMR algorithm (see Chapters 8 and 9). The crucial point in the optimal parallel string-matching algorithm presented here is to give consistent names to some segments of the pattern. The basic tool is the dictionary of basic factors (DBF, for short). However, the main drawback of the DBF is its size: it contains names for $n\log n$ objects. The solution proposed by KLP algorithm is to use a weak form of the DBF. It contains only names for all subwords of length 2^k that start at a position in the text divisible by 2^k (for each integer $k = 0, 1, \ldots, \log n$). We call this structure the *weak dictionary of basic factors* (weak DBF, for short).

Lemma 14.13 The weak DBF contains the names of $O(n)$ objects. It can be constructed in $O(\log n)$ time with $O(n/\log n)$ processors of a CRCW PRAM.

Proof The construction of the weak DBF can be carried out essentially in the same way as that of the DBF. But then the total work is linear, due to the fact that the total number of objects is linear. Hence $O(n/\log n)$ processors are sufficient. ♦

KLP algorithm reduces the string-matching problem to the suffix-prefix problem (below). That is, we need to assign names only to (some) suffixes and prefixes of two strings. The number of objects is linear, and due to this fact, an algorithm with linear total work is possible. The problem consists of essentially of building a dictionary for prefixes and suffixes. This is why the dictionary of basic factors is useful.

Let p and s be words of length n. The suffix-prefix (SP, for short) problem for words p and s of the same length is defined as follows.

SP problem Name consistently all prefixes of p and all suffixes of s together.

A linear-time sequential computation is straightforward: the SP problem for two words of length n can be solved sequentially in $O(n)$ time. This is a simple appli-

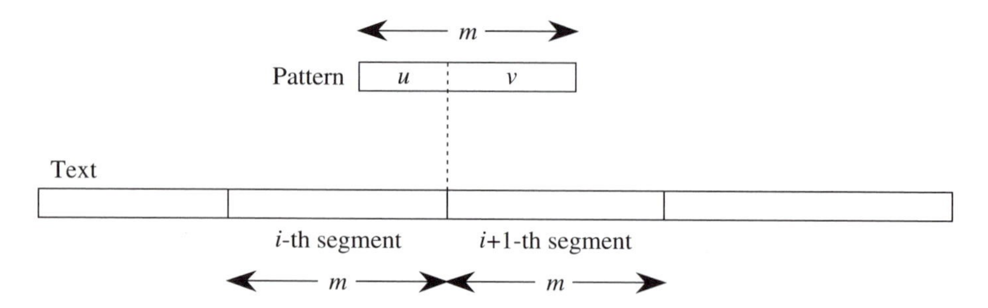

Figure 14.5: Suffix-prefix matching.

cation of Knuth-Morris-Pratt algorithm, and of the notion of failure function (see Chapter 3).

The SP problem can be solved using an optimal parallel algorithm, but this requires a nontrivial construction. It is postponed after the application to string matching presented in the next theorem. The theorem shows the significance of the SP problem. It is also presented here to acquaint the reader with the problem before constructing an optimal parallel algorithm.

Theorem 14.14 Assume that the SP problem for two strings can be solved by an optimal parallel algorithm. Then, the string-matching problem can also be solved by an optimal parallel algorithm.

Proof Let us factorize the text into (disjoint) segments of size m (length of the pattern). Then the string matching is reduced to $2n/m$ SP problems of size m. If the pattern overlaps the border between the i-th and $(i+1)$-th segment then some prefix u of the pattern should be a suffix of the i-th segment, and the associated suffix v of the pattern should be a prefix of the $(i+1)$-th segment (see Figure 14.5).

We have $O(n/m)$ SP problems of size m. They can be computed independently, each with $O(m)$ work. Altogether the algorithm works in logarithmic ($O(\log m)$) time, with a total work of $O(n)$. This completes the proof. ♦

One of the basic aspects of KLP algorithm is the reduction of the number of objects. This reduction is possible due to encoding of segments of length $\log n$ by their names of $O(1)$ size (small integers). The main tool to use is the string-matching automaton for several patterns. In what follows we assume that the input alphabet is of a constant size. This assumption is only used in the proof of Lemma 14.15 below. In fact, the lemma also holds for an unbounded alphabet, but the proof is more complicated (see Bibliographic notes). Hence, the whole suffix-prefix algorithm works essentially with the same complexity even if the assumption as to the size of the alphabet is dropped.

Lemma 14.15 Assume we have r patterns of the same length k. Then in $O(k)$ time on a CRCW PRAM

(a) we can construct the string-matching automaton G of the patterns with work $O(kr)$;

(b) if G is computed, then we can find all occurrences of the patterns in a given text of length n with $O(n)$ work.

Proof The proof of point (a) is essentially the parallel implementation of the sequential algorithm for the multi-pattern machine (see Chapter 7). The structure of this machine is based on the tree T of the prefixes of all patterns. The states of the automaton correspond to prefixes. The tree grows in a breadth-first search order. We use the algorithm for the construction of the multi-pattern automaton without use of the failure table. The transitions from a node at a given level are computed in a constant time, while transitions for all states at higher levels are computed (see Chapter 7). The bfs order is well suited to parallel construction level-by-level. At a given level we process all nodes simultaneously with rk processors. As a side effect, we have consistent names for all patterns: two patterns are equal iff they have the same names. There is a special name corresponding to all words of length k that are not equal to any of the k patterns.

The proof of point (b) is rather tricky. The constructed automaton can be used to scan the text. For each segment composed of the last k symbols read so far, write the name of the pattern equal to that segment, or the special value if no pattern matches. One automaton can sequentially process a factor of length $2k$ of the text, and give names to the last k positions (occurrences or non-occurrences of patterns) in that portion of the text. We can activate a copy of the automaton G at each position divisible by k (one can assume that k divides n). The n/k automata work simultaneously for $2k$ steps. After that, all segments of length k acquire their names. All automata traverse the same (quite large) graph representing the automaton G. The total time is $O(2k)$, and the total work is $O(n)$. ♦

For technical reasons we need to consider the following extended SP problem.

ESP problem Given strings s and $p_1, p_2, \ldots, p_r$ of the same length k, name consistently all prefixes of $p_1, p_2, \ldots, p_r$ and all suffixes of s.

A striking fact about the ESP problem is that there is an algorithm that processes prefixes and suffixes in a non symmetric fashion. In the lemma below we privilege prefixes, though one can use a reverse construction.

Lemma 14.16 (key lemma) The ESP problem for strings s and $p_1, p_2, \ldots, p_r$ of the same length k can be solved in $O(\log k)$ time, and $O(kr + k\log k)$ work.

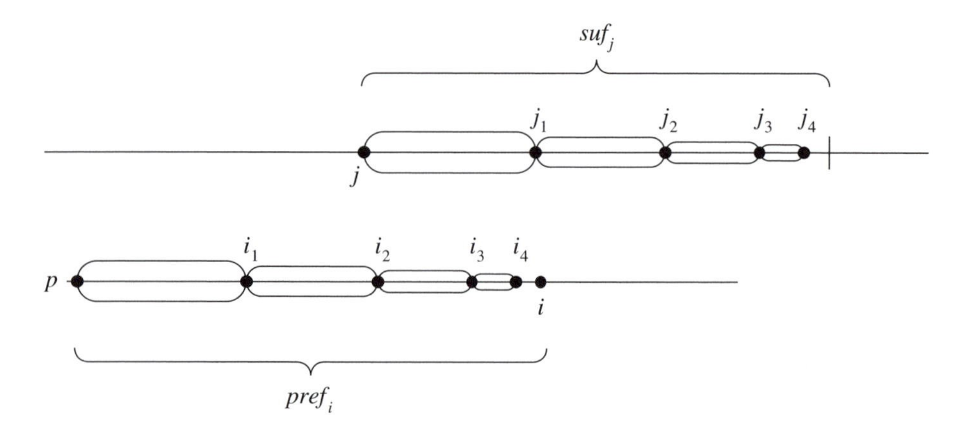

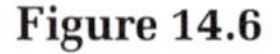
Figure 14.6

Proof Assume for simplicity that k is a power of two. Let us look for a computation on both s and one of the p_is, which is denoted by p. The same processing applies to all p_is simultaneously. Compute the complete DBF for s and the weak DBF for p together. In the weak DBF, names are given only to words of size 2^h starting at positions divisible by 2^h, for any $h \geq 0$ (see Figure 14.6). The weak DBF is of $O(k)$ size, while the complete DBF is of $O(k\log k)$ size. For all patterns, there are r weak DBFs, therefore the work spent on them is now $O(kr)$.

The algorithm proceeds in $\log k$ stages. Let us call an h-word any word in which the length is divisible by 2^h. An h-word is maximal if it cannot be extended to the right into a different h-word. The end of the maximal h-word starting at position j is denoted by $end_h(j)$. The basic property of $end_h(j)$ is that the difference $end_h(j) - end_{h+1}(j)$ is either 0 or 2^h.

In the algorithm we maintain the following invariant.

INV(h) all maximal h-words of s, and all h-prefixes of p have consistent names.

After computing the DBF for s, and the weak DBF for p, this invariant holds for $h = \log k$. In fact, the computation of the ESP problem reduces to satisfying *INV*(0). We now describe how to efficiently preserve the invariant from h+1 to h.

```
procedure STAGE(h);
{ INV(h+1) holds }
begin
   for each position j in s do in parallel begin
      if end_{h+1}(j) ≠ end_h(j) then
         combine names of subwords
         s[j..end_{h+1}(j)] and s[end_{h+1}(j)..end_h(j)];
```

```
            for each (h+1)-prefix p[1..i] of p do in parallel
                combine names of p[1..j] and p[j..j+2^h]
                to get names for p[1..j+2^h];
        end;
    { INV(h) holds }
    end.
```

The procedure can be simply extended to handle all patterns p_i simultaneously. Then, the whole algorithm has the following structure:

for h:= log$n$-1 **downto** 0 **do** *STAGE*(h).

$O(k)$ processors are obviously sufficient for processing s, one processor per each position j. Hence, $O(k\log k)$ work is spent on s. But for a given p_i of length k we spend only $O(k)$ work, since at stage h we process only (h+1)-prefixes. The total number of all h-prefixes, for h running from 0 to log$n$, is linear. Hence, the total work spent on a single p_i is $O(k)$, and it is $O(rk)$ on all p_i's. This complete the proof. ♦

Theorem 14.17 The SP problem for two words p, s of same length n can be solved on a CRCW PRAM in logarithmic time with linear work.

Proof Assume for simplicity that n is divisible by log$n$. Consider the subwords p_1, p_2, ..., $p_{\log n}$ of p of length $n - \log n$ starting inside the prefix of length log$n$ of p (see Figure 14.7).

Factorize the text s into segments of length log$n$, and call them small patterns. We can find all occurrences of the small patterns in p with $O(n)$ work, due to Lemma 14.15. The name of the small segment starting at position j in p is consistent with the name of a small pattern starting at that position.

Now we can factorize each p_i into segments of length log$n$. We name the segment consistently with names of small patterns. The name of a small segment starting at a given position of p can be found in $O(1)$ time after finding all occurrences of

Figure 14.7

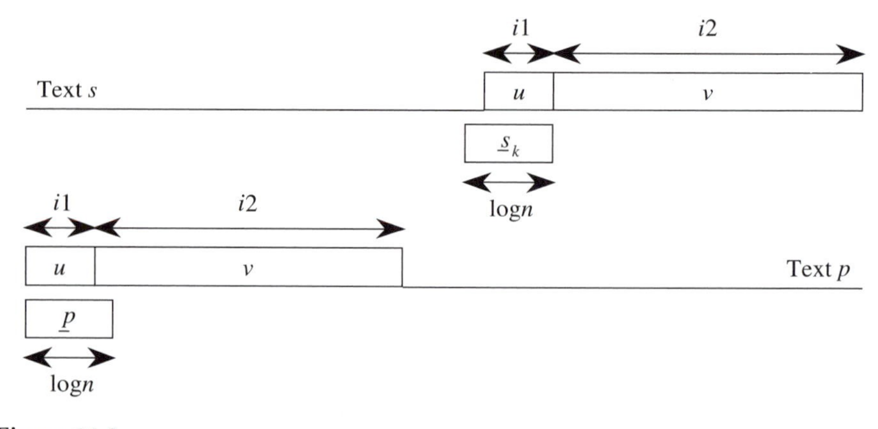

Figure 14.8

small patterns. By doing so, we compress s into a text s', and each pattern p_i into a text p_i' of length $n/\log n$.

We solve the ESP problem for s' and $p_1', p_2', p_3', \ldots, p_{\log n}'$. The time is logarithmic and the total work is linear due to Lemma 14.16. This proves the following:

Claim 1 We can compute, in logarithmic time with linear work, consistent names for all suffixes of s and all prefixes of subpatterns p_is in which the length is divisible by log$n$.

Consider now the prefix $\underline{p}$ of length log$n$ of p, and all log$n$-segments $\underline{s}_i$ of s. Assign one processor to each segment $\underline{s}_i$. This processor computes the SP problem for $\underline{p}$ and $\underline{s}_i$ sequentially in log$n$ time, due to Lemma 14.16. There are $n/\log n$ processors, hence the total work is linear. In this way we have proven:

Claim 2 We can compute optimally consistent names for all prefixes of $\underline{p}$ and all suffixes of logarithmic-size segments $\underline{s}_i$.

Now we are ready to solve the original SP problem for p and s. Take the prefix of p of a given size i, and the suffix of s of same size i. We have sufficient information for checking in constant time if the equation $p[1\ldots i] = s[n\text{-}i\text{+}1\ldots n]$ holds. We can write i as $i = i1+i2$, where $i1 < \log n$ and $i2$ is divisible by log$n$.

It is now sufficient to check names of the suffix of s of length $i2$, and prefix of p_{i1} of length $i2$ (see Figure 14.8). This can be performed in constant time, due to Claim 1, since $i2$ is divisible by log$n$. Next, we check the names of the (small) prefix of p of length $i1$ and the suffix $s[ni\text{+}1\ldots n\text{–}i2]$ of a logarithmic segment $\underline{s}_k$ of s. This can be performed in constant time due to Claim 2. We perform this operation simultaneously for all i. This completes the proof. ♦

14.4 A variation of KLP algorithm

In this section we present a variation of the KLP algorithm. Its aim is to provide a version of KLP algorithm that extends easily to the two-dimensional case. The basic parts of KLP algorithm and the dictionary algorithm (parallel version of KMR algorithm) are constructions of dictionaries that enable us to check in a constant time whether some factors of text or pattern match.

Recall that factors in which the length is a power of two are said to be *basic factors*. One can also assume that the size of the pattern is a power of two. If not, then we can search for two subpatterns that are prefix and suffix of the pattern, and in which the common size is the largest possible power of two. A basic factor f of the pattern is called a *regular basic factor* iff an occurrence of it starts in the pattern at a position divisible by the size of f. The basic property of regular factors is that a pattern of size n has $O(n)$ regular basic factors. This fact implies the key lemma, which has already been proven in Chapter 9.

Lemma 14.17 (key lemma) Consider t patterns, each of size M. There is an algorithm to locate all of these in a one-dimensional string or a two-dimensional image of size N in $O(\log M)$ time with $O(N\log M+tM)$ total work.

Proof The algorithm is derived from the parallel version of KMR algorithm. If we look more closely at the pattern-matching algorithm derived from KMR algorithm, then it is easy to see that the total work of the algorithm is proportional to the number of considered factors. We can assume w.l.o.g. that the size M of patterns is a power of two. Therefore, in patterns, only regular factors need to be processed. Hence, the work spent on one pattern is $O(M)$. Similarly, for the t patterns of size M, the work is $O(tM)$. This completes the proof. ♦

Let us fix the integer k as a parameter of order $\log m$. Lengths of texts are assumed to be multiples of k. The factors of length k of texts and patterns are called *small factors*. The small factors that have an occurrence at a position divisible by their size are called *regular small factor*. Let x be a string of length n, and let $small(x, i)$ be the small factor of x starting at position i ($1 \le i \le n$). The string $small(x, i)$ is well-defined if x is padded with k–1 special end markers at the end. Define the string $\underline{x} = \underline{x}_1\underline{x}_2\ldots\underline{x}_n$ by

$$\underline{x}_i = name(small(x, i)), \text{ for } 1 \le i \le n$$

in which $name(f)$ is the name of the regular small factor equal to f. Names are assumed to be consistent: if $small(x, i) = small(x, j)$ then $\underline{x}_i = \underline{x}_j$.

The string $\underline{x}$ is called here the *dictionary of small factors* of x. A simple way to compute such a dictionary is to use the parallel version of the Karp-Miller-Rosenberg algorithm. This gives $n\log\log n$ total work, very close to optimal. The optimality can be achieved, if the alphabet is of constant size, using a kind of the "four Russian trick," as explained below.

For simplicity, in this paragraph we consider that the alphabet is binary, and that $k = \log m/4$. There are potentially only $m^{1/2}$ binary strings of length $2k$. For each of them, we can precompute the names of all their small factors of length k. These names can be the integers having the corresponding factors as binary representations. We have sufficient processors for all the $m^{1/2}$ binary strings of length $2k$ to do that, since the total number of processors must be $O(n/\log m)$. Then, to produce $\underline{x}$ the dictionary of small factors of a string x of length n ($\geq m$), it is first factorized into segments of length $2k$. Each segment is treated independently and concurrently. One processor can encode it into a binary number in $O(\log m)$ time, looking at the precomputed table of names of its small factors, and writing down k consecutive entries of the string $\underline{x}$ in time $O(k) = O(\log m)$. Thus, the overall procedure works in $O(\log m)$ time with $O(n/\log m)$ processors. This proves the next lemma on fixed alphabets.

The breakthrough for this approach is executed by KLP algorithm: the assumption of fixed-size alphabets can be dropped. This is because only names corresponding to regular small factors need to be considered in the naming procedure for all factors (two factors which names are not equal to any regular small factor can have the same name, even if they are equal). The total length of all small regular factors is linear, and each of them has logarithmic size. The Aho-Corasick pattern machine for all regular basic factors can be constructed using an optimal algorithm, due to the linearity of their total length. But, such an approach does not work if we want to make a pattern-matching machine for all small factors.

Lemma 14.19 The dictionary of small factors can be computed in $O(\log m)$ time with linear work.

The description of the string-matching algorithm of the section is written so that it is suitable for the two-dimensional extension of Section 14.5. The algorithm for two dimensions is conceptually a natural extension of the one-dimensional case. The basic aspects of one-dimensional and two-dimensional pattern-matching algorithms are similar:

- computation of the dictionary of small factors,
- compression of strings by encoding disjoint $\log m$-size blocks by their names,
- application of the algorithm of Lemma 14.18.

Two auxiliary functions are necessary: *shift* and *compress*. Let $k = \log m$. We assume that n (length of *text*) and m (length of *pat*) are multiples of k. For $0 \leq r \leq -1$, let us denote by *shift*(*pat*, r) (see Figure 14.9) the string defined by

$$shift(pat, r) = pat[1+r \ldots m-k+r].$$

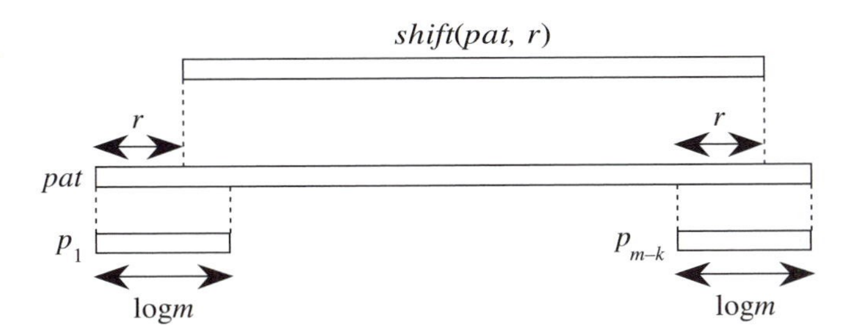

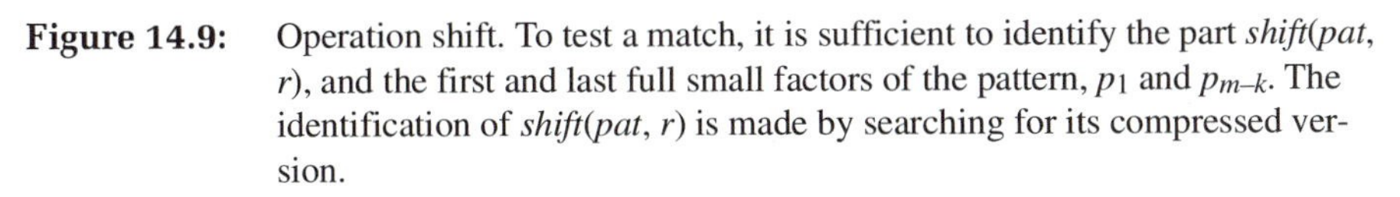

Figure 14.9: Operation shift. To test a match, it is sufficient to identify the part *shift*(*pat*, *r*), and the first and last full small factors of the pattern, p_1 and p_{m-k}. The identification of *shift*(*pat*, *r*) is made by searching for its compressed version.

For a string z let us denote by *compress*(z) the string defined by

$$compress(z) = \underline{z}_1\, \underline{z}_k\, \underline{z}_{2k} \ldots \underline{z}_{(h-1)k},$$

in which $h = |z|/k$. The string *compress*(z) contains the same information as z, but is shorter by a logarithmic reduction factor. Each letter of the new string encodes a logarithmic size block of z. Intuitively speaking *compress*(z) is a concatenation of names of consecutive small factors that compose the string z. Small factors (strings of length log$m$) are replaced by one symbol. The compression ratio is log$m$. In the following, $\underline{p}_i$ is the name of the small factor of *pat* starting at position i in *pat*, and $\underline{t}_i$ is the name of the small factor of *text* starting at position i in *text*.

The correctness of the algorithm below is based on the following obvious observation: an occurrence of *pat* occurs at position i in *text* iff the following conditions are satisfied:

(*) *compress*(*shift*(*pat*, k–(i mod k))) starts in *compress*(*text*) at position (i div k), $\underline{p}_1 = \underline{t}_{i+1}$, and $\underline{p}_{m-k} = \underline{t}_{i+m-k}$.

The condition (*) is illustrated by Figure 14.10 (see also Figure 14.9). The structure of the algorithm based on condition (*) is presented below.

```
Algorithm { parallel optimal string-matching }
begin
   build the common dictionary of small factors of pat and text;
   construct tables p and t;
   localize all patterns compress(shift(pat,0)),
        compress(shift(pat,1)),.., compress(shift(pat,k-1))
        in compress(text) by the algorithm of Lemma 14.18;
   for each position i do in parallel
        { in constant time by one processor
             due to previous localizations }
```

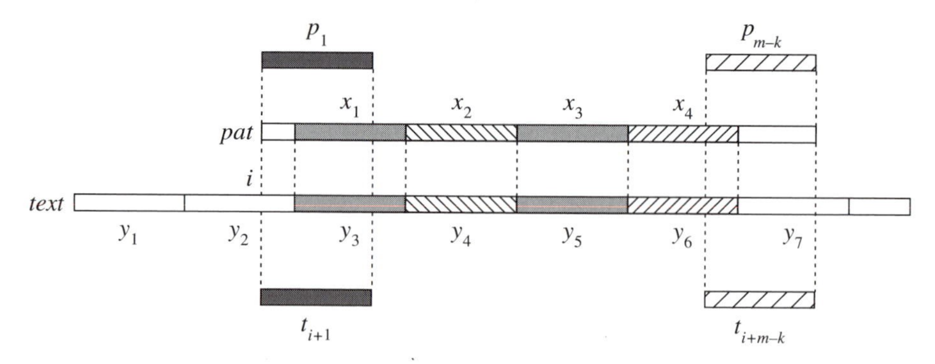

Figure 14.10: Test at position i in *text*. $x_1x_2x_3x_4 = compress(shift(pat, k-(i \bmod k)+1))$, $compress(text) = y_1y_2y_3y_4y_5y_6y_7$; *pat* occurs at position i in *text* iff $x_1x_2x_3x_4$ starts at the 3rd position of $y_1y_2y_3y_4y_5y_6y_7$, $\underline{p}_1 = \underline{t}_{i+1}$, and $\underline{p}_{m-k} = \underline{t}_{i+m-k}$.

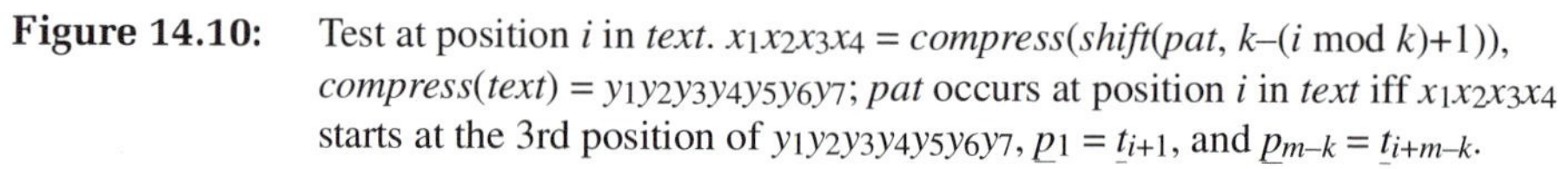
```
        if condition (*) holds for i then
            report the match at position i;
end.
```

Theorem 14.20 The above algorithm optimally solves the string-matching problem on a CRCW PRAM. It runs in $O(\log m)$ time with $O(n/\log m)$ processors (the total work is linear).

Proof The compressed text has size $n/\log m$. There are $\log m$ compressed patterns, each of size $m/\log m$. Hence, according to Lemma 14.18, the total work, when applying the algorithm of this lemma, is $O((n/\log m)\log m + \log m(m/\log m)) = O(n)$. This completes the proof. ♦

14.5 Optimal two-dimensional pattern matching

In this section, we show how the algorithm of the previous section generalizes to two-dimensional arrays, and leads to an optimal two-dimensional pattern-matching algorithm. We assume w.l.o.g. that the pattern and the text are squares: *PAT* is an $m \times m$ array, T is an $n \times n$ array. The general case of rectangular arrays can be handled in the same way. But then the shorter side of the pattern must be at least $\log m$ long, that is, the pattern must not be thin. Otherwise, the algorithm of Section 14.4 can be adapted to this specific problem.

Recall that subarrays in which the length is a power of two are said to be *basic subarrays*. We assume that m is a power of two. If not, then we can search for four (possibly overlapping) subpatterns in which the length is a power of two. We say that a basic subarray of shape $s \times s'$ is *regular* iff it starts at a position which horizontal and vertical coordinates are, respectively, multiples of s and s'. The *size* of the two-dimensional image is its area.

There is a fifth type of subarrays in addition to regular and/or basic subarrays: *thin subarrays*. It is a natural generalization of small factors of the previous section to the two-dimensional case. Thin factors are $m \times \log m$ subarrays of the pattern, and $n \times \log m$ subarrays of text. They arise if we cut the two-dimensional pattern by $m/\log m-1$ lines at distance $\log m$ from each other (see Figure 14.11).

The algorithm below provides an optimal two-dimensional matching. It is based on the fact that the key lemma of Section 14.4 (Lemma 14.18) works similarly for two-dimensional images. The compression realized, with the help of small factors in the case of strings, is executed by thin subarrays in the case of two dimensions. The text array, as well as the pattern, is cut into thin pieces on which the main part of the search is performed. We assume, for the simplicity of presentation, that we are dealing with images in which the sides have lengths divisible by m. Formally, the cut-lines are columns $\log m$, $2\log m$, ..., n-$\log m$ of the text array (see Figure 14.11). There are $n/\log m-1$ cut lines.

Let us denote by P_j the j-th column of PAT. For $0 \leq r \leq k-1$ denote by$SHIFT(PAT, r)$ the rectangle composed of columns P_{1+r}, ..., P_{m-k+r}. For a rectangle Z, denote by $COMPRESS(Z)$ the array $\underline{Z}_1\underline{Z}_k\underline{Z}_{2k} \ldots \underline{Z}_{(h-1)k}$, in which $h = |z|/k$, and $\underline{Z}_i$ is the column of names of $\log m$-size rows of the thin subarray Z_i. The rectangle $COMPRESS(Z)$ contains the same information as Z, but in a smaller form. Each column of the new rectangle encodes a thin subarray. Two-dimensional (thin) objects are reduced to one-dimensional objects. The compression ratio on sizes of arrays is $\log m$. We denote by $\underline{P}_i$ the compressed thin subarrays of PAT starting at column i, and by $\underline{T}$ the array of names of $\log m$-size factors of rows of T. The algorithm of this section implements the same idea as in the former section. The correctness of the algorithm is based on the following obvious observation (see Figure 14.11). An occurrence of PAT occurs at (i, j) in the image T iff the following condition is satisfied:

> (**) $COMPRESS(SHIFT(PAT, k-(j \bmod k)))$ occurs in $COMPRESS(T)$ at position $(i, j \text{ div } k)$, $\underline{P}_1$ occurs at position (i, j) in $\underline{T}$, and $\underline{P}_{m-k}$ occurs at position $(i, j+m-k)$ in $\underline{T}$.

The structure of the two-dimensional algorithm is essentially the same as the algorithm of Section 14.4. It yields the following result.

Theorem 14.21 Under the CRCW PRAM model, the algorithm below optimally solves the two-dimensional pattern-matching problem in $O(\log m)$ time and linear work.

Proof The proof is similar to that of Theorem 14.20. It is a simple application of Lemma 14.18. Here the size of the compressed text array is $N/\log m$ ($N = n^2$). The

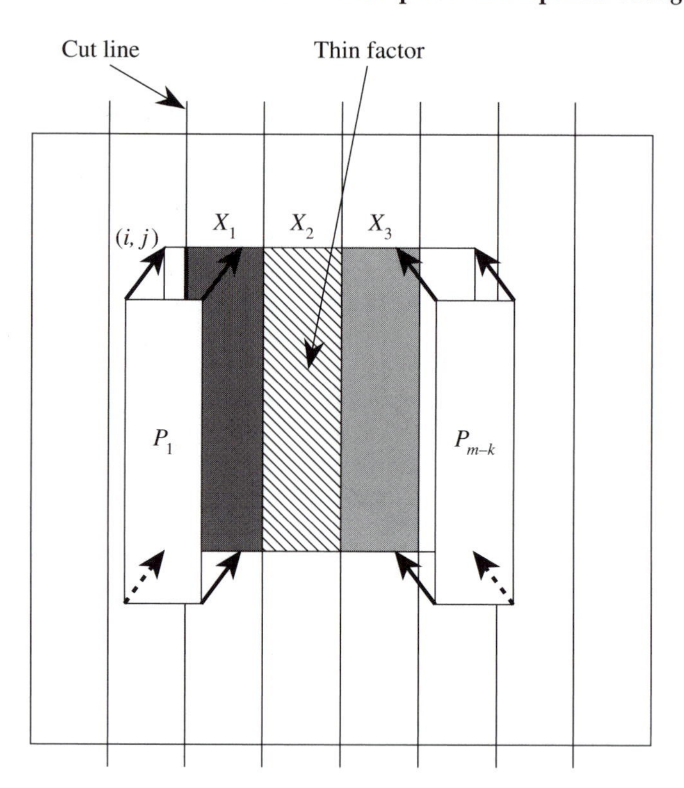

Figure 14.11: Partitioning of the pattern and the text arrays by cut-lines. We search for the compressed part of pattern image $\underline{X}_1\underline{X}_2\underline{X}_3$ in the compressed text array, and then search for the first and last thin subarrays of *PAT*, P_1 and P_{m-k}.

number of shifted patterns is still $\log m$. So, the total work is $(N/\log m)\log m + \log m(M/\log m) = O(N)$.♦

Algorithm { optimal two-dimensional pattern matching }
begin
  compute the dictionary of small subarrays together
    for rows of *PAT* and *T*;
  construct tables of names $\underline{P}$ and $\underline{T}$;
  localize patterns *COMPRESS*(*SHIFT*(*PAT*,0)),
    COMPRESS(*SHIFT*(*PAT*,1)),.., *COMPRESS*(*SHIFT*(*PAT*,*k*-1))
    in the image *COMPRESS*(*T*) by the algorithm of Lemma 14.18;
  for each *j* **do in parallel**
    find all occurrences of the first and last columns of $\underline{P}$
    in the *j*-th column of $\underline{T}$;
    { of the first and last thin factors of pattern,
      one-text/one-pattern algorithm }

```
for each position (i,j) of T do in parallel
    { in constant time by one processor for each (i,j)
        due to the information already computed }
    if condition (**) holds for (i,j) then
        report a match at position (i,j);
end.
```

14.6 Recent approach to optimal parallel computation of witness tables: The splitting technique

Quite recently a new approach was discovered for computing the witness table *WIT* of the pattern. It uses what is called the *splitting technique*. Here we sketch only the basic ideas of the technique. Several open problems have been cracked using this approach. We list the four most interesting problems that have been solved positively:

1. existence of a deterministic optimal string-matching algorithm working in $O(\log m)$ time on a CREW PRAM,
2. existence of a randomized string-matching algorithm working in constant time with a linear number of processors on a CRCW PRAM,
3. existence of an $O(\log n)$ time string-matching algorithm working on a *hypercube* computer with a linear number of processors,
4. existence of an $O(n^{1/2})$ time string-matching algorithm on a *mesh-connected* array of processors.

Theroem 14.22 Algorithms exist for each of the four points listed above.

The technique also provides a new optimal deterministic string-matching algorithm working in $O(\log\log n)$ time on a CRCW PRAM.

Recall that the CRCW PRAM is the weakest model of the PRAM with concurrent writes (whenever such writes occur the same value is written by each processor), and the CREW PRAM is the PRAM without concurrent writes. Vishkin's algorithm (see Section 14.1) works in $O(\log^2 n)$ time if implemented on a CREW PRAM. An optimal $O(\log n \log\log n)$ time algorithm for the CREW PRAM model was presented earlier by Breslauer and Galil.

The power of the splitting technique is related to the following recurrence relations:

(*) $time(n) = O(\log n) + time(n^{1/2})$,
(**) $time(n) = O(1) + time(n^{1/2})$.

Claim 1 The solutions to recurrence relations (*) and (**) satisfy, respectively:

$time(n) = O(\log n)$ and $time(n) = O(\log\log n)$.

Let P be a pattern of length m. The witness table *WIT* is computed only for the positions inside the interval *FirstHalf* = $[1 \ldots m/2]$. We say that a set S of positions is *k-regularly sparse* iff S is the set of positions i inside *FirstHalf* such that $i \bmod k = 1$. If S is regularly sparse then let *sparsity(S)* be the minimal k for which S is k-regularly sparse. Let us note

$$P^{(q)} = P(q)P(k+q)P(2k+q)P(3k+q)\ldots,$$

for $1 \leq q \leq k$. Denote by *SPLIT(P, k)* the set of strings $P^{(q)}$, $1 \leq q \leq k$.

Example *SPLIT*(*abacbdabadaa*, 3) = {*acad*, *bbba*, *adaa*}. ♦

Assume S is a k-regularly sparse set of positions. Denote by *COLLECT*(P, k) the procedure that computes values of the witness table for all positions in S, assuming that the witness tables for all strings in *SPLIT*(P, k) are known.

Claim 2 Assume the witness tables for all strings in *SPLIT*(P, k) are known. Then, *COLLECT*(P, k) can be implemented by an optimal parallel algorithm in $O(\log m)$ time on a CREW PRAM, and in $O(1)$ on a CRCW PRAM.

The next fact is more technical. Denote by *SPARSIFY*(P) the function that computes the witness table at all positions in *FirstHalf* except at a set S that is k-regularly sparse. The value returned by the function is the sparsity of S; when $k>m/2$, S is empty. In fact, the main role of the function is the sparsification of non-computed entries of the witness table.

Claim 3 *SPARSIFY*(P) can be computed by an optimal parallel algorithm in $O(\log m)$ time on a CREW PRAM, and in $O(1)$ on a CRCW PRAM. The value k of *SPARSIFY*(P) satisfies $k \geq m^{1/2}$.

The basic component of the function *SPARSIFY* is the function *FINDSUB*(P) that finds a non-periodic subword z of P of size $m^{1/2}$, or reports that there is no such subword. A similar construction is used in the sieve algorithm of Section 14.2. It is easy to check whether the prefix z' of size $m^{1/2}$ is non-periodic or not (we have a quadratic number of processors with respect to $m^{1/2}$); if z' is periodic we find the continuation of the periodicity and take the last subword of size $m^{1/2}$. The computed segment z can be preprocessed (its witness table is computed). Then, all occurrences of z are found, and based on these the sparsification is performed.

We only present how to compute witness tables in $O(\log m)$ time using $O(m\log m)$ processors. The number of processors can be further reduced by a logarithmic factor, which makes the algorithm optimal. The algorithm is illustrated as the following procedure *Compute_by_Splitting*.

According to the recurrence relation (*), the time for computing the witness table using the procedure *Compute_by_Splitting* is $O(\log m)$ on a CREW PRAM, and $O(\log\log m)$ on a CRCW PRAM. Implementations of the subprocedures *SPARSIFY* and *COLLECT* on a hypercube and on a mesh-connected computer give the results stated in points 3 and 4 earlier.

```
procedure Compute_by_Splitting(P);
begin
    k := 1;
    while k ≤ m/2 do begin
        k := SPARSIFY(P);
        (P^(1), P^(2), ..., P^(k)) := SPLIT(P, k);
        for each q, 1 ≤ q ≤ k do in parallel
            Compute_by_Splitting(P^(q));
        COLLECT(P, k);
    end;
end;
```

The constant-time randomized algorithm (point 2) is much more complicated to design. After achieving a large sparsification, the algorithm stops further calls, and begins a special randomized iteration. Indeed, it is more convenient to consider an iterative algorithm. The definition of *SPARSIFY*(P) needs to be slightly changed: the new version sparsifies the set S of non-computed entries of the witness talbe assuming that S is already sparse. The basic point is that the sparsity grows according to the inequality:

$$k' \geq k.(m/k)^{1/2},$$

in which k is the old sparsity, and k' is the new sparsity of the set S of non-computed entries. The randomization occurs when sparsity $\geq m^{7/8}$. This is achieved after at most three deterministic iterations.

The constant-time string matching also requires a quite technical (though very interesting) construction of several deterministic samples in constant time. But this is outside the scope of this book. We refer the reader to the Bibliographic notes for details.

Bibliographic notes

The first optimal parallel algorithm for string matching was presented by Galil in [Ga 85]. The algorithm is optimal only for alphabets of constant size. Vishkin improved on the notion of the slow duels of Galil, and described the more powerful concept of (fast) duels that leads to an optimal algorithm independently of the size of the alphabet [Vi 85]. The optimal parallel string-matching algorithm working in $O(\log^2 n)$ time on a CREW PRAM is also from Vishkin [Vi 85].

The idea of witnesses and duels is used by Vishkin in [Vi 91] in the string matching by sampling. The concept of deterministic sampling is very powerful. It has been used by Galil to design a constant-time optimal parallel searching algorithm (the preprocessing is not included). This result was an improvement upon the $O(\log^* n)$ result of Vishkin, though $O(\log^* n)$ can also be treated practically as a constant.

If preprocessing is included, the lower bound for the parallel time of string matching is $O(\log\log n)$. It has been proven by Breslauer and Galil [BG 90], who also gave an optimal $O(\log\log n)$ time algorithm. The algorithm uses a scheme similar to an optimal algorithm for finding the maximum of n integers by Shiloach and Vishkin [SV 81]. Recently it has been shown that it is possible to combine a $O(\log\log n)$ time preprocessing with the constant-time optimal parallel search of Galil (see [C-R 93b]).

Section 14.3 is adapted from the suffix-prefix approach of Kedem, Landau, and Palem [KLP 89]. The counterpart to the relative simplicity of their algorithm is the use of a large auxiliary space (though space $n^{1+\varepsilon}$ is sufficient, for any $\varepsilon>0$, with the help of some arithmetic tricks). Algorithms of Galil and Vishkin require only linear space, but their algorithms are much more sophisticated. The assumption as to the size of alphabet in Section 14.3 can be dropped (see [KLP 89]). We do not know how to convert the KLP suffix-prefix algorithm into an optimal fast parallel string-matching algorithm working on a CREW PRAM. Such an algorithm working in $O(\log^2 n)$ time was given by Vishkin (see [Vi 85]).

The "four Russian trick" of encoding small segments by numbers is a classical method. It is used in particular in [Ga 85], and in [ML 85] for string problems.

The optimal $O(\log n)$ time parallel two-dimensional pattern matching is from [CR 92]. An optimal searching algorithm running in $O(\log\log n)$ time is presented in [ABF 92b]. The splitting technique of Section 14.6 is adapted from [C-R 93c]. Consequences of the technique can also be found in this paper.

Selected references

[BG 90] Breslauer, D., & Galil Z., "An optimal O(*loglogn*) time parallel string-matching algorithm," *SIAM J.Comput* 196 (1990): 1051–1058.

[C-R 93c] Cole, R, Crochemore, M, Galil, Z., Gasieniec, L., Hariharan, R., Muthukrishnan, S., Park, K., & Rytter, W., "Optimally fast parallel algorithms for preprocessing and pattern matching in one and two dimensions," in: (*Proc. 34th IEEE Symposium on Foundations of Computer Science* 1993): 248–258.

[CR 92] Crochemore, M., & Rytter, W., "Note on two-dimensional pattern matching by optimal parallel algorithms," in (*Parallel Image Analysis*, A. Nakamura, M. Nivat, A. Saoudi, P.S.P. Wang, K. Inoue, editors, LNCS 654, Springer-Verlag, 1992): 100–112.

[Ga 92] Galil, Z., "A constant-time optimal parallel string-matching algorithm," in (*Proc. 24th ACM Symp. on Theory of Computing*, 1992): 69–76.

[KLP 89] Kedem, Z.M., Landau, G.M., & Palem, K.V., "Optimal parallel suffix-prefix matching algorithm and applications," in (*Proc. ACM Symposium on Parallel Algorithms*, Association for Computing Machinery, New York, 1989): 388–398.

[Vi 85] Vishkin, U., "Optimal parallel pattern matching in strings," *Information and Control* 67 (1985): 91–113.

Miscellaneous

This chapter presents several interesting questions about strings that have not already been considered in previous chapters. They are: *string-matching by hashing* and its extension to two-dimensional pattern matching, *tree-pattern matching* as an example of pattern-matching questions on trees, *shortest common superstrings*, *cyclic equality of words* and *Lyndon factorization of words*, *unique decipherability* problem of codes, *equality of words over partially commutative alphabets*, and breaking paragraphs into lines. The treatment of problems is not always handled in full detail.

15.1 String matching by hashing: Karp-Rabin algorithm

The concept of hashing and fingerprinting is very successful from a practical point of view. When we need to compare two objects x and y, we can look at their "fingerprints" given by $hash(x)$, $hash(y)$. If the fingerprints of two objects are equal, there is a strong likelihood that they are really the same object, and we can then apply a more thorough test of equality (if necessary). The two basic properties of fingerprints are:

- efficiently computable,
- highly discriminating: it is unlikely to have both $x \neq y$ and $hash(x) = hash(y)$.

The idea of hashing is utilized in the Karp-Rabin string-matching algorithm. The fingerprint FP of the pattern (of length m) is computed first. Then, for each posi-

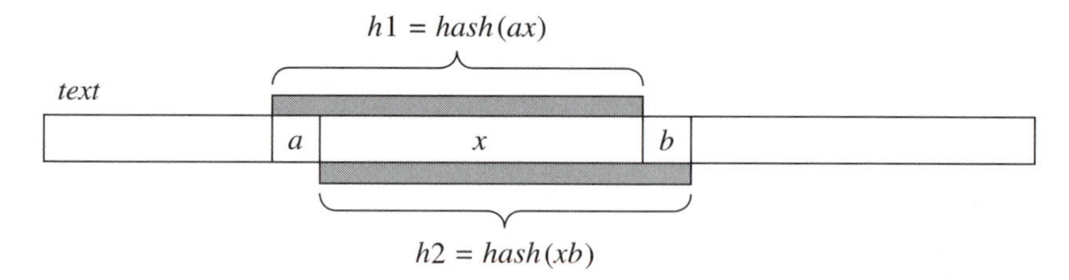

Figure 15.1: $h2$ should be easily computable from $h1$; $h2 = f(a, b, h1)$, where f is easy to compute.

tion i on the text, the fingerprint FT of $text[i+1 \ldots i+m]$ is computed. If ever $FT = FP$, we check directly to see if the equation $pat = text[i+1 \ldots i+m]$ really holds.

Going from a position i on the text to the next position $i+1$ efficiently requires another property of hashing functions for this specific problem (see Figure 15.1):

- hash($text[i+1 \ldots i+m]$) should be easily computable from $hash(text[i \ldots i+m-1])$.

Assume for simplicity that the alphabet is $\{0,1\}$. Then each string x of length m can be treated as a binary integer. If m is large, the number becomes too large to fit into a unique memory cell. It is convenient then to take as a fingerprint the value "x mod Q," in which Q is a prime number as large as possible (for example, the largest prime number which fits into a memory cell). The fingerprint function is then

$$hash(x) = [x]_2 \bmod Q,$$

where $[u]_2$ is the number which binary representation is the string u of length m. All string arguments of $hash$ are of length m. Let $g = 2^{m-1} \bmod Q$. Then, the function f (see Figure 15.1) can be computed by the formula:

$$f(a, b, h) = 2(h-ag)+b.$$

Proceeding in this way, the third basic property of fingerprints is satisfied: f is easily computable. This is implemented in the algorithm below.

```
Algorithm Karp-Rabin; { string-matching by hashing }
begin
   FP := [pat[1..m]]2 mod Q;   g := 2^(m-1) mod Q;
   FT := [text[1..m]]2 mod Q;
   for i := 0 to n-m do begin
      if FT = FP { small probability } then
         check equality pat = text[i+1..i+m]
         applying symbol by symbol comparisons, { cost = O(m) }
         and report a possible match;
```

```
            FT := f(text[i+1], text[i+m+1], FT);
        end;
    end.
```

The worst-case complexity of the algorithm is quadratic. But it could be difficult to find interesting input data causing the algorithm to make a quadratic number of comparisons (the non-interesting example is that in which *pat* and *text* consist only of repetitions of the same symbol). On the average, the algorithm is fast, but the best time complexity is still linear. This is to be compared with the lower bound of string matching on the average (which is $O(n\log m/m)$), and the best time complexity of Boyer-Moore type of algorithms (which is $O(n/m)$). String matching by hashing produces a straightforward $O(\log n)$ randomized optimal parallel algorithm because the process is reduced to prefix computations.

One can also apply other hashing functions that satisfy the three basic properties above. The original Karp-Rabin algorithm chooses the prime number randomly.

Essentially, the idea of hashing can also be used to solve the problem of finding repetitions in strings and arrays (looking for repetitions of fingerprints). The algorithm below is an extension of Karp-Rabin algorithm to two-dimensional pattern matching. Let m be the number of rows of the pattern array. Fingerprints are computed for columns of the pattern, and for factors of length m of columns of the text array. The problem then reduces to ordinary string matching on the alphabet of fingerprints. Since this alphabet is large, an algorithm in which the performance is independent of the alphabet is actually required.

```
Algorithm two-dimensional pattern matching by hashing;
{ PAT is an m × m' array, T is an n × n' array }
begin
    pat := hash(P1)...hash(Pm'), where Pj is the j-th column of PAT;
    text := hash(T1)...hash(Tn'), where Tj is the prefix of length m
                                            of the j-th column of T;
    for i := 0 to n-m do begin
        if pat occurs in text at position j then
            check if PAT occurs in T at position (i, j)
            applying symbol by symbol comparisons, { cost = O(mm') }
            and report a possible match;
        if i ≠ n-m then      { shift one row down }
            for j := 1 to n' do
                text[j] := f(T[i+1, j], T[i+m+1, j], text[j]);
    end;
end.
```

15.2 Efficient tree-pattern matching

The pattern-matching problem for trees consists of finding all occurrences of the pattern tree P in the host tree T. The sizes (number of nodes) of P and T are, respec-

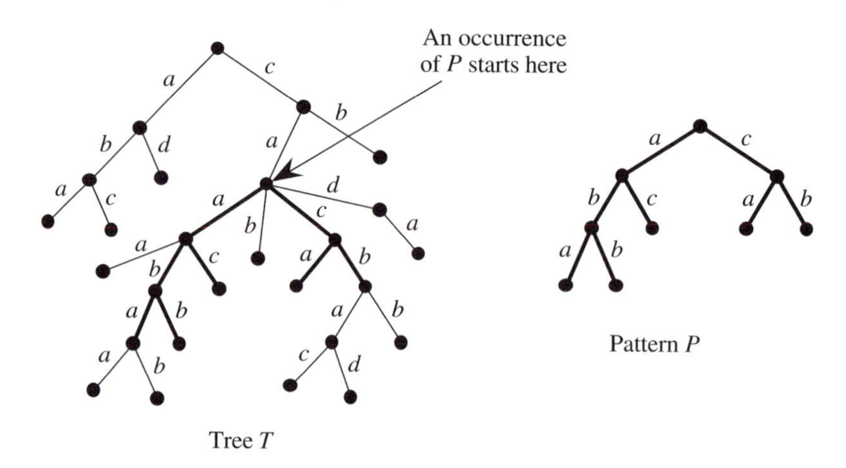

Figure 15.2: An occurrence of pattern P in the tree T.

tively, m and n. An example of occurrence of a pattern P in a tree T is shown in Figure 15.2.

We consider rooted trees in which the edges are labeled by symbols. We assume that the edges leading to the sons of the same node have distinct labels. If the tree is ordered (the sons of each node are linearly ordered) then this assumption can be dropped, because then the number of a son is an implicit part of the label.

Let $p(v)$ be the path from the root of P to the node v in P. The previous assumption allows us to identify this path with its label (the string of labels of edges along the path). The paths $p(v)$s are called node-paths here. We say that a path (string) p occurs at node w if there is a path down the tree starting at w and labeled by p. Then, we say that pattern P occurs at node w in T iff paths $p(v)$ occur at w, for all leaves v of P. The usual string-matching problem for a single pattern is a special case of the tree-pattern-matching problem. In this case P and T are single paths. If the pattern is a single path and T is an arbitrary tree, then, the problem is easy to carry out. Just build the KMP string-matching automaton G for P, and traverse the tree T in a DFS or BFS order computing the occurrences of P along the paths. The total time is linear. In fact, the same procedure can be applied if the set of all $p(v)$s is suffix-free (a set of strings is suffix-free if no string is a suffix of another one in the set).

Lemma 15.1 (easy tree-pattern matching) If the set of all node-paths corresponding to leaves of P is suffix-free, then we can find all occurrences of P in T in linear time.

Proof Construct the multi-pattern string-matching automaton G, in which the set of string patterns is given by all $p(v)$s. Then, traverse T using G, and, for each node w in which an occurrence of $p(v)$ for some leaf v ends, record it in the ancestor u of w at

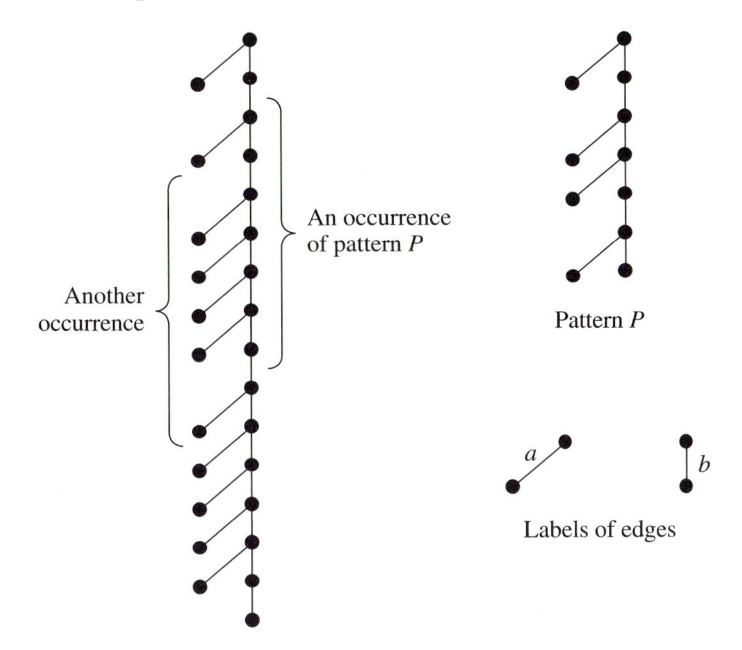

Figure 15.3: The tree-pattern matching corresponds here to string matching with don't cares, where *text* = 1010111101111100, and *pat* = 1∅11∅1∅.

distance $height_P(v)$ from w. There is a linear number of such recordings due to the fact that no $p(v)$ can end on a same node. Afterward, we check for each node u if it has the record for each leaf v of P. If yes, then u is the beginning of an occurrence of P in T. Altogether this takes linear time. ♦

We say that a pattern is a simple *caterpillar* if it consists of one path, called the *backbone*, and edges leading from the backbone to leaves (see Figure 15.3). The edges of the backbone are labeled by a, and all other edges are labeled by b.

Lemma 15.2 (easy tree-pattern matching) Let P and T be simple caterpillars. All occurrences of P in T can found in $O(n\log^2 n)$ time.

Proof The problem reduces to the string matching with don't care symbols, which has the required complexity (in fact even slightly smaller) (see Chapter 11). We place 1 on a given node of the backbone if there is a b-edge outgoing that node, otherwise we place 0. The pattern is processed similarly, except that instead of zeros we place don't care symbols ∅ (see Figure 15.3). This symbol matches any other symbol. In this way, P and T are substituted by two strings. The string matching with don't care symbols is then applied. This completes the proof. ♦

Lemma 15.3 (easy tree-pattern matching) Assume that the pattern tree P contains a suffix-free set S of size k of node-paths. Then, all occurrences of P in T can be found in $O(nm/k + m\log m)$ time.

Proof First we show how to find such a set S, when it is known to exist. S is found by constructing the suffix tree for the set of all reversed node-paths with added end marker #. Afterward, edges labeled with # are removed. This guarantees that there is a node for each reversed node-path. This suffix tree can be constructed in $O(m\log m)$ time adapting the construction of suffix trees for a single string. The path $p(v)$ is not a suffix of $p(v')$ if the node corresponding to $p(v)^R$ in the constructed suffix tree is not an ancestor of node corresponding to $p(v')^R$. The assumption of the lemma implies that there are at least k leaves in the suffix tree. Choose any k leaves. They correspond to a suffix-free set S of node-paths of P. In this way, S is built.

Then, identify S with the sample subpattern of P consisting only of the node-paths of P that are in S. We can find all occurrences of the sample S in linear time, due to Lemma 15.1. It can be checked that there are at most n/k occurrences of S in T. For each of the occurrences of the sample, we check a possible occurrence of the whole pattern P using a "naive" $O(m)$-time algorithm. This takes $O(nm/k)$ time. Globally, the matching takes $O(nm/k + m\log m)$ time as announced in the statement. ♦

Lemma 15.4 (easy tree-pattern matching) Assume P is of height h. Then, all occurrences of P in T can found in $O(nh)$ time.

Proof The naive algorithm actually works in that time. The proof is left to the reader. ♦

The aim is to design an algorithm working in $O(nm^{0.5})$ time. Let $l = m^{0.5}$. If we have $k \geq l$ in Lemma 15.3, or $h < 3l$ in Lemma 15.4, then we have an algorithm with the required complexity. In any case, the subpattern consisting of node-paths of length at most $3l$ can be matched separately, using Lemma 15.4. Therefore, we are left with hard patterns, which do not satisfy assumptions of Lemmas 15.3 or 15.4. The pattern P is said to be *hard* iff it satisfies:

(*) in each set of at least l node-paths there are two strings such that one is the suffix of the other,

(**) the depth of all leaves is at least $3l$.

The key to the efficient algorithm is the high periodicity of node-paths of a hard pattern tree. For a leaf v of P, consider the bottom l+1 ancestors of v. Then, there are two distinct ancestors u, w such that $p(u)$ is a proper suffix of $p(w)$. The segment p of size $|p(w)|-|p(u)|$ is a period of $p(w)$, and the latter word results by cutting off the path t from v to w. This path t is called the tail.

Let us take the lowest nodes u, w. Then p is primitive. The pair (t, p) is uniquely determined by v. It is called the tail-period of $p(v)$ and of v. We leave to the reader the proof that there are at most l distinct tail-period pairs, and that they can be found all in $O(ml)$ time (by a version of KMP algorithm). This proves the following lemma.

Lemma 15.5 Assume that P is hard. Then each node-path of P has a period $p \leq l$ $(= m^{0.5})$, after cutting off a part t of size at most l at the bottom. The word p is primitive. There are $O(p)$ distinct tail-period pairs (t, p), and they can all be found in $O(ml)$ time.

The main result of this section stated as follows yields an upper bound to the tree-pattern-matching problem.

Theorem 15.6 The tree-pattern-matching problem can be solved in $O(nm^{0.5} \log^2 n)$ time.

Proof Let $P' = P(t, p)$ be the part of pattern P formed by all node-paths with given characteristics (t, p). It is sufficient to show how to find all occurrences of P' in T in linear time. There are at most $|p|$ leaves of P', hence it is sufficient to prove that all occurrences of each node-path of P in T can be found in $O(n/|p|)$ time. This looks impossible, because $n/|p|$ is much smaller than n, but the entire tree T of size n is to be searched. However, the key point is that we can make a suitable preprocessing that takes linear time, and which can be used later for all node-paths of P'. After the preprocessing, the tree T is compressed into a smaller structure of $O(n/|p|)$ size. We have to look carefully at the structure of node-paths of P' of periodicity p, together with the tails t outgoing; they all have the same characteristic (t, p).

The tree P' consists of a small top tree and a set of branches, which results by cutting off that tree. These branches are called main branches. Each such branch starts with the full occurrence of the period p. At some points, where an occurrence of p ends, an occurrence of the tail t can start (see Figure 15.4). There are at most p main branches. All path-nodes of the small tree are proper suffixes of the period p. We leave to the reader the exercise of finding all occurrences in T of the top tree in $O(n)$ time. It is adequate for now to find occurrences of all branches in $O(n)$ time.

First we preprocess the tree T. We find all occurrences of the period p, and of the tail t in T in linear time due to Lemma 15.1. Each subpath of T from a node v to some node w corresponding to an occurrence of p is replaced by a special edge (v, w) labeled a. After that, we remove isolated special edges (one-edge biconnected components), as each main branch contains at least two ps. Then, for each node w, if an occurrence of the tail t starts at w, we add a special edge (w, w') labeled b, where w' is a newly created leaf. Let us consider the graph T' consisting of such special edges. The constructed graph is a forest, since the tail is not a continuation of p (p is primitive).

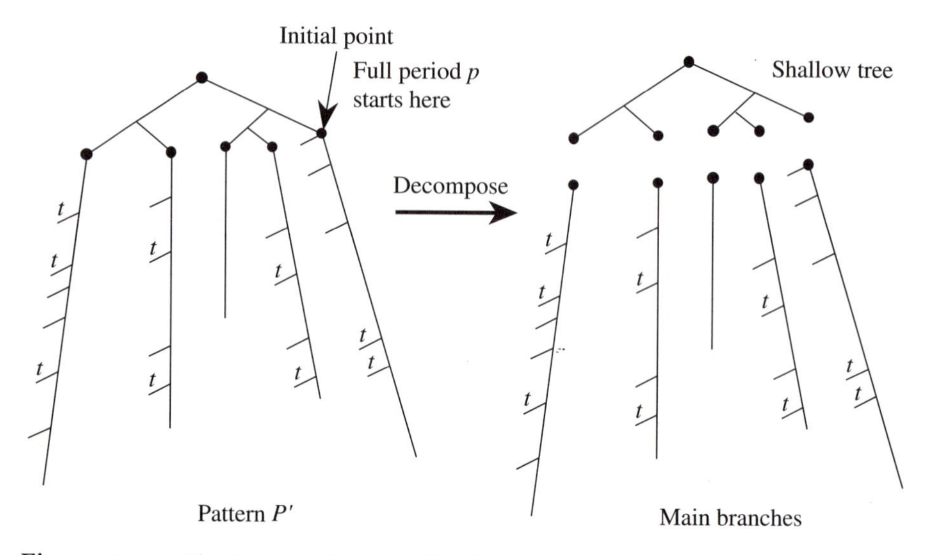

Figure 15.4: The structure of the tree $P' = P(t, p)$, and its decomposition.

The forest T' consists of disjoint caterpillars. We leave as an exercise showing that the total size of T' is $O(n/p)$.

Let us compress each main branch of P' in a manner similar as to how we transformed T into T' (see Figure 15.5). The basic point is that after compression each main branch becomes a simple caterpillar. It is easy to see that it is sufficient to find all occurrences of such caterpillars in T'.

Now we can use Lemma 15.2. Each compressed main branch (a simple caterpillar) can be matched against the set T' of compressed caterpillars in $O(n/p.\log^2 n)$ time, due to Lemma 15.2. There are $O(p)$ branches. Hence, the total time is $O(n)$ for finding all occurrences of $P(t, p)$. But there are only $O(m^{0.5})$ such trees. Therefore, the total time needed for the tree-pattern matching is $O(nm^{0.5}\log^2 n)$. This completes the proof. ♦

15.3 Shortest common superstrings: Gallant algorithm

The shortest common superstring problem is defined as follows: Given a finite set of strings R find a shortest text w such that $Fac(w) \supseteq R$. The size of the problem is the total size of all words in R.

The problem is known to be NP-complete. So the natural question is to find a polynomial approximation of the exact algorithm. The aim of Gallant's method is to compute an approximate solution.

Without loss of generality, we assume (throughout this section) that R is a factor-free set, this means that no word in R is a factor of another word in R. Otherwise,

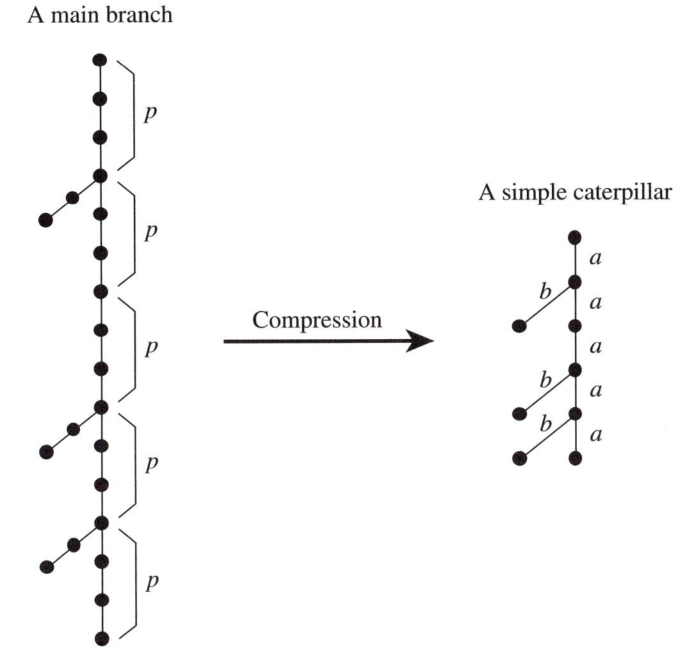

Figure 15.5: The compression of a main branch into a simple caterpillar.

if u is a factor of v (u, $v \in R$), a solution for R-$\{u\}$ is a solution for R. For two words x, y define *Overlap*(x, y) as the longest prefix of y which is a suffix of x. If v = *Overlap*(x, y), then x, y are of the form

$$x = u_1v,\ y = vu_2.$$

Let us define $x © y$ as the word u_1vu_2 ($= u_1y = xu_2$). Observe that the shortest superstring of two words u, v is $u © v$, or $v © u$. Since the set R is factor-free, the operation © has the following properties:

(*) operation © is associative (on R),

(**) the minimal shortest superstring for R is of the form $x_1©x_2©x_3©\ldots©x_k$ in which $x_1, x_2, x_3, \ldots, x_k$ is a permutation of all words of the set R.

```
function Gallant(R); { greedy approach to SCS }
begin
   if R consists of one word w then
      return w
   else begin
      find two words x, y such that |Overlap(x,y)| is maximal;
      S := R - {x, y} ∪ {x © y};
```

```
        return Gallant(S);
    end;
end;
```

The above algorithm is quite effective in terms of compression. The superstring represents (in a certain sense) all subwords of R. Let n be the sum of lengths of all words in R. Let w_{min} be a shortest common superstring, and let w_{Gal} be the output of Gallant algorithm. Note that $|w_{min}| \leq |w_{Gal}| \leq n$. The difference $n - |w_{min}|$ is the size of compression. The better the compression, the smaller the shortest superstring. The following lemma states that the compression reached by Gallant algorithm is at least half the optimal value (see Bibliographic notes).

Lemma 15.7 $n - |w_{Gal}| \geq (n - |w_{min}|)/2$.

Example Take the following example set $R = \{w_1, w_2, w_3, w_4, w_5\}$, in which

$$w_1 = egiakh,\ w_2 = fbdiac,\ w_3 = hbgegi,\ w_4 = iacbd,\ w_5 = bdiach.$$

Then, the shortest superstring has length 15. However, Gallant algorithm produces the following result

$$\begin{aligned} w_{Gal} &= \text{Gallant}(w_1, w_2 © w_5, w_3, w_4) \\ &= \text{Gallant}(w_3 © w_1, w_2 © w_5, w_4) \\ &= \text{Gallant}(w_3 © w_1 © w_2 © w_5, w_4) \\ &= w_3 © w_1 © w_2 © w_5 © w_4. \end{aligned}$$

Its size is

$$n - Overlap(w_3, w_1) - Overlap(w_1, w_2) - Overlap(w_2, w_5) - Overlap(w_5, w_4) = 29 - 3 - 0 - 5 - 0 = 21.$$

In this case, $n - |w_{Gal}| = 8$, and $n - |w_{min}| = 14$. ♦

An alternative approach to Gallant algorithm is to find a permutation $x_1, x_2, x_3, \ldots, x_k$ of all words of R, such that $x_1©x_2©x_3©\ldots©x_k$ is of minimal size. But, this produces exactly a shortest superstring (property **). Translated into graph notation (with nodes x_is, linked by edges weighted by lengths of overlaps) the problem reduces to the Traveling Salesman problem, which is also NP-complete. Heuristics for this latter problem can be used for the shortest superstring problem.

The complexity of Gallant algorithm depends on the implementation. Obviously the basic operation is computing overlaps. It is easy to see that for two given strings u, v the overlap is the size of the border of the word $v\#u$. Hence, methods from Chapter 3 can be used here. This leads to an $O(nk)$ implementation of Gallant method. The best known implementations of the method work in $O(n\log n)$ time, using sophisticated data structures.

15.4 Cyclic equality of words, and Lyndon factorization

A *conjugate* or *rotation* of a word u of length n is any word of the form $u[k+1\ldots n]u[1\ldots k]$, denoted by $u^{(k)}$ (note that $u^{(0)} = u^{(n)} = u$).

Let u, w be two words of the same length n. They are said to be *cyclic-equivalent* (or conjugate) iff $u^{(i)} = w^{(j)}$ for some i, j. In other words, words u and w have the same set of conjugates. If words u and w are written as circles, they are cyclic-equivalent if the circles coincide after appropriate rotations.

There are several linear-time algorithms for testing the cyclic-equivalence of two words. The simplest one is to apply any string matching to pattern $pat = u$ and text $text = ww$ (words u and w are conjugate iff pat occurs in $text$). Another algorithm is to find maximal suffixes of uu and ww. Let u' (resp. w') be the prefix of length n of the maximal suffix of u (resp. w). Denote $u' = maxconj(u)$, $w' = maxconj(w)$. The word u' (resp. w') is a maximal conjugate of u (resp. w). Then, it is sufficient to check if $maxconj(u) = maxconj(w)$.

We have chosen this problem because there is a simpler interesting algorithm, working in linear time and constant space simultaneously, that deserves presentation.

```
Algorithm { checks cyclic equality of u, w of common length n }
begin
   x := ww, y := uu ;
   i := 0; j := 0;
   while (i < n) and (j < n) do begin
      k := 1;
      while x[i+k] = y[j+k] do k := k+1;
      if k > n then return true;
      if x[i+k] > y[j+k] then i := i+k else j := j+k;
      { invariant }
   end;
   return false;
end.
```

Let $D(w)$ and $D(u)$ be defined by

$$D(w) = \{k\colon 1 \le k \le n \text{ and } w^{(k)} > u^{(j)} \text{ for some } j\},$$
$$D(u) = \{k\colon 1 \le k \le n \text{ and } u^{(k)} > w^{(j)} \text{ for some } j\}.$$

We use the following simple fact:

if $D(w) = [1\ldots n]$, or $D(u) = [1\ldots n]$, then words u, w are not cyclic equivalent.

Now the correctness of the algorithm follows from preserving the invariant:

$D(w) \supseteq [1\ldots i]$, and $D(u) \supseteq [1\ldots j]$.

The number of symbol comparisons is linear. We leave to the reader the estimation of this number. The largest number of comparisons is for words u = 111..1201 and w = 1111...120.

There is a strong connection between the preceding problem and the Lyndon factorization of words which is considered for the rest of the section. The factorization is related to Lyndon words defined according to a lexicographic ordering on words on a fixed alphabet A.

Let u and v be two words from A^*. We say that u is *strongly smaller* than v, denoted by $u << v$, if u and v can be written wau' and wbv', respectively, with w, u', v' words, and a, b letters such that $a < b$. The lexicographic ordering induced on A^* by the ordering of letters is also denoted by $\leq$, and is defined by

$u \leq v$ iff ($u << v$ or u prefix of v).

A Lyndon word is a non-empty word u such that $u < u^{(k)}$ for $0 < k < |u|$. Therefore, a Lyndon word is smaller than its conjugates (in a nontrivial sense). It is equivalent to say (but not entirely straightforward) that the non-empty word u is smaller than (according to the lexicographic ordering) its proper non-empty suffixes. This is the reason why the algorithm below is also closely related to the computation of the maximum suffix of a word (see algorithm *Maxsuf* in Chapter 13). It is useful to note that a Lyndon word u is border-free, which means that $period(u) = |u|$.

A Lyndon factorization of a word x is a sequence of Lyndon words $(u_1, u_2, \ldots, u_k)$, for $k \geq 0$, such that both $x = u_1u_2\ldots u_k$, and $u_1 \geq u_2 \geq \ldots \geq u_k$. In fact, there is a unique Lyndon factorization of a given word, as stated by the next theorem. Since letters are Lyndon words, any non-empty word is a composition of Lyndon words. Regarding Lemma 15.9, the Lyndon factorization is such a factorization having the minimum number of elements.

Theorem 15.8 Any word has a unique Lyndon factorization.

```
Algorithm Lyndon_Fact(x);
{ computes the Lyndon factorization of x of length n }
{ operates in linear time and constant space }
begin
  x[n+1] := '#'; { # is smaller than all other letters }
  ms := 0; j := 1; k := 1; p := 1;
  while (j + k ≤ n+1) do begin
     a' := x[ms+k]; a := x[j+k];
     if (a' < a) then begin
        j := j+k; k := 1; p := j-ms;
     end else if (a' = a) then
```

```
            if (k ≠ p) then
                k := k+1
            else begin
                j := j+p; k := 1;
            end
        else {a' > a} begin
            repeat
                writeln(x[ms+1..ms+p]); ms := ms+p;
            until ms = j;
            j := ms+1; k := 1; p := 1;
        end;
    end;
end.
```

Example Let A be $\{a, b\}$ with the usual ordering ($a < b$). The set of Lyndon words on A contains

$b, a, ab, aab, abb, aaab, aabb, abbb, \ldots.$

It also contains the words *aababbababbb*, or *aababbaabbabab*, for example. The Lyndon factorization of *babababbabaababaabaa* is

$(b, abababb, ab, aabab, aab, a, a)$. ♦

The algorithm *Lyndon_Fact* above produces the Lyndon factorization of its input string. We do not formally prove its correctness, but we do indicate how the algorithm works, and what the main properties useful for proving it are.

Figure 15.6 shows the variables used in the algorithm. The invariant of the main loop is the following. The prefix $x[1\ldots j+k-1]$ of x is factorized as $u_1u_2\ldots u_gu_{g+1}\ldots u_fv$, with

- $u_1, u_2, \ldots, u_g, u_{g+1}, \ldots, u_f$ are Lyndon words,
- $u_1 \geq u_2 \geq \ldots \geq u_g > u_{g+1} = \ldots = u_f$,
- v is a proper prefix (possibly empty) of u_f,
- $u_g >> u_{g+1}\ldots u_fv$.

The variable ms is such that $x[1\ldots ms] = u_1u_2\ldots u_g$. In other words, ms is the position of the factor u_{g+1} in x. The Lyndon factorization of prefix $x[1\ldots ms]$ of x is already computed when ms has this value. Moreover, $u_1, u_2, \ldots, u_g$ are the g first elements of the Lyndon factorization of the whole word x. The variable j stores the position of v in x, that is, such that $x[1\ldots j] = u_1u_2\ldots u_f$. The value of variable p is the common length of $u_{g+1}, \ldots, u_f$. It is also the period of $u_{g+1}\ldots u_fv$. Finally, $k = |v|+1$.

At this current step of the algorithm, the next letter a of x is compared with the corresponding letter a' in u_{g+1} (letter following the prefix v of u_{g+1}). Three cases arise

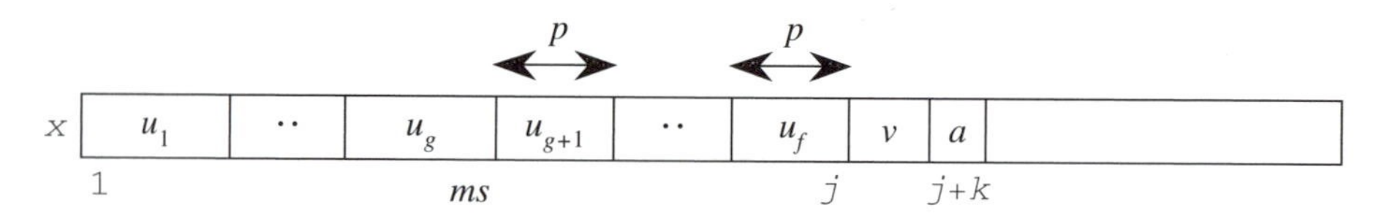

Figure 15.6: Variables in *Lyndon_Fact* algorithm.

from the comparison of letters. If they coincide, then the same periodicity (as that of $u_{g+1}\ldots u_f v$) continues, and the algorithm essentially has to maintain the third property. If a is smaller than a', u_{g+1}, …, u_f become elements of the Lyndon factorization of x, and the entire factorization process restarts at the beginning of v. In the third situation, a' is smaller than a. A very surprising fact is that all properties imply then that the word $u_{g+1}\ldots u_f va$ is a Lyndon word. Since va' is a prefix of the Lyndon word u_f, and $a' < a$, by Lemma 15.10, va is a Lyndon word. Moreover, we have $u_f < va$ (and even $u_f << va$). Thus, by applying Lemma 15.9 successively, we get that $u_f va$, $u_{f-1}u_f va$, …, and finally $u_{g+1}\ldots u_{f-1}u_f va$ are Lyndon words (all greater than u_{g+1}). Therefore, the proof of correctness is essentially a consequence of the two next lemmas. The first one provides an important property of Lyndon words that can be used to generate them in lexicographic order.

Lemma 15.9 If u and v are Lyndon words such that $u < v$, then uv is a Lyndon word and $u < uv < v$.

Proof We prove that proper non-empty suffixes z of uv are greater than uv itself.

First, consider the case $z = v$. We have to prove $uv < v$. This is obvious when $u << v$. Otherwise, u is a proper prefix of v, that is, $v = uw$ for a non-empty word w. Since v is a Lyndon word, we have $v < w$, that implies $uv < uw$ as required.

If z is a proper suffix of v, we have $uv < v < z$, because v is a Lyndon word, and due to the above result.

Finally, let z be a suffix of uv of the form wv, with w a proper non-empty suffix of u. Since u is a Lyndon word, we have $u < w$, and even $u << w$ because u cannot be a prefix of w. Thus, $uv < wv$. ♦

Lemma 15.10 If vb is a prefix of a Lyndon word, and $b < c$ (b, c letters), then vc is a Lyndon word.

Proof Let u be the Lyndon word having prefix vb.

We prove that proper non-empty suffixes of vc are greater than vc itself. These suffixes are of the form zc. For some word w, zbw is a suffix of u. Since u is a Lyndon word, $u < zbw$.

If zb is not a prefix of u, we have indeed $u << zb$, and thus also $v << zb$, because v is a prefix of u not shorter than zb. Therefore, $vc < zb < zc$.

If zb is a prefix of u, it is also a prefix of v, and we have directly $v < zc$. ♦

15.5 Unique decipherability problem

A set of words H is said to be a *uniquely-decipherable code* if words that are compositions of words of H have only one factorization according to H. The unique decipherability problem consists of testing whether a set of words satisfies the condition. The n size of the problem, when H is a finite set is the total length of all elements of H; in particular, the cardinality of H is also bounded by n. Note that we can consider that $\varepsilon \notin H$, because otherwise H is not uniquely decipherable and the problem is solved.

Another way to set up the problem is to consider a *coding function* h, or substitution, from B^* to A^* (B and A are two finite alphabets). The function is a morphism (i.e., it satisfies both properties: $h(\varepsilon) = \varepsilon$, $h(uv) = h(u)h(v)$ for all u, v), and the set H, called the code, is $\{h(a)\colon a \in B\}$. The elements of H are called *code words*. Then, asking whether h is a one-to-one function is equivalent to the unique decipherability for H, provided all $h(a)$s are pairwise distinct. Coding functions related to data compression algorithms are considered in Chapter 10. We can assume that all code words $h(a)$ are non-empty and pairwise distinct. If not, then obviously the function is not one-to-one and the problem is solved.

We translate the unique decipherability condition for H into a problem on a graph G that is now defined. The nodes of G are suffixes of the code words (including the empty word). There is an edge in G from u to v iff $v = u//x$ or $v = x//u$ for some code word x. The operation $y//z$ is defined only if z is a prefix of y, that is, if $y = zw$ for some word w, and the result is precisely this word w.

A set of initial nodes, *Init*, is defined for the graph G. Initial nodes are those of the form $x//y$ in which x, y are two (distinct) code words. Let us call the empty word the *sink*. Then, it is easy to prove the following fact:

> The code H is uniquely decipherable iff there is no path in G from an initial node to the sink.

Example Let $H = \{ab, abba, baaabad, aa, badcc, cc, dccbad, badba\}$. The corresponding graph G is presented in Figure 15.7.

This code is not uniquely decipherable because there is a path from ba to the sink. The word ba is an initial node because $ba = abba//ab$. The path is:

$$ba \rightarrow aabad \rightarrow bad \rightarrow cc \rightarrow \varepsilon.$$

We have $baaabad//ba = aabad$; $aabad//aa = bad$; $badcc//bad = cc$; $cc//cc = \varepsilon$. The path corresponds to two factorizations of the word *baaabadcc*:

baaabad.cc and *ba.aa.badcc.*

♦

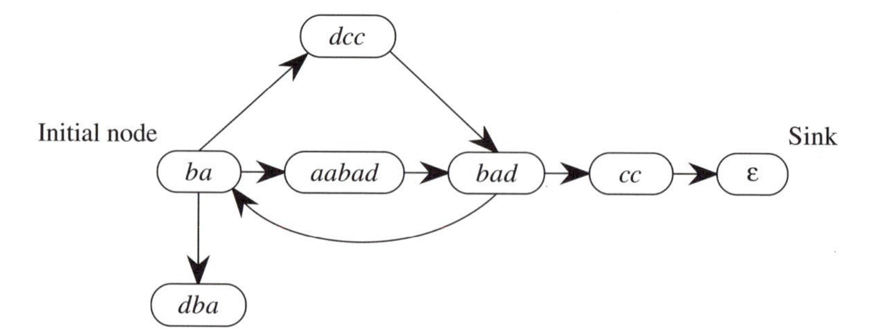

Figure 15.7: Graph G for the example code.

The size of graph G is $O(n^2)$, and we can search a path from an initial node to the sink in time proportional to the size of the graph by any standard algorithm. We show that the construction of G can also be accomplished within the same time bound. If we can answer questions like "is $x//y$ defined?" and "is $y//x$ defined?" in constant time, then, there is at most a quadratic number of such questions, and we are done. The question "is $y//x$ defined?" is equivalent to "does the pattern x start in the text y at position $|y|-|x|$?." Therefore, it is related to string matching. Hence, it is sufficient for each code word y treated as a text, to make string matchings with respect to all other code words x treated as patterns. For a fixed code word y (as a text) this takes $O(n+|y|)$ time, since n is the total size of all code words. Altogether this takes $O(n^2)$ time if we sum over all ys. This proves the following.

Theorem 15.11 The unique decipherability problem can be solved in $O(n^2)$ time.

A more precise estimation of the same algorithm shows that it works in $O(nk)$ time, where k is the number of code words. By applying some data structures, the space complexity can also be improved for some instances of the problem. The close relationship between the unique decipherability problem and accessibility problem in graphs is quite inherent, particularly if space complexity is considered. Indeed, the problems are mutually reducible using additional constant memory of a random access machine (or $\log n$ deterministic space of a Turing machine).

15.6* Equality of words over partially commutative alphabets

Assume that we have a symmetric binary relation C over the alphabet A. The relation is called the *commutativity relation*. Relation C induces an equivalence on A^*. Two words, x and y over A, are said to be equivalent, denoted by $x \approx y$, iff one of them can be obtained from the other by commuting certain symbols in the word several times. Formally, we define $ubav \approx uabv$ when $(a, b) \in C$ (symbols a, b commute), and we get an equivalence relation by closing this relation under reflexivity and transitivity. We consider the complexity of the following problem: check the equivalence of two strings x, y of the same length n.

At first glance, the problem looks like an NP-complete problem. The straightforward algorithm is to generate all strings equivalent to x and check if one of them is y. However, the equivalence class of x can contain an exponential number of words (for example, when symbols a, b commute and $x = a^m b^m$), leading to an exponential-time algorithm. Fortunately, there is a very simple algorithm with linear-time complexity in the case of fixed-size alphabets. This is due to a specific combinatorial property of words over partially commutative alphabets.

Denote by $h_{a,b}$ the morphism that erases from its input word all symbols except occurrences of a and b. Let $\#_a(x)$ be the number of occurrences of symbol a in x.

Lemma 15.12 Words x and y are equivalent, $x \approx y$, iff the following two conditions hold:

(1) $\#_a(x) = \#_a(y)$ for each symbol a of the alphabet,

(2) $h_{a,b}(x) = h_{a,b}(y)$ for each pair of distinct non-commuting symbols a, b.

As a simple corollary we have the following theorem.

Theorem 15.13 The equality of two words over a partially commutative alphabet of fixed size can be checked in linear time.

We soon discuss what happens when the size k of the alphabet is not treated as a constant. The lemma implies an $O(k^2 n)$ algorithm, because the number of distinct pairs of letters is quadratic. However, instead of taking morphisms leaving only two symbols in texts (erasing all others), we can consider morphisms leaving a set of mutually non-commuting symbols. Denote by h_X the morphism that erases in the text all symbols except symbols in set X. Let Π be a family of subsets of the alphabet such that each two symbols contained in the same set do not commute, and each two non-commuting symbols are contained in some set from Π together. Then condition (2) of Lemma 15.12 is equivalent to $h_X(x) = h_X(y)$ for each X in Π. The cardinality of Π can be much smaller than the number of all non-commuting pairs. For example, consider the following relation C over alphabet $\{1, 2, 3, 4, 5, 6\}$:

$$(i, j) \in C \text{ iff } |i-j] = 1.$$

There are 10 pairs of non-commuting symbols, while we can take Π of cardinality 5,

$$\Pi = \{\{1, 3, 5\}, \{2, 4, 6\}, \{3, 6\}, \{4, 1, 6\}, \{5, 2\}\}.$$

Hence, we have only to check equality of 5 pairs of words rather than 10.

But there are cases when this approach does not help. Take the alphabet $\{1, 2, \ldots, 2r\}$ with commutativity relation:

$$(i, j) \in C \text{ iff } i, j \leq r \text{ or } i, j > r.$$

The size of the alphabet here is $k = 2r$. The graph of C consists of two disjoint cliques, each of size r. There is a quadratic number of non-commuting pairs, and no set Π helps, because for each three distinct symbols two of them commute. This suggests that the algorithm should have complexity of order $k^2 n$. However, there is a simple $O(kn)$ time algorithm for checking $x \approx y$ in this case. Let $\#_1(z)$ be a vector of size r in which the i-th component is the number of occurrences of the i-th symbol in z, and let $\#_2(z)$ be a vector of size r in which the i-th component is the number of occurrences of the $(r+i)$-th symbol in z. For each word x, let $H(x)$ be the string obtained from x by replacing each factor z between two consecutive symbols from $\{1, 2, \ldots, r\}$ by $\#_2(z)$. Then, each word z (over $\{1, 2, \ldots, r\}$) between two consecutive newly inserted vectors is replaced by $\#_1(z)$. $H(x)$ can be computed in $O(nk)$ time. Finally, it is easy to see that $x \approx y$ iff $H(x) = H(y)$.

It is typical in problems related to partially commutative alphabets that their complexity depends on the structure of the commutativity relation graph.

For many textual problems the introduction of partial commutativity of the alphabet changes their complexity dramatically. The typical example is the unique decipherability problem. For partially commutative alphabets with three letters, the problem has still an efficient (polynomial time) solution. However, for partially commutative alphabets with four letters there does not exist any algorithm for this problem at all. In other words, the problem is undecidable.

We will soon prove that for three-letter alphabets the unique decipherability problem is still algorithmically tractable. This can be seen as follows. The unique decipherability problem is easily reducible to the disjointness problem of two languages L_1, L_2 accepted by finite automata A_1, A_2. Consider the following two commutativity relations on the alphabet $\{a, b, c\}$ presented graphically by

$R1$: $a \quad b—c$,
$R2$: $a—c—b$.

Let $Cl(L)$ be the closure of language L with respect to the relation $\approx$.

$$Cl(L) = \{x: x \approx y \text{ for some } y \text{ in } L\}.$$

Take the following morphism h defined by $h(a) = a$, $h(b) = b$, $h(c) = \varepsilon$. The only action of the morphism is to erase all letters c in its input text. Now, it can be easily proved, in the case of relations R_1 and R_2, that, knowing automata A_1, A_2, we can construct a non-deterministic pushdown automaton A accepting the language

$$L = h(Cl(L_1) \cap Cl(L_2)).$$

Therefore, the languages L_1, L_2 are disjoint iff $L = \emptyset$. Finally, the emptiness problem for non-deterministic pushdown automata is solvable in polynomial time.

We describe briefly the construction of automaton A for the commutativity relation R_2. The pushdown store of A acts as a counter (guessed number of occurrences of symbol c). We can view A as a composition of A_1 and A_2. Whenever A reads the symbol a or b, then A_1 and A_2 go to next states, and their next states form a pair that is a next state of A. At any time, A can non-deterministically assume in an ε-move the input symbol c for one of the machines A_1, A_2 (without advancing its input head); then, the simulated machine makes a move as if the reading symbol were c. If the machine A_1 is chosen, the counter is incremented by one; on the contrary, if A_2 is chosen, the counter is decremented by one. The machine A accepts, if there is a computation that ends in accepting states (for both A_1 and A_2), and if the final value of the counter is zero. Using the description of A_1, A_2 the construction of A can be executed efficiently.

The case of the relation R_1 is similar. Other possible relations for three-letter alphabets are easy to deal with. It is interesting to note that if we take a relation R on four letters similar to R_2, then the problem becomes more complex. Let the commutativity relation R be defined by a—b—c—d. The decidability (or undecidability) of the unique decipherability problem with relation R is still an open problem. But, if we add the pair a—d to the relation R, then, it is known that the unique decipherability problem becomes undecidable.

15.7 Hirschberg-Larmore algorithm for breaking paragraphs into lines

In this section, we describe an application of text manipulation to text editing. It is the problem of breaking a paragraph optimally into lines. The algorithm may be seen as another application of the notion of the failure function, introduced in Chapter 3 for the KMP string-matching algorithm.

The problem of breaking a paragraph is defined as follows. We are given a *paragraph* (a sequence) of n words (in the usual sense) $x_1, \ldots, x_n$, and bounds *lmin*, *lmax* on lengths of lines. The i-th word of the paragraph has length w_i. A *line* is an interval $[i\ldots j]$ of consecutive words x_ks ($i \le k \le j$). The length of line $[i\ldots j]$, denoted by $line(i, j)$, is the total length of its words, that is, $w_i + w_{i+1} + \ldots + w_j$. Bounds *lmin* and *lmax* are related to the smallest and largest length of lines respectively. The optimal length of a line is *lmax*. Moreover, the length of the line $[i\ldots j]$ is said to be *legal* iff $lmin \le line(i, j) \le lmax$. Let us denote the corresponding predicate by $legal(i, j)$.

For a legal line, its penalty is defined as $penalty(i, j) = C.(lmax-line(i, j))$, for some constant C. The problem of breaking a paragraph consists of finding a sequence of integers i_1 (=1), $i_2, i_3, \ldots, i_k$ (=n) such that both lines $[i_1, i_2]$, $[i_2+1, i_3]$, $\ldots$, $[i_{k-1}+1, i_k]$ have legal lengths, and the total penalty (sum of penalties of lines) is minimum. Integers $i_1, i_2, i_3, \ldots, i_k$ are called the *breaking-points* of the paragraph.

We assume that there is no penalty (or zero penalty) for the first line, if its length does not exceed *lmax*. It is as if we treated the paragraph starting from the end. In the usual definition of the problem the last line is not penalized for being too short. Reversing the order of words in paragraphs leads to an algorithm that is even more similar to KMP algorithm, since indices are processed from left to right.

Let *break*[*i*] be the right most breaking-point preceding *i* in an optimal breaking of the subparagraph [1...*i*]. When *break*[*i*] is computed for all values of *i* (from 1 to *n*), the problem is solved: the sequence of breaking-point can be recovered by iterating *break* from *n*.

We first design a brute-force breaking algorithm that uses quadratic time. The informal scheme of such a naive algorithm is given below. It uses the table *f*: *f*[*i*] is the total penalty of breaking into lines the subparagraph [1...*i*]. The scheme assumes that *f*[*i*] is initialized to 0 for all integers *i* such that *line*[1, *i*] ≤ *lmax*, and *f*[*i*] is initialized to infinity for all other values of *i*.

```
for i := 1 to n do begin
    j := an integer which minimizes f[j]+penalty(j+1, i);
    break[i] := j;
    f[i] := f[j]+penalty(j+1, i);
end;
```

The value of *j* is computed by scanning the interval [*first*[*i*], ..., *last*[*i*]], in which *first*[*i*] is the smallest integer *k* for which *legal*(*k*, *i*) holds, and in which, similarly, *last*[*i*] is the largest such *k* less than *i*. The interval [*first*[*i*], ..., *last*[*i*]] is called the *legal interval* of i. All values *first*[*i*], *last*[*i*], *line*(1, *i*) can be precomputed. Therefore, the above scheme yields an $O(n^2)$ time algorithm. The next theorem shows that this can be improved upon considerably.

Theorem 15.14 The problem of optimally breaking a paragraph can be solved in $O(n)$ time.

Proof Define the function $g[j] = f[j] + C.line(j, n)$. The crucial point is the following property:

> a value of *j* that minimizes *f*[*j*]+*penalty*(*j*+1, *i*) also minimizes *g*[*j*] in the legal interval for *j*.

That the difference between expressions depends only on *i* can be checked using simple arithmetics. Another important property is the monotonicity of breaking-points:

> $i' < i$ implies $break[i'] \le break[i]$.

Hence, the value of *j* is to be found in the interval [max(*first*[*i*], *break*[*i*–1]), *last*[*i*]]. Define *Next*[*j*] to be the first position $k \le i$ to the right of *j* such that $g[k] \le g[j]$. When looking for the minimal value of *g*[*j*] in a legal interval, we can initialize *j* to the beginning of the interval, and compute successive positions by iterating *Next*

$j_1 = Next[j]$, $j_2 = Next[j_1]$, ...,

until the value is undefined, or until it goes outside the interval. In this way, *Next* works as the failure table of KMP algorithm. This produces the following algorithm.

```
Algorithm HL; { breaking a paragraph of n words into lines }
begin
   precompute values first [i], last [i], line(1, i) for all i ≤ n;
   let k be the maximal index such that line[1, k] ≤ lmax;
   initialize f [i] to 0, break [i] to 0 for all i ≤ k,
         and compute corresponding values g[i] and Next [i] for j ≤ k;
   j := 0;
   for i := k+1 to n do begin
         { invariant: break [i] < i, first [i] < last [i] < i }
         { legal interval is [first [i],...,last [i]] }
         if j < first [i] then j := first [i];
         while Next [j] defined and Next [j] ≤ last [i] do
         { j is not rightmost minimal }      j := Next [j];
         break [i] := j; f [i] := f [j]+penalty(j+1, i);
         g[i] := f [i]+C.line(1, i);
         Update_table_next;
   end;
   return table break, which gives an optimal breaking;
end.
```

The algorithm works in linear time as per a similar argument as that used in the analysis of the KMP algorithm, if the total cost of updating the table *Next* is linear. Let j_1, j_2, ..., j_r be the increasing sequence of indices j for which $Next[j]$ is not defined at a given stage of the algorithm. Then the values of $g[j]$ are strictly increasing for this sequence. We keep the sequence $j_1, j_2, \ldots, j_r$ on a stack S, the last element at the top. The following procedure is applied.

```
procedure Update_table_next;
begin
   repeat
         pop the top indices j of the stack S;
         Next [j] := i;
   until g[j] < g[i];
   push i onto S;
end;
```

The total complexity of computing table *Next* is linear, since each position is popped from S at most once. This completes the proof of the theorem. ♦

We refer the reader to [HL 87a] for more details on the algorithm, and for extensions to more sophisticated penalty functions. For example, we can assume that optimal lengths of lines are properly within the interval [*lmin*, *lmax*], and that the penalty of a line is a linear function of the distance from optimal length. The penalty can also be a nonlinear function. The linear time complexity works as long as the penalty function is concave.

Bibliographic notes

String matching by hashing was first considered by Harrison [Ha 71]. A complete analysis is presented by Karp and Rabin in [KR 87]. The same idea (using hashing) is applied to finding repetitions in [Ra 85]. An adaptation of Karp-Rabin algorithm to two-dimensional pattern matching has been designed by Feng and Takaoka [FT 89].

The contents of Section 15.2 is adapted from the tree-pattern-matching algorithm of Dubiner, Galil, and Magen [DGM 90].

The approximation of the SCS problem by Gallant may be found in [GMS 80]. An efficient implementation has been designed by Tarhio and Ukkonen [TU 88]. Stronger methods are developed in [Tu 89], providing an $O(n\log n)$-time algorithm, but requiring quite complicated data structures.

The book by Lothaire [Lo 83] contains motivations for factorization problems from the point of view of combinatorics on texts. The Lyndon factorization algorithm of Section 15.4 is from Duval [Du 83]. An algorithm for the canonization of circular strings (computing the smallest conjugate) based on KMP string-matching algorithm is presented by Booth in [Bo 80]. The fastest known algorithm is by Shiloach [Sh 81]. The relationship between Lyndon factorization and smallest conjugate of a word in considered by Apostolico and Crochemore in [AC 91].

The algorithm for testing unique decipherability of a code is usually attributed to Sardinas and Paterson (see [Lo 83]). By using specific data structures, Apostolico and Giancarlo have improved on the space complexity of the algorithm [AG 84]. The literature on the unique decipherability problem is quite rich (see, for example, [BP 85]). The problem is complete in the class of non-deterministic $\log n$-space computations (see [Ry 86]).

An interesting exposition on algorithmic problems related to partially commutative alphabets is presented by Perrin in [Pe 85]. The algorithmic properties of the unique decipherability problem for partially commutative alphabets were considered by Chrobak and Rytter in [CR 87b]. For the emptiness problem for non-deterministic pushdown automata in polynomial time the reader can refer to [HU 79]. String matching over partially commutative alphabets is solved by Hashiguchi and Yamada in [HY 92].

The application of failure functions to the problem of breaking a paragraph into lines is from Hirschberg and Larmore [HL 87a].

Selected references

[AG 84] Apostolico, A., & Giancarlo, R., "Pattern matching machine implementation of a fast test for unique decipherability," *Inf. Process. Lett.* 18 (1984): 155–158.

[DGM 90] Dubiner, M., Galil, Z., & Magen, E., "Faster tree-pattern matching," in (*Proceedings of 31st FOCS*, 1990): 145–150.

[Du 83] Duval, J-P., "Factorizing words over an ordered alphabet," *J. Algorithms* 4 (1983): 363–381.

[GMS 80] Gallant, J., Maier, D., & Storer, J.A., "On finding minimal length superstrings," *J. Comput. Syst. Sci.* 20 (1980): 50–58.

[HY 92] Hashiguchi, K., & Yamada, K., "Two recognizable string-matching problems over free partially commutative monoids," *Theoret. Comput. Sci.* 92 (1992): 77–86.

[HL 87a] Hirschberg, D.S., & Larmore. L.L., "New applications of failure functions," *J. ACM* 34 (1987): 616–625.

[KR 87] Karp, R.M., & Rabin, M.O., "Efficient randomized pattern-matching algorithms," *IBM J. Res.Dev.* 31 (1987): 249–260.

[Sh 81] Shiloach Y., "Fast canonization of circular strings," *J. Algorithms* 2 (1981): 107–121.

[TU 88] Tarhio, J., & Ukkonen, E., "A greedy approximation algorithm for constructing shortest common superstrings," *Theoret. Comput. Sci.* 57 (1988): 131–146.

Exercises

1. Let x be a non-empty word. Let u be the shortest word such that x is a prefix of ux. Show that $|u| = period(x)$.

Prove that $period(x^2) = |x|$ iff x is *primitive* (i.e., cannot be written u^k for $k > 1$) iff x^2 contains only two occurrences of x.

2. Show that any non-empty word x can be uniquely written as u^k, where u is a primitive word and $k > 0$. The word u is called the *root* of x and k is called its *exponent*. Show that the set $\{y \in A^* : xy = yx\}$ is equal to u^* where u is the root of x.

Two words y and z are said to be *conjugate* if they can be written $y = uv$ and $z = vu$. Show that this defines an equivalence relation on A^*. Show that two words are conjugate iff their roots are conjugate. Show that the equivalence class of x has cardinality $|x|/k$ where k is the exponent of x.

3. Prove the following properties of Fibonacci words Fib_k (for large enough k):

- $Fib_kFib_{k+1} = c(Fib_{k+1}Fib_k)$, where c is the operation of exchanging the last two letters of the text,
- Fib_k is an "almost" palindrome, i.e., after cutting the last two letters it becomes a palindrome.

Find a morphism h, defined from $\{a, b\}^*$ to itself, such that $Fib_k = h^{k-2}(a)$ for $k > 1$.

4. *Show that if the non-empty word u^2 is factor of a Fibonacci word, then $|u|$ is a Fibonacci number. Deduce that Fibonacci words have no non-empty factor of the form u^4.

Prove that a Fibonacci word of length n contains $O(\log n)$ squares (of primitive words).

{First statement is by Séébold; for the second, see [Cr 81]}

5. A set $C \subseteq A^*$ is a uniquely decipherable code iff for any two sequences of elements of C, $u = (u_1, u_2, \ldots, u_k)$ and $v = (v_1, v_2, \ldots, v_k)$, the equation $u_1u_2\ldots u_k = v_1v_2\ldots v_k$ implies $u = v$.

Let $B \subseteq A^*$. Show that the minimal generator of B^* is $C = (B^*-\{\varepsilon\})^2-(B^*-\{\varepsilon\})$. Show that C is a uniquely decipherable code iff, for all $w, y, z \in A^*$, $y, z, yw, wz \in B^* \Rightarrow w \in B^*$.

Show that, for any $B \subseteq A^*$, there exists a unique uniquely decipherable code C such that $B^* \subseteq C^*$ and that, for any uniquely decipherable code C' satisfying $B^* \subseteq C'^*$, we have $C^* \subseteq C'^*$ (i.e., C^* is minimal for inclusion). The set C is called the *uniquely decipherable code induced by B*.

Propose an algorithm to compute the uniquely decipherable code induced by B.

{see [Lo 83], or [BP 85] for further studies on the subject}

6. **Defect theorem.* Let B be a subset of A^* that is not a uniquely decipherable code. Show that the code C induced by B (see previous exercise) satisfies $|C| < |B|$.

Deduce from the statement:

- the two-word set $\{x, y\}$ is a uniquely decipherable code iff $xy \neq yx$,
- two words y and z satisfy $yz = zy$ iff they are powers of a same word,
- the same conclusion holds iff $y^i = z^j$ for two integers i and j.

7. A word w is called *border-free* if its only borders are itself and the empty word, or equivalently *period*$(w) = |w|$. Assume that the word x has a shortest non-empty border u. Show that u is border-free, and that either $x = u$ or $x = uvu$ for some word v.

Let $B(x) = \{u \in A^* : u$ is prefix of x, and u is border-free$\}$. Show that the minimal generator of $B(x)^*$ is a uniquely decipherable code $C(x)$.

Let $x = u_1u_2\ldots u_k$ (with u_is in $C(x)$) be the unique factorization of x according to $C(x)$. For example, the decomposition of *abacaba* is (*abac*, *ab*, *a*). Show that u_k is the shortest border-free prefix of x, and that u_1 is the longest border-free prefix of x. Propose a linear-time algorithm to compute the decomposition $(u_1, u_2, \ldots, u_k)$ of x.

{Hint: use the failure function of the MP algorithm.}

8. Let $Maxsuf(x)$ be the maximal suffix of the word x ($\neq \varepsilon$) with respect to lexicographic ordering. Let $u = Maxsuf(x)$, and $va = Maxsuf(xa)$, for some non-empty x and letter a. Prove that v is a border of u.

Let $(u_1, u_2, \ldots, u_k)$ be the sequence of proper borders of $Maxsuf(x)$ in decreasing order of lengths. Define letters a_is and words v_is by

$$Maxsuf(x) = u_1a_1v_1 = u_2a_2v_2 = \ldots = u_ka_kv_k.$$

Show that $a_1 \leq a_2 \leq \ldots \leq a_k$.

For a letter $a \leq a_1$, show that $Maxsuf(xa) = Maxsuf(x)a$. Otherwise (when $a > a_1$), show that $Maxsuf(xa) = u_ga_g$ for the largest integer g such that $a_g \leq a$.

Develop a full proof of algorithm *Maxsuf* of Chapter 13.

9. An *overlap* is a word w that satisfies $period(w) < |w|/2$. A *simple overlap* is a word of the form $auaua$ for a letter a and a word u. Show that a word x contains an overlap factor iff it contains a simple overlap factor.

A word that contains no factor that is an overlap (or a simple overlap) is said to be *overlap-free*.

Let x be an overlap-free word on the alphabet $\{a, b\}$. Show that if $|x| \geq 7$, x has a unique factorization as gud with $g, d \in \{\varepsilon, a, b, aa, bb\}$ and $u \in \{ab, ba\}^*$.

Let h be the (Thue-Morse) morphism defined by $h(a) = ab$ and $h(b) = ba$. For the word x, show that, for some integer r and word u, $x = g_1g_2\ldots g_r h^{r-1}(u)d_r\ldots d_2d_1$, with $|u| < 7$ and for all $i = 1, \ldots, r$ $g_i, d_i \in \{\varepsilon, h^{i-1}(a), h^{i-1}(b), h^{i-1}(aa), h^{i-1}(bb)\}$.

*Design a linear-time algorithm to test the existence of an overlap factor in a given text.

{see [Kf 86] for the algorithm; see also [RS 85]}

10. Let w be a non-empty word. Assume that w has a strict border u, and that u has a strict border v. Then, show that $|w| > |u| + |v| + 1$.

Prove that $delay(|x|) \leq \log_\Phi(|x|+1)$ (Lemma 3.4), where $\Phi = (1+\sqrt{5})/2$. Prove that the delay is $O(\log|x|)$ for Fibonacci patterns.

{Hint: Fibonacci words satisfy the inequality, for large enough k, $|Fib_k| = F_k \geq \Phi^{k-2}$.}

11. Design a linear-time algorithm computing the table *t_Bord* used by MPS algorithm (Chapter 3).

12. *Write real-time versions of KMP and MPS algorithms on a random access machine.

{see [Ga 81]}

13. *Give a linear-time algorithm (independent of the alphabet) that, for each position i in the word *text*, finds the longest suffix ending at x and belonging to the language $\{x, y, z\}^*$, where x, y, and z are given texts. The size of the problem is the total size of x, y, z, and *text*.

{see [NC 92]}

14. Assume that $a^k b$ ($a, b \in A$, $a \neq b$, $k > 0$) is a prefix of pattern x. Consider the following variant of KMP algorithm, which uses the same failure table: the scan is done from left to right on the suffix $x_{k+1}...x_m$ of x; if the suffix occurs in *text* at the current position, we check to see if it is preceded by a^k and report a possible match.

Write the above algorithm in full detail. Prove that the search makes at most $1.5.|text|$ comparisons.

{idea in [AC 91]}

*Show that considering other interrupted periodicities in addition to the first one as above leads to a searching algorithm making at most $4|text|/3$.

For finding all occurrences of the pattern *aba* in a text of length n, prove that any algorithm has to make at least $4n/3–c$ comparisons, where c is a constant.

{see [CGG 90]; other upper bounds are in [CH 92] and [ZP 92].}

15. Let $k \geq 2$. A k-hrp of a string x is a primitive word u such that u^k is prefix of x (sections 13.2 and 13.3)

Let $k > 2$. Show that if u and v are k-hrps of x with $|u| < |v|$, then $(k\text{-}1)|u| < |v|$. Deduce that x has at most $\log_{k-1}|x|$ k-hrps.

*Let $k = 2$. Show that if u, v, and w are 2-hrps of x with $|u| < |v| < |w|$, then $|u| + |v| \leq |w|$. Deduce that x has at most $\log_{\emptyset}|x|$ 2-hrps ($\emptyset = (1+\sqrt{5})/2$). What are the words reaching the worst case? How many 2-hrps have Fibonacci words and their prefixes?

{see [CR 93]}

16. A language $L \subseteq A^*–\{\varepsilon\}$ is said to have the *left cancellation property* if $u, uv \in L$ implies $v \in L$.

Let $P \subseteq A^*–\{\varepsilon\}$ be a language such that P^* has the left cancellation property. Assume that there is algorithm that, given a string x of length n, can find a prefix u of x, $u \in P$, or prove that no such prefix exists, in $O(n)$ time. Then, prove that it can be tested if $x \in P^*$ in $O(n)$ time.

Show that the set *Pal* of even palindromes has the left cancellation property. Propose a linear-time algorithm for the recognition of *Pal**.

*Write a real-time algorithm for the recognition of even palindromes.

{see [KMP 77]}

17. Design a linear-time version of KMP algorithm for a Turing machine.

{see [FP 74]}

18. Prove that in the search phase of BM algorithm any symbol at a given position in the text can be inspected at most a logarithmic number of times.

19. Give a worst-case pattern for Turbo_BM algorithm (search phase).

20. Give an example of text for which the uncompacted position tree (each edge is labeled by a single symbol) has quadratic size.

21. Implement the three suffix tree constructions of Chapter 5. Compare them according to time and space actually required.

Implement the suffix array method of Chapter 5. Compare with above implementations.

22. Design a linear-time algorithm for the computation of suffix trees using only the table *sext* of Section 6.4 (without tables *link* and *test*).

23. Let T be a tree in which the edges are labeled with symbols. Assume that the edges outgoing the same node are labeled with distinct symbols. Let $SUF(T)$ be the set of all suffixes of the paths of T. Construct an $O(n\log n)$ time algorithm that builds the compressed tree representing all strings in $SUF(T)$.

24. Let T be a tree in which the edges are labeled with symbols. Let $REV(T)$ be the set of all reverses of paths of T. Construct an $O(n\log n)$-time algorithm that builds the compressed tree representing all strings in $REV(T)$.

25. What is the complexity of the problems of the two preceding exercises if, instead of a tree, we have a directed acyclic graph? Note that the resulting tree can be very large (investigate how large). Denote its size by s. Can you obtain $O(s)$ time algorithms?

26. Consider a directed acyclic graph with one root and one sink. Assume that edges outgoing a same node have distinct labels. The set of patterns consists of all labels of paths from the root to the sink. Assume the total size of the graph is m. Can you construct an $O(m+n)$-time algorithm for pattern searching in a text of length n?

27. Let $twice(u)$ be the size of the largest suffix of u that occurs in x at least twice. Design a linear-time algorithm that, for each prefix u of a given word x, finds the value of $twice(u)$ (assume that the size of the alphabet is fixed).

Compute the function $twice(x)$ for a single word x in linear time, and simultaneously in constant space.

28. Let us define the function $k\text{-}times(x)$. Its value is the size of the largest prefix of x that occurs in x at least k times. Compute $k\text{-}times(x)$ in linear time. The parameter k can be a function of n (the alphabet is fixed).

29. Let P be a family of disjoint subsets of $\{1, 2, \ldots, m\}$ satisfying the non-overlap property: if two subsets from P have a common element then one of them is included in the other. Prove that P has less than $2m$ elements. How this is related to the size of suffix dawgs?

Assume that $|pat| > 2$. Prove that the number of states of the suffix dawg $dawg(pat)$ is at most $2|pat|-1$, and that the bound is reached only if pat belongs to ab^*, where a and b are distinct letters.

Assume that $|pat| > 3$. Prove that the number of edges of the suffix dawg $dawg(pat)$ is at most $3|pat|-4$, and that the bound is reached only if pat belongs to ab^*c, where a, b, and c are letters with $a \neq b$ and $b \neq c$.

{see [BBEHCS 85] and [Cr 86]}

30. Give exact bounds, in term of m, of the size of the minimal deterministic automaton—*factor automaton*—accepting the set of all factors of a text of length m. Note that this automaton can be smaller than the minimal deterministic automaton accepting suffixes of the same text.

On a fixed alphabet, design a linear on-line algorithm that builds the factor automaton of a text.

{see [BBEHCS 85] and [Cr 86]}

31. Propose an algorithm to compute a suffix dawg for a finite set of patterns.

{see [BBEHM 87]}

32. *Let us call a *compact dawg* a dawg in which the edges are labeled by strings and no node has only one outgoing edge, except possibly the initial state. Design a linear-time algorithm for building the compact suffix dawg of a text.

33. Let $G = DAWG(pat)$, and $G' = DAWG(pat^R)$. Prove that there is a one-to-one correspondence between essential nodes of G and G' (internal nodes of out-degree larger than one).

34. Design efficient algorithms to compute, for a given text w, the number of distinct

- primitive factors of w,
- squares of primitive words, factors of w,
- squares, factors of w,
- even palindromes, factors of w.

35. Design an efficient algorithm to test if a given word is a "squarestar" (i.e., a product of non-empty squares).

Design an efficient algorithm to test if a given word is a "cubestar" (i.e., product of non-empty cubes).

36. Implement and compare Aho-Corasick and Multi_BDM multi-pattern-matching algorithms (Sections 7.1 and 7.2).

37. Show that the smallest deterministic automaton accepting all words over alphabet $\{a, b\}$ in which the 7-th letter from the end is a, has 128 states.
Generalize the example.

38. Design an efficient algorithm to compute the number of all distinct subsequences of the word. What is the maximal value of this number for a text of length 10?

39. On a fixed alphabet, design an algorithm that builds the minimal automaton accepting the set of subsequences of a text. What are the time and space complexities?
Design algorithms that use this kind of automaton to find a longest common subsequence of two words, and a shortest word that is subsequence of only one of two words.

{see [HC 86]; an unpublished algorithm for the last problem is by Simon as an implementation of Theorem 6.2.11 in [Lo 83] Chapter 6.}

40. Construct 2dpdas accepting the languages:

- $L_{\text{prefpal}} = \{ww^R u$: w is non-empty$\}$,
- $L_{\text{prefpal3}} = \{ww^R uu^R vv^R z$: w, u, v are non-empty words$\}$,
- $L_{\text{pal2}} = \{ww^R uu^R$: w, u are non-empty words$\}$.

41. Construct 2dpdas accepting the languages $\{a^n b^m: m = 2^n\}$, $\{a^n b^m: m = n^4\}$, and $\{a^n b^m: m = \log^* n\}$.

42. *Construct a 2dpda accepting the language $L = \{1^n$: n is a square of an integer$\}$.

{see [Mo 85]}

43. Write a version of the Karp-Miller-Rosenberg algorithm for the two-dimensional case (and generally for k dimensions, where k is a constant) working in $O(N\log N)$, where N is the total size of the input.

44. Prove the combinatorial properties required for the correctness of the linear-time constant-space algorithm for *righttest* (Section 8.2). Give a formal proof of algorithm *righttest*.
Prove that *righttest*(u, v) is computed in $O(|v|)$ time by the algorithm.

{see [ML 79] or [ML 84]}

45. Prove that the *f*-factorization of texts introduced in Section 8.2 can be computed in linear time using suffix trees or suffix dawgs.
According to the *f*-factorization $(v_1, v_2, \ldots, v_m)$ of *text*, prove that this word contains a square iff, for some k, $pos(v_k)+|v_k| \geq |v_1 v_2 \ldots v_{k-1}|$ (self-overlapping of v_k), or *lefttest*(v_{k-1}, v_k) or *righttest*(v_{k-1}, v_k), or *righttest*$(v_1 v_2 \ldots v_{k-2}, v_{k-1} v_k)$.

{see [Cr 86]}

46. Estimate the number of comparisons made by Manacher algorithm (Section 8.3).

Propose a similar algorithm for computing the radii of odd palindromes, and running in linear time.

{see [Ma 75]}

47. Write an efficient algorithm computing the number of all distinct (as strings) palindromes contained in a given text.

48. Give a full proof of the Lemma 8.14: $parse1(text) \in \{first1(text), 2first1(text)-1, 2first1(text)+1\}$.

49. Compute tables of all longest palindromes starting or ending at positions $i = 1, 2, \ldots, n$ of a word of length n by a linear-time algorithm.

50. * Design an efficient algorithm checking membership of a given text in PAL_k.

{Linear time for this problem is an open question.}

51. Given n subintervals of $[1\ldots n]$, design a parallel algorithm that computes which subintervals are maximal (in the sense of inclusion) in $O(\log n)$ time with n processors.

{see [GR 88]}

52. Given a linked list of n elements stored in an array of size n, design a parallel algorithm that computes for each element its rank in the list in $O(\log n)$ time with n processors. Does the algorithm have optimal speed-up?

{see [Ma 89]}

53. Prove Lemma 9.2.

{see [GR 88]}

54. *Implement the Ziv-Lempel compression algorithm in parallel: $O(\log n)$ time with $O(n)$ processors of a CRCW PRAM.

{Hint: use suffix trees, see [CR 91a].}

55. Design an efficient (serial) algorithm computing the edit distance where the costs of one-symbol manipulations depend on the symbols involved.

56. Propose a complete algorithm for the computation of a longest common subsequence of two texts, and not only its length (Chapter 11).

57. Show how to reduce in linear time the string-matching problem with don't care symbols to the tree-pattern-matching problem.

58. Consider the pair of words (*abb*, *baa*). A replacement in a word is a replacement of an occurrence of *abb* by *baa*, or a replacement of an occurrence of *baa* by *abb*. Two words are equivalent iff one can be obtained from the other by a sequence of such replacements. Construct a linear-time and constant-space algorithm that checks whether two words are equivalent. What happens for the pair of strings (*aa*, *bb*)? Does the complexity change for the pair (a^k, b^j)?

59. Implement efficiently the operations *CLASS, SPLIT, UNION* for intervals of integers used by Hunt-Szymanski algorithm (Section 11.2).

{Hint: use B-trees, see [AHU 83].}

60. Estimate the number of symbol comparisons made by the algorithm checking the cyclic-equivalence of two words in linear time and constant space (Chapter 15). Show that the maximum number of comparisons this algorithm makes for texts u=111..1201 and w=1111...120.

61. Prove the Lyndon Factorization Theorem (Theorem 15.8).

Show that if u and v are Lyndon words with $u < v$, then, for any $k, l > 0$, $u^k v^l$ is a Lyndon word. [Hint: use Lemma 15.9.]

Show that w is a Lyndon word of length greater than 1 iff $w = uv$ for two Lyndon words u and v satisfying $u < v$. Propose a linear-time algorithm to compute the factorization uv of w.

{see [Lo 83]}

62. Prove Lemma 15.12.

Bibliography

[Ab 89] Abrahamson, D.M., "Generalized string-matching," *SIAM J. Comput.* 16 (1989): 77–83.

[Ah 80] Aho, A.V., "Pattern matching in strings," in (Book, editor, *Formal Language Theory—Perspectives and Open Problems*, Academic Press, Orlando, Florida 1980): 325–347.

[Ah 90] Aho, A.V., "Algorithms for finding patterns in strings," in (J. van Leeuwen, editor, *Handbook of Theoretical Computer Science*, vol A, *Algorithms and complexity*, Elsevier, Amsterdam, 1990): 255–300.

[AC 75] Aho, A.V., & Corasick, M., "Efficient string matching: An aid to bibliographic search," *Comm. ACM* 18 (1975): 333–340.

[AHU 76] Aho, A.V., Hirschberg, D.S., & Ullman, J.D., "Bounds on the complexity of the longest common subsequence problem," *J. ACM* 23 (1976): 1–12.

[AHU 74] Aho, A.V., Hopcroft, J.E., & Ullman, J.D., *The Design and Analysis of Computer Algorithms*, Addison-Wesley, Reading, Mass., 1974.

[AHU 83] Aho, A.V., Hopcroft, J. E., & Ullman, J.D., *Data Structures and Algorithms*, Addison-Wesley, Reading, Mass., 1983.

[AKW 88] Aho, A.V., Kernighan, B.W. & Weinberger, P.J., *The AWK Programming Language*, Addison-Wesley, Reading, Mass., 1988.

[ASU 86] Aho, A.V., Sethi, R., & Ullman, J.D., *Compilers—Principles, Techniques and Tools*, Addison-Wesley, Reading, Mass., 1986.

[AD 86] Allison, L., & Dix, T.I., "A bit string longest common subsequence algorithm," *Inf. Process. Lett.* 23 (1986): 305–310.

[AB 92] Amir, A., & Benson, G., "Two-dimensional periodicity and its application," in (*Proc. Symp. On Discrete Algorithms*, 1992): 440–452.

[ABF 92a] Amir, A., Benson, G., & Farach, M., "Alphabet-independent two-dimensional matching," in (*Proc. 24th ACM Symp. on Theory Of Computing*, 1992): 59–68.

[ABF 92b] Amir, A., Benson, G., & Farach, M., "Optimal parallel two-dimensional pattern matching," 1992, manuscript.

[AF 91] Amir, A., & Farach M., "Efficient two-dimensional approximate matching of non-rectangular figures," in (*Proc. Symp. On Discrete Algorithms*, 1991): 212–223.

[AL 91] Amir, A., & Landau, G.M., "Fast parallel and serial multidimensional approximate array matching," *Theoret. Comput. Sci.* 81 (1991): 97–115.

[ALV 92] Amir, A., Landau, G.M., & Vishkin, U., "Efficient pattern matching with scaling," *J. Algorithms* 13 (1992): 2–32.

[An 80] Angluin, D., "Finding patterns common to a set of strings," *J. Comput. Syst. Sci.* 21 (1980): 46–62.

[Ap 85] Apostolico, A., "The myriad virtues of suffix trees," in: [AG 85]: 85–96.

[Ap 86] Apostolico, A., "Improving the worst-case performance of the Hunt-Szymanski strategy for the longest common subsequence of two strings," *Inf. Process. Lett.* 23 (1986): 63–69.

[Ap 87] Apostolico, A., "Remark on the Hsu-Du new algorithm for the longest common subsequence problem," *Inf. Process. Lett.* 25 (1987): 235–236.

[Ap 92] Apostolico, A., "Fast parallel detection of squares in strings," *Algorithmica* 8 (1992): 285–319.

[AALF 88] Apostolico, A., Atallah, M.J., Larmore, L.L., & McFaddin, H.S., "Efficient parallel algorithms for string editing and related problems," Report CSD-TR-724, Purdue University, 1988.

[ABG 92] Apostolico, A., Breslauer, D., & Galil, Z., "Optimal parallel algorithms for periods, palindromes and squares," in: (*ICALP'92*, LNCS, Springer-Verlag, 1992).

[ABG 92] Apostolico, A., Browne, S., & Guerra, C., "Fast linear-space computations of longest common subsequences," *Theoret. Comput. Sci.* 92 (1992): 3–17.

[AC 91] Apostolico, A., & Crochemore, M., "Optimal canonization of all substrings of a string," *Information and Computation* 95, 1 (1991): 76–95.

[AC 90] Apostolico, A., & Crochemore, M., "Fast parallel Lyndon factorization and applications," *Information and Computation*, (1990). To appear.

[AG 85] Apostolico, A., & Galil, Z., editors, *Combinatorial Algorithms on Words*, NATO Advanced Science Institutes, Series F, vol 12, Springer-Verlag, Berlin, 1985.

[AG 84] Apostolico, A., & Giancarlo, R., "Pattern-matching machine implementation of a fast test for unique decipherability," *Inf. Process. Lett.* 18 (1984): 155–158.

[AG 86] Apostolico, A., & Giancarlo, R., "The Boyer-Moore-Galil string-searching strategies revisited," *SIAM J.Comput.* 15 (1986): 98–105.

[AG 87] Apostolico, A., & Guerra C., "The longest common subsequence problem revisited," *Algorithmica* 2 (1987): 315–336.

[AILSV 88] Apostolico, A., Iliopoulos, C., Landau, G.M., Schieber, B., & Vishkin, U., "Parallel construction of a suffix tree with applications," *Algorithmica* 3 (1988): 347–365.

[AP 83] Apostolico, A., & Preparata, F.P., "Optimal off-line detection of repetitions in a string," *Theoret. Comput. Sci.* 22 (1983): 297–315.

[AP 85] Apostolico, A., Preparata, F.P., "Structural properties of the string statistics problem," *J. Comput. Syst. Sci.* 31 (1985): 394–411.

[AKLMT 89] Atallah, M.J., Kosaraju, S.R., Larmore, L.L., Miller, G.L., & Teng, S-H., "Constructing trees in parallel," Report CSD-TR-883, Purdue University, 1989.

[Ba 88] Baase, S., *Computer Algorithms—Introduction to Design and Analysis*, Addison-Wesley, Reading, Mass., 1988. 2nd edition.

[Ba 89] Baeza-Yates, R.A., "Improved string searching," *Software—Practice and Experience* 19 (1989) 257–271.

[BCG 93] Baeza-Yates, R.A., Choffrut, C., & Gonnet, G.H., "On Boyer-Moore automata," *Algorithmica* (1993). To appear.

[BG 89] Baeza-Yates, R.A., & Gonnet, G.H., "Efficient text searching of regular expressions," in: (Ausiello, Dezani-Ciancaglini, Ronchi Della Rocca, editors, *Automata, Languages and Programming*, Lecture Notes in Computer Science 372, Springer-Verlag, Berlin, 1989): 46–62.

[BG 92] Baeza-Yates, R.A., & Gonnet, G.H., "A new approach to text searching," *Comm. ACM* 35, 10 (1992): 74–82.

[BGR 90] Baeza-Yates, R.A., Gonnet, G.H., & Régnier, M., "Analysis of Boyer-Moore type string searching algorithms," in: (*Proc. of 1st ACM-SIAM Symposium on Discrete Algorithms*, American Mathematical Society, Providence, 1990): 328–343.

[BR 90] Baeza-Yates, R.A., & Régnier, M., "Fast algorithms for two-dimensional and multiple pattern matching," in (R. Karlsson, J. Gilbert, editors, *Proc. 2nd*

Scandinavian Workshop in Algorithmic Theory, Lecture Notes in Computer Science 447, Springer-Verlag, Berlin, 1990): 332–347.

[BR 92] Baeza-Yates, R.A., & Régnier, M., "Average running time of the Boyer-Moore-Horspool algorithm," *Theoret. Comput. Sci.* 92 (1992): 19–31.

[Ba 93] Baker, B., "A theory of parameterized pattern matching: Algorithms and applications," in: (*STOC'93*, 1993): 71–80.

[Ba 78] Baker, T.P., "A technique for extending rapid exact-match string matching to arrays of more than one dimension," *SIAM J.Comput.* 7 (1978): 533–541.

[Ba 81] Barth, G., "An alternative for the implementation of the Knuth-Morris-Pratt algorithm," *Inf. Process. Lett.* 13 (1981): 134–137.

[Ba 84] Barth, G., "An analytical comparison of two string-searching algorithms," *Inf. Process. Lett.* 18 (1984): 249–256.

[Ba 85] Barth, G., "Relating the average-case cost of the brute-force and the Knuth-Morris-Pratt string-matching algorithm," in [AG 85]: 45–58.

[BEM 79] Bean, D., Ehrenfeucht, A., & McNulty, G., "Avoidable patterns in strings of symbols," *Pacific Journal of Math.* 85 (1979): 261–294.

[BBC 92] Beauquier, D., Berstel, J., & Chrétienne, P., *Éléments d'Algorithmique*, Masson, Paris, 1992.

[BCW 90] Bell, T.C., Cleary, J.G., & Witten, I.H., *Text Compression*, Prentice Hall, Englewood Cliffs, New Jersey, 1990.

[BSTW 86] Bentley, J.L., Sleator, D.D., Tarjan, R.E., & Wei V.K., "A locally adaptive data compression scheme," *Commun. ACM* 29, 4 (1986): 320–330.

[BBGSV 89] Berkman, O., Breslauer, D., Galil, Z., Schieber, B., & Vishkin, U., "Highly parallelizable problems," in: (*Proc. 21st ACM Symposium on Theory of Computing*, Association for Computing Machinery, New York, 1989): 309–319.

[BP 85] Berstel, J., & Perrin, D., *Theory of Codes*, Academic Press, Orlando, Florida, 1985.

[Bi 77] Bird, R.S., "Two-dimensional pattern matching," *Inf. Process. Lett.* 6 (1977) 168–170.

[BR 87] Bishop, M.J. & Rawlings, C.J., *Nucleic Acid and Protein Sequence Analysis: A Practical Approach*, IRL Press Limited, Oxford, England, 1987.

[BBEHM 83] Blumer, A., Blumer, J., Ehrenfeucht, A., Haussler, D., & McConnell, R., "Linear size finite automata for the set of all subwords of a word an outline of results," *Bull. Europ. Assoc. Theoret. Comput. Sci.* 21 (1983): 12–20.

[BBEHCS 85] Blumer, A., Blumer, J., Ehrenfeucht, A., Haussler, Chen, M.T., & Seiferas, J., "The smallest automaton recognizing the subwords of a text," *Theoret. Comput. Sci.* 40 (1985): 31–55.

[BBEHM 87] Blumer, A., Blumer, J., Ehrenfeucht, A., Haussler, D., & McConnell, R., "Complete inverted files for efficient text retrieval and analysis," *J. ACM* 34 (1987): 578–595.

[Bl 87] Blumer, J.A., "How much is that dawg in the window? A moving window algorithm for the directed acyclic word graph," *J. Algorithms* 8 (1987): 451–469.

[Bo 80] Booth, K., "Lexicographically least circular strings," *Inf. Process. Lett.* 10 (1980): 240–242.

[BM 77] Boyer, R.S., & Moore, J.S., "A fast string-searching algorithm," *Comm. ACM* 20 (1977): 762–772.

[BG 90] Breslauer, D., & Galil, Z., "An optimal O(loglog$n$)-time parallel string-matching algorithm," *SIAM J.Comput* 19, 6 (1990): 1051–1058.

[BCT 93] Breslauer, D., Colussi, L., & Toniolo, L., "Tight comparison bounds for the string prefix-matching problem," *Inf. Process. Lett.* 47 (1993): 51–57.

[Ca 90] Capocelli, R., editor, *Sequences: Combinatorics, Compression, Security and Transmission*, Springer-Verlag, New-York, 1990.

[CSV 93] Capocelli, R., & de Santis, A., Vaccaro, U., editors, *Sequences II*, Springer-Verlag, New-York, 1993.

[CL 90] Chang, W.I., & Lawler, E.L., "Approximate string matching in sublinear expected time," in: (*FOCS'90*): 116–124.

[CS 85] Chen, M.T., & Seiferas, J., "Efficient and elegant subword tree construction," in: [AG 85]: 97–107.

[Ch 90] Choffrut, C., "An optimal algorithm for building the Boyer-Moore automaton," *Bull. Europ. Assoc. Theoret. Comput. Sci.* 40 (1990): 217–225.

[CR 87a] Chrobak, M., & Rytter, W., "Remarks on string matching and one-way multihead automata," *Inf. Process. Lett.* 24 (1987): 325–329.

[CR 87b] Chrobak, M., & Rytter, W., "Unique decipherability for partially commutative alphabets," *Fundamenta Informaticae* (1987): 323–336.

[CS 75] Chvatal, V., & Sankoff, D., "Longest common subsequence of two random sequences," *J. Appl. Prob.* 12 (1975): 306–315.

[Co 88] Cole, R., "Parallel merge sort," *SIAM J.Comput.* 17 (1988): 770–785.

[Co 90] Cole, R., "A new proof of the linearity of the Boyer-Moore string-matching algorithm," in: (*Maryland Theoretical Science Day*, 1990).

[Co 91] Cole, R., "Tight bounds on the complexity of the Boyer-Moore algorithm," in: (*Proceedings of the Second Annual ACM-SIAM Symposium on Discrete Algorithms*, 1991): 224–233, and *SIAM J.Comput.* to appear.

[C-R 93b] Cole, R., Crochemore, M., Galil, Z., Gasieniec, L., Hariharan, R., Muthukrishnan, S., Park, K., & Rytter, W., "Optimally fast parallel algorithms for preprocessing and pattern matching in one and two dimensions," in: (*FOCS'93*, 1993): 248–258.

[CH 92] Cole, R., & Hariharan, R., "Tighter bound on the exact complexity of string matching," in: (*FOCS'92*): 600–609.

[CGG 90] Colussi, L., Galil, Z., & Giancarlo, R., "On the exact complexity of string matching," in (*Proc. 31st Symposium on Foundations of Computer Science*, IEEE, 1990): 135–143.

[Co 79] Commentz-Walter, B., "A string-matching algorithm fast on the average," in: (*Automata, Languages and Programming*, Lecture Notes in Computer Science, Springer-Verlag, Berlin, 1979): 118–132.

[CD 89] Consel, C., & Danvy, O., "Partial evaluation of pattern matching in strings," *Inf. Process. Lett.* 30 (1989): 79–86.

[Co 72] Cook, S.A., "Linear-time simulation of deterministic two-way pushdown automata," *Information Processing* 71 (1972): 75–80.

[CH 84] Cormack, G.V., & Horspool, R.N.S., "Algorithms for adaptive Huffman codes," *Inf. Process. Lett.* 18 (1984): 159–165.

[CLR 89] Cormen, T.H., Leirserson, C.E., & Rivest, R.L., *Introduction to Algorithms*, The MIT Press, Cambridge, Mass., 1989.

[Cr 81] Crochemore, M., "An optimal algorithm for computing the repetitions in a word," *Inf. Process. Lett.* 12 (1981): 244–250.

[Cr 83] Crochemore, M., "Recherche linéaire d'un carré dans un mot," *C. R. Acad. Sc. Paris*, t. 296 (1983) Série 1, 781–784.

[Cr 85] Crochemore, M., "Optimal factor transducers," in: [AG 85]: 31–43.

[Cr 86] Crochemore, M., "Transducers and repetitions," *Theoret. Comput. Sci.* 45 (1986): 63–86.

[Cr 87] Crochemore, M., "Longest common factor of two words," in: (Ehrig, Kowalski, Levi & Montanari, editors, *TAPSOFT'87*, vol 1, Springer-Verlag, Berlin, 1987): 26–36.

[Cr 88] Crochemore, M., "String matching with constraints," in: (M. Chytil, L. Janiga, & V. Koubek, editors, *Mathematical Foundations of Computer Science 1988*, Lecture Notes in Computer Science 324, Springer-Verlag, Berlin, 1988): 44–58.

[Cr 89a] Crochemore, M., "Data compression with substitution," in: [GP 89]: 1–16.

[Cr 89b] Crochemore, M., "Automata and algorithms," in: (J-E. Pin, editor, *Formal Properties of Finite Automata and Applications*, Lecture Notes in Computer Science 386, Springer-Verlag, Berlin, 1989): 166–175.

[Cr 89c] Crochemore, M., "String matching and periods," *Bul. Euro. Assoc. Theor. Comput. Sci.* 39 (1989): 149–153.

[Cr 92] Crochemore, M., "String matching on ordered alphabets," *Theoret. Comput. Sci.* 92 (1992): 33–47.

[C-R 92] Crochemore, M., Czumaj, A., Gasieniec, L., Jarominek, S., Lecroq, T., Plandowski, W., & Rytter, W., "Speeding up two string-matching algorithms," in: (A. Finkel and M. Jantzen, editors, *9th Annual Symposium on Theoretical Aspects of Computer Science*, Springer-Verlag, Berlin, 1992): 589–600.

[C-R 93a] Crochemore, M., Czumaj, A., Gasieniec, L., Jarominek, S., Lecroq, T., Plandowski, W., & Rytter, W., "Fast multi-pattern matching," Rapport I.G.M. 93–3, Université de Marne la Vallée, 1993.

[CGPR 93] Crochemore, M., Gasieniec, L., & Rytter, W., "Two-dimensional pattern matching by sampling," *Inform. Process. Lett. 96* (1993): 159–162.

[CN 90] Crochemore, M., & Néraud, J., "Unitary monoid with two generators: An algorithmic point of view," in: (A. Arnold, editor, *CAAP'90*, Lecture Notes in Computer Science 431, Springer-Verlag, Berlin, 1990): 117–131.

[CP 91] Crochemore, M., & Perrin, D., "Two-way string matching," *J. ACM* 38, 3 (1991): 651–675.

[CR 90] Crochemore, M., & Rytter, W., "Parallel construction of minimal suffix and factor automata," *Inform. Process. Lett.* 35 (1990): 121–128.

[CR 91a] Crochemore, M., & Rytter, W., "Efficient parallel algorithms to test square-freeness and factorize strings," *Inf. Process. Lett.* 38 (1991): 57–60.

[CR 91c] Crochemore, M., & Rytter, W., "Usefulness of the Karp-Miller-Rosenberg algorithm in parallel computations on strings and arrays," *Theoret. Comput. Sci.* 88 (1991): 59–82.

[CR 92] Crochemore, M., & Rytter, W., "Note on two-dimensional pattern matching by optimal parallel algorithms," in: (*Parallel Image Analysis*, A. Nakamura, M. Nivat, A. Saoudi, P.S.P. Wang, K. Inoue, editors, LNCS 654, Springer-Verlag, 1992): 100–112.

[CR 94] Crochemore, M., & Rytter, W., "Cubes, squares and time-space efficient string searching," *Algorithmica* (1994). To appear.

[De 79] Deken, J., "Some limit results for longest common subsequences," *Discrete Math.* 26 (1979): 17–31.

[DGM 90] Dubiner, M., Galil, Z., & Magen, E., "Faster tree-pattern matching," in: (*Proceedings of 31st FOCS*, 1990):145–150.

[Du 79] Duval, J-P., "Périodes et répétitions des mots du monoïde libre," *Theoret. Comput. Sci.* 9 (1979): 17–26.

[Du 82] Duval, J-P., "Relationship between the period of a finite word and the length of its unbordered segments," *Discrete Math.* 40 (1982): 31–44.

[Du 83] Duval, J-P., "Factorizing words over an ordered alphabet," *J. Algorithms* 4 (1983): 363–381.

[EV 88] Eilam-Tzoreff, T., & Vishkin, U., "Matching patterns in strings subject to multi-linear transformations," *Theoret. Comput. Sci.* 60 (1988): 231–254.

[EGGI 92] Eppstein, D., Galil, Z., Giancarlo, R., & Italiano, G., "Sparse dynamic programming I," *J. ACM* 39 (1990): 519–545.

[Fa 73] Faller, N., "An adaptive system for data compression," in: (*Record of the 7th Asilomar Conference on Circuits, Systems, and Computers*, 1973): 593–597.

[FT 87] Feng, Z.R., & Takaoka, T., "On improving the average case of the Boyer-Moore string-matching algorithm," *J. Inf. Process.* 10, 3 (1987): 173–177.

[FT 89] Feng, Z.R., & Takaoka, T., "A technique for two-dimensional pattern matching," *Comm. ACM* 32 (1989): 1110–1120.

[FW 65] Fine, N.J., & Wilf, H.S., "Uniqueness theorems for periodic functions," *Proc. Amer. Math. Soc.* 16 (1965): 109–114.

[FP 74] Fischer, M.J., & Paterson, M.S., "String matching and other products," in: (R.M. Karp, editor, *Proc. of SIAM-AMS Conference on Complexity of Computation*, American Mathematical Society, Providence, R.I., 1974) 113–125.

[Fr 75] Fredman, M.L., "On computing the length of longest increasing subsequences," *Discrete Math.* 11 (1975): 29–35.

[Ga 76] Galil, Z., "Two fast simulations which imply fast string-matching and palindrome-recognition algorithms," *Inf. Process. Lett.* 4, 4 (1976): 85–87.

[Ga 77] Galil, Z., "Some open problems in the theory of computations as questions about two-way deterministic pushdown automaton languages," *Math. Syst. Theory* 10 (1977): 211–228.

[Ga 78] Galil, Z., "Palindrome recognition in real time by a multitape Turing machine," *J. Comput. Syst. Sci.* 16 (1978): 140–157.

[Ga 79] Galil, Z., "On improving the worst case-running time of the Boyer-Moore string-searching algorithm," *Comm. ACM* 22 (1979): 505–508.

[Ga 81] Galil, Z., "String matching in real time," *J. ACM* 28 (1981): 134–149.

[Ga 85a] Galil, Z., "Open problems in stringology," in: [AG 85]: 1–12.

[Ga 85b] Galil, Z., "Optimal parallel algorithm for string matching," *Information and Control* 67 (1985): 144–157.

[Ga 92] Galil, Z., "A constant-time optimal parallel string-matching algorithm," in: (*Proc. 24th ACM Symp. on Theory Of Computing*, 1992): 69–76.

[GG 87] Galil, Z., & Giancarlo, R., "Parallel string matching with k mismatches," *Theoret. Comput. Sci.* 51 (1987) 341–348.

[GG 88] Galil, Z., & Giancarlo, R., "Data structures and algorithms for approximate string matching," *J. Complexity* 4 (1988): 33–72.

[GG 89] Galil, Z., & Giancarlo, R., "Speeding up dynamic programming with applications to molecular biology," *Theoret. Comput. Sci.* 64 (1989): 107–118.

[GP 89] Galil, Z., & Park, K., "An improved algorithm for approximate string matching," in: (Ausiello, Dezani-Ciancaglini, Ronchi Della Rocca, editors, *Automata, Languages and Programming*, Lecture Notes in Computer Science 372, Springer-Verlag, Berlin, 1989): 394–404.

[GP 92a] Galil, Z., & Park, K., "Dynamic programming with convexity, concavity and sparsity," *Theoret. Comput. Sci.* 92 (1992): 49–76.

[GP 92b] Galil, Z., & Park, K., "Truly alphabet-independent two-dimensional matching," in: (*Proc. 33rd Annual IEEE Symposium on the Foundations of Computer Science*, 1992): 247–256.

[GR 92] Galil, Z., & Rabani, Y., "On the sparse complexity of some algorithms for sequence comparison," *Theoret. Comput. Sci.* 95, 2 (1992): 231–244.

[GS 78] Galil, Z., & Seiferas, J., "A linear-time on-line recognition algorithm for `Palstars,'" *J. ACM* 25 (1978): 102–111.

[GS 80] Galil, Z., & Seiferas, J., "Saving space in fast string matching," *SIAM J.Comput.* 9 (1980): 417–438.

[GS 81] Galil, Z., & Seiferas, J., "Linear-time string matching using only a fixed number of local storage locations," *Theoret. Comput. Sci.* 13 (1981): 331–336.

[GS 83] Galil, Z., & Seiferas, J., "Time-space optimal string matching," *J. Comput. Syst. Sci.* 26 (1983): 280–294.

[Ga 78] Gallager, R.G., "Variations on a theme by Huffman," *I.E.E.E. Trans. Inform. Theory* IT 24, 6 (1978): 668–674.

[GMS 80] Gallant, J., Maier, D., & Storer, J.A., "On finding minimal length superstrings," *J. Comput. Syst. Sci.* 20 (1980): 50–58.

[GJ 79] Garey, M.R., & Johnson, D.S., *Computers and Intractability: A Guide to the Theory of NP-Completeness*, W.H. Freeman, New York, 1979.

[Gi 93] Giancarlo, R., "The suffix tree of a square matrix, with applications," in: (*Proc. Symp. On Discrete Algorithms*, 1993).

[GR 86] Gibbons, A., & Rytter, W., "On the decidability of some problems about rational subsets of free partially commutative monoids," *Theoret. Comput. Sci.* 48 (1986): 329–337.

[GR 88] Gibbons, A., & Rytter, W., *Efficient Parallel Algorithms*, Cambridge University Press, Cambridge, U.K., 1988.

[GR 89] Gibbons, A., & Rytter, W., "Optimal parallel algorithm for dynamic expression evaluation and application to context-free recognition," *Information and Computation* 81, 1 (1989): 32–45.

[GB 91] Gonnet, G.H., & Baeza-Yates, R., *Handbook of Algorithms and Data Structures*, Addison-Wesley, Reading, Mass., 1991.

[Gr 91] Gross, M., "Constructing lexicon-grammars," in: (B.T.S. Atkins & A. Zampolli editors, *Computational Approaches to the Lexicon*, Oxford University Press, 1991).

[GP 89] Gross, M., & Perrin, D., editors, *Electronic Dictionaries and Automata in Computational Linguistics*, Lecture Notes in Computer Science 377, Springer-Verlag, Berlin, 1989.

[GO 80] Guibas, L.J., & Odlyzko, A.M., "A new proof of the linearity of the Boyer-Moore string-searching algorithm," *SIAM J.Comput.* 9 (1980): 672–682.

[GO 81a] Guibas, L.J., & Odlyzko, A.M., "Periods in strings," *J. Comb. Th. A* 30 (1981): 19–42.

[GO 81b] Guibas, L.J., & Odlyzko, A.M., "String overlaps, pattern matching and nontransitive games," *J. Comb. Th. A* 30 (1981): 183–208.

[HD 80] Hall, P.A.V., & Dowling, G.R., "Approximate string matching," *ACM Comput.Surv.* 12 (1980): 381–402.

[Ha 93] Hancart, C., "On Simon's string-searching algorithm," *Inf. Process. Lett.* 47 (1993): 95–99.

[HPS 92] Hansel, G., Perrin, D., & Simon, I., "Compression and entropy," in: (*STACS 92*, A. Finkel and M. Jantzen editors., LNCS 577, Springer-Verlag, Berlin, 1992): 515–528.

[HT 84] Harel, D., & Tarjan, R.E., "Fast algorithms for finding nearest common ancestors," *SIAM J.Comput.* 13 (1984): 338–355.

[Ha 71] Harrison, M.C., "Implementation of the substring test by hashing," *Comm. ACM* 14, 12 (1971): 777–779.

[HR 85] Hartman, A., & Rodeh, M., "Optimal parsing of strings," in: [AG 85]: 155–167.

[HY 92] Hashiguchi, K., & Yamada, K., "Two recognizable string-matching problems over free partially commutative monoids," *Theoret. Comput. Sci.* 92 (1992): 77–86.

[HC 86] Hébrard, J-J., & Crochemore, M., "Calcul de la distance par les sous-mots," *R.A.I.R.O. Informatique Théorique* 20 (1986): 441–456.

[He 87] Held, G., *Data Compression—Techniques and Applications, Hardware and Software Considerations*, John Wiley & Sons, New York, NY, 1987. 2nd edition.

[Hi 75] Hirschberg, D.S., "A linear-space algorithm for computing maximal common subsequences," *Comm. ACM* 18, 6 (1975): 341–343.

[Hi 77] Hirschberg, D.S., "Algorithms for the longest common subsequence problem," *J. ACM* 24 (1977): 664–675.

[Hi 78] Hirschberg, D.S., "An information theoretic lower bound for the longest common subsequence problem," *Inf. Process. Lett.* 7 (1978): 40–41.

[HL 87a] Hirschberg, D.S., & Larmore, L.L., "New applications of failure functions," *J. ACM* 34 (1987): 616–625.

[HL 87b] Hirschberg, D.S., & Larmore, L.L., "The set LCS problem," *Algorithmica* 2 (1987): 91–95.

[HO 82] Hoffman, C.M., & O'Donnell, M.J., "Pattern matching in trees," *J. ACM* 29, 1 (1982): 68–95.

[HU 79] Hopcroft, J.E., & Ullman, J.D., *Introduction to Automata, Languages and Computations*, Addison-Wesley, Reading, Mass., 1979.

[Ho 80] Horspool, R.N., "Practical fast searching in strings," *Software—Practice and Experience* 10 (1980): 501–506.

[HD 84] Hsu, W.J., & Du, M.W., "New algorithms for the LCS problem," *J. Comput. Syst. Sci.* 29 (1984): 133–152.

[Hu 51] Huffman, D.A., "A method for the construction of minimum redundancy codes," *Proceedings of the I.R.E.* 40 (1951): 1098–1101.

[Hu 88] Hume, A., "A tale of two greps," *Software—Practice and Experience* 18 (1988): 1063–1072.

[HS 91] Hume, A., & Sunday, D.M., "Fast string searching," *Software—Practice and Experience* 21, 11 (1991): 1221–1248.

[HS 77] Hunt, J.W., & Szymanski, T.G., "A fast algorithm for computing longest common subsequences," *Comm. ACM* 20 (1977): 350–353.

[IS 93] Idury, R., & Schäffer, A., "Multiple matching of rectangular patterns," in: (*STOC'93*, 1993): 81–90.

[IS 92] Iliopoulos, C.S., & Smyth, W.F., "Optimal algorithms for computing the canonical form of a circular string," *Theoret. Comput. Sci.* 92 (1992): 87–105.

[Já 92] Jájá, J., *An Introduction to Parallel Algorithms*, Addison-Wesley, Reading, Mass., 1992.

[Ja 92] Jantke, K., "Polynomial time inference of general pattern languages," in: (*STACS'92*, Lecture Notes in Computer Science 166, 1992): 314–325.

[Jo 77] Jones, N.D., "A note on linear-time simulation of deterministic two-way pushdown automata," *Inf. Process. Lett.* 6, 4 (1977): 110–112.

[KMR 72] Karp, R.M., Miller, R.E., & Rosenberg, A.L., "Rapid identification of repeated patterns in strings, arrays and trees," in: (*Proc. 4th ACM Symposium on Theory of Computing*, Association for Computing Machinery, New York, 1972): 125–136.

[KR 87] Karp, R.M., & Rabin, M.O., "Efficient randomized pattern-matching algorithms," *IBM J. Res.Dev.* 31 (1987): 249–260.

[KLP 89] Kedem, Z.M., Landau, G.M., & Palem, K.V., "Optimal parallel suffix-prefix matching algorithm and applications," in: (*Proc. ACM Symposium on Parallel Algorithms*, Association for Computing Machinery, New York, 1989): 388–398.

[KP 89] Kedem, Z.M., & Palem, K.V., "Optimal parallel algorithms for forest and term matching," *Theoret. Comput. Sci.* (1989).

[Kf 88] Kfoury, A.J., "A linear-time algorithm to decide whether a word is overlap-free," *Theoret. Comput. Sci.* 22 (1988): 135–145.

[KS 92] Kim, J.Y., & Shawe-Taylor, J., "An approximate string-matching algorithm," *Theoret. Comput. Sci.* 92 (1992): 107–117.

[Kl 56] Kleene, S.C., "Representation of events in nerve nets and finite automata," in: (Shannon and McCarthy editors, *Automata studies*, Princeton University Press, 1956): 3–40.

[Kn 85] Knuth, D.E., "Dynamic Huffman coding," *J. Algorithms* 6 (1985): 163–180.

[KP 71] Knuth, D.E., & Pratt, V.R., "Automata theory can be useful," Report, Stanford University, 1971.

[KMP 77] Knuth, D.E., Morris Jr, J.H., & Pratt, V.R., "Fast pattern matching in strings," *SIAM J.Comput.* 6 (1977): 323–350.

[KH 87] Ko, Ker-I, & Hua, Chin-Ming, "A note on the two-variable pattern-finding problem," *J. Comput. Syst. Sci.* 34 (1987): 75–86.

[Ko 89] Kosaraju, S.R., "Efficient tree-pattern matching," *FOCS'89* (1989): 178–183.

[LSV 87] Landau, G.M., Schieber, B., & Vishkin, U., "Parallel construction of a suffix tree," in: (*Automata, Languages and Programming*, Lecture Notes in Computer Science, Springer-Verlag, Berlin, 1987): 314–325.

[LV 86a] Landau, G.M., & Vishkin, U., "Introducing efficient parallelism into approximate string matching," *STOC* 18 (1986): 220–230.

[LV 86b] Landau, G.M., & Vishkin, U., "Efficient string matching with *k* mismatches," *Theoret. Comput. Sci.* 43 (1986): 239–249.

[LV 88] Landau, G.M., & Vishkin, U., "Fast string matching with *k* differences," *J. Comput. Syst. Sci.* 37 (1988): 63–78.

[LV 89] Landau, G.M., & Vishkin, U., "Fast parallel and serial approximate string matching," *J. Algorithms* 10 (1989): 158–169.

[Le 92] Lecroq, T., "A variation on the Boyer-Moore algorithm," *Theoret. Comput. Sci.* 92 (1992): 119–144.

[LZ 76] Lempel, A., & Ziv, J., "On the complexity of finite sequences," *IEEE Trans. Inform.Theory* IT 22, 1 (1976): 75–81.

[Li 81] Liu, L., "On string pattern matching: A new model with a polynomial time algorithm," *SIAM J.Comput.* 10 (1981): 118–139.

[Lo 83] Lothaire, M., *Combinatorics on Words*, Addison-Wesley, Reading, Mass., 1983.

[LW 75] Lowrance, R., & Wagner, R.A., "An extension of the string-to-string correction problem," *J. ACM 25* (1975): 177–183.

[LS 62] Lyndon, R.C., & Schützenberger, M.P., "The equation $a^M = b^N c^P$ in a free group," *Michigan Math. J.* 9 (1962): 422–432.

[Ma 77] Maier, D., "The complexity of some problems on subsequences and supersequences," *J. ACM* 25 (1977): 322–336.

[MS 77] Maier, D., & Storer, J.A., "A note on the complexity of superstring problem," Report 233, Computer Science Lab., Princeton University, 1977.

[Ma 89] Main, M.G., "Detecting left most maximal periodicities," *Discrete Applied Math.* 25 (1989): 145–153.

[ML 79] Main, M.G., & Lorentz, R.J., "An $O(n \log n)$ algorithm for finding repetition in a string," TR CS-79–056, Washington State University, Pullman, 1979.

[ML 84] Main, M.G., & Lorentz, R.J., "An $O(n \log n)$ algorithm for finding all repetitions in a string," *J. Algorithms* (1984): 422–432.

[ML 85] Main, M.G., & Lorentz, R.J., "Linear-time recognition of square-free strings," in: [AG 85]: 271–278.

[MR 80] Majster, M.E., & Ryser, A., "Efficient on-line construction and correction of position trees," *SIAM J.Comput.* 9, 4 (1980): 785–807.

[Ma 75] Manacher, G., "A new linear-time on-line algorithm for finding the smallest initial palindrome of the string," *J. ACM* 22 (1975): 346–351.

[Ma 76] Manacher, G., "An application of pattern matching to a problem in geometrical complexity," *Inf. Process. Lett.* 5 (1976): 6–7.

[Ma 89] Manber, U., *Introduction to Algorithms*, Addison-Wesley, Reading, Mass., 1989.

[MM 90] Manber, U., & Myers, E., "Suffix arrays: A new method for on-line string searches," in: (*Proc. of 1st ACM-SIAM Symposium on Discrete Algorithms*, American Mathematical Society, Providence, R.I., 1990): 319–327.

[MP 80] Masek, W.J., & Paterson, M.S., "A faster algorithm computing string edit distances," *J. Comput. Syst. Sci.* 20, 1 (1980): 18–31.

[McC 76] McCreight, E.M., "A space-economical suffix tree construction algorithm," *J. ACM* 23, 2 (1976): 262–272.

[Mo 84] Monien, B., "Deterministic two-way one-head pushdown automata are very powerful," *Inf. Process. Lett.* 18, 3 (1984): 239–242.

[MP 70] Morris Jr, J.H., & Pratt, V.R., "A linear pattern-matching algorithm," Report 40, University of California, Berkeley, 1970.

[Mo 68] Morrison, D.R., "PATRICIA—practical algorithm to retrieve information coded in alphanumeric," *J. ACM* 15 (1968): 514–534.

[My 86] Myers, E.W., "An *O(ND)* difference algorithm and its variations," *Algorithmica* 1 (1986): 251–266.

[MM 89] Myers, E.W., & Miller, W., "Approximate matching of regular expressions," *Bull. Math. Biol.* 51 (1989): 5–37.

[NKY 82] Nakatsu, N., Kambayashi, Y., & Yajima, S., "A longest common subsequence algorithm suitable for similar text strings," *Acta Informatica* 18 (1982): 171–179.

[NW 70] Needleman, S.B., & Wunsch, C.D., "A general method applicable to the search for similarities in the amino-acid sequence of two proteins," *Journal of Molecular Biology 48* (1970): 443–453.

[Ne 92] Nelson, M., *The Data Compression Book*, M&T Publishing, Inc., Redwood City, Calif., 1992.

[NC 92] Néraud, J., & Crochemore, M., "A string matching interpretation of the equation $x^m y^n = z^p$," *Theoret. Comput. Sci.* 92 (1992): 145–164.

[Pe 85] Perrin, D., "Words over a partially commutative alphabet," in: [AG 85]: 329–340.

[Pe 90] Perrin, D., "Finite automata," in: (J. van Leeuwen, editor, *Handbook of Theoretical Computer Science*, vol B, *Formal Models and Semantics*, Elsevier, Amsterdam, 1990): 1–57.

[Qu 92] Quong, R.W., "Fast average-case pattern matching by multiplexing sparse tables," *Theoret. Comput. Sci.* 92 (1992): 165–179.

[RS 59] Rabin, M.O., & Scott, D., "Finite automata and their decision problems," *IBM J. Research and Development* 3 (1959): 114–125. Reprinted in: (E.F. Moore editor, *Sequential Machines: Selected Papers*, Addison-Wesley, Reading, Mass., 1964): 63–91.

[Ra 85] Rabin, M.O., "Discovering repetitions in strings," in: [AG 85]: 279–288.

[Ré 89] Régnier, M., "Knuth-Morris-Pratt algorithm: An analysis," in: (*MFCS'89*, Lecture Notes in Computer Science 379, Springer-Verlag, Berlin, 1989): 431–444.

[RR 93] Régnier, M., & Rostami, L., "A unifying look at d-dimensional periodicities and space coverings," in: (*Combinatorial Pattern Matching*, Lecture Notes in Computer Science 684, Springer-Verlag, Berlin, 1993): 215–227.

[RS 85] Restivo, A., & Salemi, S., "Some decision results on nonrepetitive words," in: [AG 85]: 289–295.

[Re 92] Revuz, D., "Minimization of acyclic deterministic automata in linear time," *Theoret. Comput. Sci.* 92 (1992): 181–189.

[Ri 77] Rivest, R.L., "On the worst-case behavior of string-searching algorithms," *SIAM J.Comput.* 6, 4 (1977): 669–674.

[RT 85] Robert, Y., & Tchuente, M., "A systolic array for the longest common subsequence problem," *Inf. Process. Lett.* 21 (1885): 191–198.

[RPE 81] Rodeh, M., Pratt, V.R., & Even, S., "Linear algorithm for data compression via string matching," *J. ACM* 28 (1981): 16–24.

[Ry 80] Rytter, W., "A correct preprocessing algorithm for Boyer-Moore string searching," *SIAM J.Comput.* 9 (1980): 509–512.

[Ry 85] Rytter, W., "The complexity of two-way pushdown automata and recursive programs," in: [AG 85]: 341–356.

[Ry 86] Rytter, W., "The space complexity of the unique decipherability problem," *Inf. Process. Lett.* 23 (1986): 1–3.

[Ry 88] Rytter, W., "On efficient computations of costs of paths of a grid graph," *Inf. Process. Lett.* 29 (1988): 71–74.

[Ry 89] Rytter, W., "On the parallel transformations of regular expressions to non-deterministic finite automata," *Inf. Process. Lett.* 31 (1989): 103–109.

[RD 90] Rytter, W., & Diks, K., "On optimal parallel computations for sequences of brackets," in: [Ca 90]: 92–105.

[Sa 89] Salton, G., *Automatic Text Processing*, Addison-Wesley, Reading, Mass., 1989.

[SK 83] Sankoff, D., & Kruskal, J.B., *Time Warps, String Edits and Macromolecules: The Theory and Practice of Sequence Comparison*, Addison-Wesley, Reading, Mass., 1983.

[Sc 88] Schaback, R., "On the expected sublinearity of the Boyer-Moore string-searching algorithm," *SIAM J.Comput.* 17 (1988): 648–658.

[SV 88] Schieber, B., & Vishkin, U., "On finding lowest common ancestors: simplification and parallelization," *SIAM J.Comput.* 17 (1988): 1253–1262.

[Se 88] Sedgewick, R., *Algorithms*, Addison-Wesley, Reading, Mass., 1988. 2nd edition.

[SG 77] Seiferas, J., & Galil, Z., "Real-time recognition of substring repetition and reversal," *Math. Syst. Theory* 11 (1977): 111–146.

[Se 77] Selkow, S.M., "The tree-to-tree editing problem," *Inf. Process. Lett.* 6 (1977): 184–186.

[Se 74a] Sellers, P.H., "An algorithm for the distance between two finite sequences," *J. Comb. Th. A* 16 (1974): 253–258.

[Se 74b] Sellers, P.H., "On the theory and computation of evolutionary distances," *SIAM J. Appl. Math.* 26 (1974): 787–793.

[Se 80] Sellers, P.H., "The theory and computation of evolutionary distances: Pattern recognition," *J. Algorithms* 1 (1980): 359–373.

[Se 85] Semba, I., "An efficient string-searching algorithm," *J. Inf. Process.* 8 (1985): 101–109.

[Sh 81] Shiloach, Y., "Fast canonization of circular strings," *J. Algorithms* 2 (1981): 107–121.

[SV 81] Shiloach, Y., & Vishkin, U., "Finding a maximum, merging and sorting in parallel computation model," *J. Algorithms* 2 (1981): 88–102.

[Sl 83] Slisenko, A.O., "Detection of periodicities and string matching in real time," *J. Sov. Math.* 22, 3 (1983): 1326–1387.

[Sm 82] Smit, G. de V., "A comparison of three string-matching algorithms," *Software—Practice and Experience* 12 (1982): 57–66.

[Sm 90] Smith, P.D., *An Introduction to Text Processing*, The MIT Press, Cambridge, Mass., 1990.

[Sp 86] Spehner, J-C., "La reconaissance des facteurs d'un mot dans un texte," *Theoret. Comput. Sci.* 48 (1986): 35–52.

[St 77] Storer, J.A., "NP-completeness results concerning data compression," Report 234, Princeton University, 1977.

[SS 78] Storer, J.A., & Szymanski, T.G., "The macro model for data compression," in: (*Proc. 10th ACM Symposium on Theory of Computing*, Association for Computing Machinery, New York, 1978): 30–39.

[St 88] Storer, J.A., *Data Compression: Methods and Theory*, Computer Science Press, Rockville, MD, 1988.

[Su 90] Sunday, D.M., "A very fast substring search algorithm," *Comm. ACM* 33, 8 (1990): 132–142.

[TU 88] Tarhio, J., & Ukkonen, E., "A greedy approximation algorithm for constructing shortest common superstrings," *Theoret. Comput. Sci.* 57 (1988): 131–146.

[TU 90] Tarhio, J., & Ukkonen, E., "Boyer-Moore approach to approximate string matching," in: (*Proc. 2nd Scandinavian Workshop in Algorithmic Theory*, Lecture Notes in Computer Science 447, Springer-Verlag, Berlin, 1990): 348–359.

[Th 68] Thompson, K., "Regular expression search algorithm," *Comm. ACM* 11 (1968): 419–422.

[Ti 84] Tichy, W.F., "The string-to-string correction problem with block moves," *ACM Trans. Comput. Syst.* 2 (1984): 309–321.

[Uk 85a] Ukkonen, E., "Finding approximate patterns in strings," *J. Algorithms* 6 (1985): 132–137.

[Uk 85b] Ukkonen, E., "Algorithms for approximate string matching," *Information and Control* 64 (1985) 100–118.

[Uk 92] Ukkonen, E., "Approximate string matching with q-grams and maximal matches," *Theoret. Comput. Sci.* 92 (1992): 191–211.

[Uk 92] Ukkonen, E., "Constructing suffix trees on-line in linear time," in: (*IFIP*'92): 484–492.

[UW 93] Ukkonen, E., & Wood, D., "Approximate string matching with suffix automata," *Algorithmica* (1993). To appear.

[Vi 85] Vishkin, U., "Optimal parallel pattern matching in strings," *Information and Control* 67 (1985): 91–113.

[Vi 91] Vishkin, U., "Deterministic sampling—A new technique for fast pattern matching," *SIAM J.Comput.* 20, 1 (1991): 22–40.

[Vi 87] Vitter, J.S., "Design and analysis of dynamic Huffman codes," *J. ACM* 34 (1987): 825–845.

[Wa 74] Wagner, R.A., "Order-n correction for regular languages," *Comm. ACM* 17, 6 (1974): 265–268.

[Wa 75] Wagner, R.A., "On the complexity of extended string-to-string correction problem," in: (*Proc. 7th ACM Symposium on Theory of Computing*, Association for Computing Machinery, New York, 1975): 218–223.

[WF 74] Wagner, R.A., & Fischer, M.J., "The string-to-string correction problem," *J. ACM* 21 (1974): 168–178.

[Wa 89] Waterman, M.S., *Mathematical Methods for DNA Sequences*, CRC Press, Boca Raton, Fla., 1989.

[We 73] Weiner, P., "Linear pattern-matching algorithms," in: (*Proc. 14th IEEE Annual Symposium on Switching and Automata Theory*, Institute of Electrical Electronics Engineers, New York, 1973): 1–11.

[We 84] Welch, T.A., "A technique for high-performance data compression," *IEEE Computer* 17,6 (1984): 8–19.

[WNC 87] Witten, I.H., Neal, R.M., & Cleary, J.G., "Arithmetic coding for data compression," *Commun. ACM* 30, 6 (1987): 520–540.

[WC 76] Wong, C.K., & Chandra, A.K., "Bounds for the string-editing problem," *J. ACM* 23 (1976): 13–16.

[WM 92] Wu, S., & Manber, U., "Fast text searching allowing errors," *Comm. ACM* 35, 10 (1992): 83–91.

[Ya 79] Yao, A.C., "The complexity of pattern matching for a random string," *SIAM J.Comput.* 8 (1979): 368–387.

[Zi 92] Zipstein, M., "Data compression with factor automata," *Theoret. Comput. Sci.* 92 (1992): 213–221.

[ZC 89] Zipstein, M., & Crochemore, M., "Transducteurs arithmétiques," Rapport L.I.T.P. 89–12, Université Paris 7, 1989.

[ZL 77] Ziv, J., & Lempel, A., "A universal algorithm for sequential data compression," *IEEE Trans. Inform. Theory* 23 (1977): 337–343.

[ZL 78] Ziv, J., & Lempel, A., "Compression of individual sequences via variable length coding," *IEEE Trans. Inform. Theory* 24 (1978): 530–536.

[ZP 92] Zwick, U., & Paterson, M.S., "Lower bounds for string-matching in the sequential comparison model," manuscript, 1992.

Index